Linear and Nonlinear Waves in Microstructured Solids

Linear and Nonlinear Waves in Microstructured Solids

Homogenization and Asymptotic Approaches

Igor V. Andrianov
(igor.andrianov@gmail.com)

Jan Awrejcewicz
(jan.awrejcewicz@p.lodz.pl)

Vladyslav Danishevskyy
(vladyslav.danishevskyy@gmail.com)

CRC Press is an imprint of the
Taylor & Francis Group, an **informa** business

First edition published 2021
by CRC Press
6000 Broken Sound Parkway NW, Suite 300, Boca Raton, FL 33487-2742

and by CRC Press
2 Park Square, Milton Park, Abingdon, Oxon, OX14 4RN

CRC Press is an imprint of Taylor & Francis Group, LLC

Library of Congress Cataloging-in-Publication Data

ISBN: [9780367704124] (hbk)
ISBN: [9781003146162] (ebk)

Typeset in Nimbus font
by KnowledgeWorks Global Ltd.

Dedication

This is a modest contribution dedicated to the work and virtue of Professor Leonid I. Manevitch (02.04.1938 - 20.08.2020), a prominent figure in Dynamics and Mechanics of Composite Materials

Contents

Preface

Wave propagation in heterogeneous materials and structures is characterized by a number of significant phenomena, which can never be observed in homogeneous solids. In recent years, intensive studies have been devoted to phononic band gaps, negative refraction, dynamic anisotropy and waves focusing, acoustic diodes, acoustically invisible cloaks, waves localization in structures with defects. All these remarkable effects and properties have a great practical importance for a large variety of applications in physics, engineering, biomechanics, and many other areas. Composite materials, metamaterials and acoustic crystals, consisting of several components with different physical and mechanical properties, are widely used in modern civil and mechanical engineering, aircraft and rocketry industries. Examples of composites are reinforced plastics, concrete, laminated wood, plywood and many other materials. In recent years, technologies for creating nanocomposites have been intensively developed, in which reinforcing particles (fullerenes, nanotubes) have a characteristic size from 1 nm to 100 nm. Unlike ordinary substances, in the nanoparticles most of the atoms are placed on the surface, which leads to their extremely high stiffness and strength. Due to this, nanocomposites are characterized by special mechanical properties that significantly exceed the properties of traditional materials.

From the point of view of mechanics, the main feature of modeling and calculation of composite materials is to take into account the heterogeneity of their internal structure. Therefore, models and methods of the theory of composites can also be used to study spatially heterogeneous structures, such as beams with attached masses, layered, ribbed, perforated plates and shells, pile foundations, and many others.

The propagation of elastic strain waves in composite materials is accompanied multiple reflections and refractions of the signal on the nonhomogeneities of the internal structure, which lead to dispersion of the wave field and scattering of mechanical energy. The influence of nonlinearity, on the contrary, leads to the concentration of energy and to the generation of higher harmonics. In the case of a balance between the effects of dispersion and nonlinearity, the formation of stationary nonlinear waves (periodic or solitary) is possible, which propagate without changing the shape and speed.

The properties of linear and nonlinear strain waves propagating in an nonhomogeneous medium substantially depend on the characteristics of the microstructure. Studying the processes of formation and propagation of waves, one can get detailed information about the internal structure and defects of the composite material.

Localized nonlinear waves accumulate a significant amount of mechanical energy, which can lead to local destruction of structures. Therefore, the consideration of nonlinear effects is necessary when developing new, refined criteria for dynamic strength.

The special properties of nonlinear waves (such as dispersion, localization, the dependence of the shape and propagation velocity on the amplitude) can be used to create various types of acoustic instruments and devices: vibration dampers, filters, receivers, transmitters and waveguides.

Linear and nonlinear elastic waves in solids have been investigated by many authors. In most studies, homogeneous materials and structures were considered when the dispersion phenomenon is caused by geometric factors (the presence of bending forces, reflection of waves from the side surfaces of the waveguide, etc.). The propagation of strain waves in composite materials has been much less studied. Many approaches are phenomenological in nature; therefore, the numerical values of the coefficients in the wave equations are often not defined. It is assumed that they can be found experimentally. However, for many real materials, the exact values of the phenomenological parameters remain unknown.

Therefore, the critical problem is the development of new methods for modeling linear and nonlinear waves in composite materials and finding solutions that would allow taking into account the influence of the internal structure of the medium on the processes of wave propagation at the macrolevel.

This monograph can be treated as extension/continuation of our previous monograph [19]. In the mentioned book we considered the static problems of the theory of composites using asymptotic approaches. In the present book these approaches are generalized to the dynamic case. The same asymptotic methods are employed, i.e. the homogenization approach, singular and regular asymptotics, as well as one- and multi-point Padé approximations.

In the first chapter of the book, we illustrate and discuss the mathematical models and the methods devoted to study elastic waves propagating in nonlinear and nonhomogeneous materials.

First, a brief literature overview is presented highlighting various effects of the wave evolution including nonlinearity, dispersion and dissipation features. The nonlinearly exhibited by a solid body may include geometric, physical and structural components. Then the state-of-the art of the analytical/numerical approaches is presented with an emphasis to scale effects studied with the help of the asymptotic homogenization method. Namely, it is known that when the wavelength of a travelling wave signal becomes comparable with characteristic size of heterogeneities, reflections and refractions of the wave may generate polarization, dispersion and attenuation effects on both micro and macro level.

In addition benefits given by asymptotic methods and with the homogenization method are outlined.

The second chapter is aimed on the study of waves in layered materials from perspective of a linear analysis. First, 1D dynamic problem with a help of the Bloch-Floquet method is considered. The use of higher-order homogenization method is described. Next, the solution to the exact dispersion equation is compared with that obtained through the Bloch-Floquet method. Two numerical examples are presented.

The similar like methodology is presented in Chapter 3 where waves in fiber composites are investigated in frame of linear statement. First, the 2D dynamical problem is considered. Then the method of higher-order homogenization is employed to solve the problem. The comparison of solutions yielded by the Bloch-Floquet method and by the standard Fourier series approximation is carried out based on numerical investigations. Moreover, the shear waves dispersion in cylindrically structured cancellous viscoelastic bones are modeled and analyzed.

Longitudinal waves in layered composite materials exhibiting both physical and geometric nonlinearity are analyzed in Chapter 4. In beginning, the fundamental relations of nonlinear theory of elasticity are introduced. Then, a solution to the macroscopic wave equation is derived.

Furthermore, analytical solution for stationary waves propagation constant velocity and form is obtained including the case of soft and hard nonlinearity. Numerical examples validating the obtained solutions are provided.

Chapter 5 is devoted to the antiplane shear waves in fiber composite with an account of a structural nonlinearity. Boundary value problem for imperfect bounding conditions is defined first. Then the microscopic wave equation is solved using the method of homogenization. Analytical solution governing stationary wave propagation is derived including soft and hard nonlinearity effects.

Formation and propagation of localized nonlinear waves in composite materials are considered in Chapter 6. The model of longitudinal waves propagating in a layered composite in a direction perpendicular to the layers arrangement is derived and solved using the pseudo-spectral method and the Fourier-Padé approximation. Moreover, the numerical analysis of formation and propagation of non-stationary nonlinear waves is carried out including three cases: (i) generation of a single localized wave; (ii) generation of a train of localized waves; (iii) scattering of an initial impulse of tension.

Localization of vibrations in 1D linear and nonlinear lattices based on the developed discrete and continuous model is analyzed in Chapter 7. First, a monatomic lattice with a perturbed mass is studied based on lumped-mass approximation. Then the same problem is solved based on the continuous approximation of mass distribution. Then, the same methodology is employed to study of a diatomic lattice. The problem of vibration of a lattice on the support with a defect is solved. Moreover, the problem governed by a set of second-order ODEs with the Duffing-type nonlinearity governing vibration of a lattice is solved. The effects of nonlinearity on pass bands and stop bands are outlined.

Chapter 8 deals with spatial localization of linear elastic waves in composite materials with defects. We begin with a simple problem of an elastic composite material consisting of two alternating layers which is solved by the transfer-matrix method. Then, the problem of localization in a layered composite material based on lattice approach is solved. Moreover, antiplane shear waves in a fiber 2D composite are analyzed.

The aim of the Chapter 9 is to study how viscous damping influences mode coupling in nonlinear vibrations of microstructured solids. Natural longitudinal vibrations of a layered composite are considered. The macroscopic dynamic equation is obtained by asymptotic homogenization. The input continuous problem is analyzed using a spatial discretization procedure. An asymptotic solution is developed by the method of multiple time scales and the fourth-order Runge-Kutta method is employed for numerical simulations. Internal resonances and energy transfers between the vibrating modes are predicted and analyzed. The conditions for possible truncation of the original infinite system are discussed. The obtained numerical and analytical results are in good agreement.

Chapter 10 focuses on continuous models derived from a discrete microstructure. Various continualization procedures that approximately take into account the nonlocal interaction between variables of the discrete media are analyzed. Theories of elasticity with couple-stresses are also under consideration. Continualizations of FPU lattice are analyzed. Relations between molecular dynamics simulations and continualization as well as between continualization and discretization are also described.

In Chapter 11 we study various variants of Verhulst-like ODE and ordinary difference equations. Usually Verhulst ODE serves as an example of a deterministic system and discrete logistic equation is a classic example of a simple system with very complicated (chaotic) behavior. We present examples of deterministic discretization and chaotic continualization. Continualization procedure is based on Padé approximants. To correctly characterize the dynamics of obtained ODE we measured such characteristic parameters of chaotic dynamical systems as the Lyapunov exponents and the Lyapunov dimensions.

It is assumed that the reader has knowledge of basic calculus as well as the elementary properties of ODEs and PDEs, strength of materials and theory of elasticity and oscillation theory. Our previous monograph contains a voluminous tutorial, which we do not repeat in this book but recommends for a quick acquaintance with the applied technique.

Further development of the book described in this monograph can go in the following directions:

1. The solution of specific tasks important for practice, for example, the study waves in viscoelastic composites with imperfect bonding between components, wave propagation in layered composites with degraded matrices at locations of imperfect bonding, etc. [30, 31].
2. Calculation of dynamics of composite materials of finite size [21, 24]. Perhaps, it is more efficient to apply spectral or pseudo-spectral methods from the very beginning to study these systems. In this case, methods for overcoming of the Wilbraham-Gibbs phenomenon require further development. In particular, previous results show that the combination of Padé approximants and filter function improve converges of used series better.
3. The use of short-wave homogenization methods [126, 130, 133] seems promising.
4. Most of the real composites possess random microstructures. For analysis of this composites a lot of homogenization techniques can be found in literature such as ensemble averaging [459]. Example calculations demonstrate that "effective modulus" and "effective density" are operators, nonlocal in space and time [313, 460, 461].

Hence, in order to obtain the exact homogenized solution, one needs, generally speaking, an infinite set of correlation functions, which are intended to provide a statistical description of the microgeometry of the composite. Naturally, in many cases such a description is not available. On the other hand, various methods give a possibility to estimate bounds for the effective properties using only a limiting amount of microstructural information. As it were shown, in many cases a regular composite possesses the extreme effective properties among the corresponding random structures [75, 76, 252, 477]. Therefore, a solution for the perfectly regular composite can be considered as bounds on the effective coefficient of random ones, at least for the shaking-geometry random structures [33].

The limitation of the analytical approaches motivated their usage of the finite element model of the unit cell. The numerical homogenization not being bound to certain geometries for the structure of the unit cell [7, 312]. Numerical solutions are important not only on their own, but they also give possibillity to evaluate the applicability of analytical solutions and refine them. At the same time, analytic solutions can be used as benchmark results for numerical ones. They also are impotant for solving of optimization problems and as the first approximation for numerical approaches.

We would like to acknowledge A.L. Kalamkarov, V.I. Malyi, L.I. Manevitch, V.V. Mityushev, A.V. Porubov, G.A. Starushenko, H. Topol, D. Weichert for the helpful and inspiring exchanges of ideas, fruitful collaboration, stimulating discussions, encouraging remarks and thoughtful criticism we have received over the years.

In addition, let us acknowledge that some results presented in this monograph have been supported by (i) the German Research Foundation (Deutsche Forschungsgemeinschaft), grant no. WE736/30-1 (I.V. Andrianov), (ii) Alexander von Humboldt Foundation (grant no.3.4-Fokoop-UKR/1070297) and European Union Horizon 2020 Research and Innovation Programme (Maria Sklodowska-Curie grant no. 655177) (V.V. Danishevskyy), and (iii) Polish Natonal Science Centre under the grant OPUS14 No. 2017/27/B/ST8/01330 (J. Awrejcewicz).

We are aware of the fact that the book may contain controversial statements, too personal or one-sided arguments, inaccuracies and typographical errors. Any of remarks, comments and criticisms regarding the book are appreciated.

Aachen, Germany — *Igor V. Andrianov*
Lodz, Poland — *Jan Awrejcewicz*
Dnipro, Ukraine — *Vladyslav V. Danishevskyy*

Permissions

Figures 4.3–4.8 is reprinted from the paper "Dynamic homogenization and wave propagation in a nonlinear 1D composite material" authored by I. Andrianov, V. Danishevs'kyy, O. Ryzhkov, D. Weichert, Wave Motion, vol. 50, pages 271–281, 2013. Copyright 4913751111215 (2020), with permission from Elsevier.

Figures 5.4–5.12 is reprinted from the paper "Anti-plane shear waves in a fiber-reinforced composite with a non-linear imperfect interface" authored by V. Danishevs'kyy, J. Kaplunov, G. Rogerson, International Journal of Non-linear Mechanics, vol. 76, pages 223–232, 2015. Copyright 4915481459476 (2020), with permission from Elsevier.

Figures 6.1–6.9 is reprinted from the paper "Numerical study of formation of solitary strain waves in a nonlinear elastic layered composite material" authored by I.V. Andrianov, V.V. Danishevs'kyy, O.I. Ryzhkov, D. Weichert, Wave Motion, vol. 51, pages 405–417, 2014. Copyright 4913760036534 (2020), with permission from Elsevier.

Text and figures of Chapter 7 are reprinted from the paper "Vibration localization in one-dimensional linear and nonlinear lattices: discrete and continuum models" authored by I.V. Andrianov, V. Danishevs'kyy, A.L. Kalamkarov, Nonlinear Dynamics, vol. 72 (1-2), pages 37–48, 2013. Copyright 4915400154694 (2020), with permission from Springer Nature.

Text and figures of Chapter 8 are reprinted from the paper "Spatial localization of linear elastic waves in composite materials with defects" authored by I.V. Andrianov, V.V. Danishevs'kyy, Ie.A. Kushnierov, ZAMM, vol. 94, pages 1001–1010, 2014. Copyright 4914180424490 (2020), with permission from John Wiley and Sons.

Text and figures of Chapter 9 are reprinted from the paper "Internal resonances and modes interactions in non-linear vibrations of viscoelastic heterogeneous solids" authored by I.V. Andrianov, V.V. Danishevskyy, G. Rogerson, vol. 433, pages 55–64, 2018, Journal of Sound and Vibration. Copyright 4914141333640 (2020), with permission from Elsevier.

1 Models and Methods to Study Elastic Waves in Nonlinear and Nonhomogeneous Materials

In this chapter problems of wave propagation in heterogeneous materials and structures and existing approaches for their effective solutions are analyzed. The main features which must be taken into account are dispersion, dissipation and nonlinearity. They are caused by polarization, reflections and refraction at the microinhomoneinities, as well as geometric, physical and structural nonlinearities. The existing methods of analysis are mainly based on the homogenization and Bloch-Floquet approaches. General ideas of higher- order homogenization are described based on a simple case study. The papers and books dealing with analytical and numerical solutions are also reviewed. Relations between numerical and analytical solutions are analyzed. Advantages of analytical approaches are outlined. General results yielded by the homogenization theory are presented and discussed.

1.1 BRIEF LITERATURE OVERVIEW

Composite materials are materials consisting of several components with different physical and mechanical characteristics. By the special choice of their volume fractions and geometric shapes, it is possible to create new structural materials with valuable properties like those having high strength and stiffness with low weight, materials with improved heat and electrical conductivity, materials characterized by resistance to aggressive environments, etc.

Consider the characteristic scale l of the internal structure of the composite (the size of the reinforcing particle or grain) is small in comparison to the size of the structure. At the same time, it is assumed that l is much larger than an interatomic distance. In this case, the physical behavior of the composite material can be described using equations of continuum mechanics.

Literature dedicated to the statics of composite materials is thoroughly analyzed in [19]. In this monograph, we focus on the dynamic problems. Generally, in the problems of dynamics, one of the manifestations of the scale effect stands for the dispersion of waves in composites caused by local reflections and refractions of the signal on the nonhomogeneities of the medium. Boutin [89–91], Boutin and

Auriault [92], Bakhvalov and Eglit [53, 54], Fish and Chen [120, 170, 171] using the higher-order homogenization method obtained solutions suitable for long waves analysis ($\eta << 1$). Dispersion in inhomogeneous media can also be described on the basis of the theory of effective stiffness [4, 144, 145, 206, 425] and theory of mixtures [59–61, 93, 203, 391]. Detailed reviews of higher-order microstructural theories are given by Ting [431], Cattani and Rushchitsky [112], Rushchitsky [392] as well as various self-consistent methods are described in [225–228]. Parnell [347,348], Parnell and Abrahams [349–351] investigated the effect of initial stresses on wave propagation in composite materials. Waves scattering and diffraction on single medium nonhomogeneities (cracks, holes, inclusions) were considered in [57].

When the wavelength decreases and becomes commensurate with the size of the internal structure, the pass and stop bands are detected in the composite. If the frequency falls into the stop band, a standing wave arises in the material, the group velocity of which is zero. In the latter case, the signal amplitude at the macro level decays exponentially. Thus, the composite plays the role of a selective wave filter. The results of experimental studies of pass and stop bands for acoustic waves in inhomogeneous media are presented in the Wolf's monograph [462] and in references [284, 320, 427, 433].

A solution for short waves can be obtained using the Bloch representation [82], which is an interpretation of the Floquet theorem [176] for differential equations with periodic coefficients. According to this method, the solution is sought in the form of the product of the harmonic wave $\exp(i\mu x)\exp(i\omega t)$ and some modulating function $F(x)$, and due to the periodicity of the composite structure $F(x)$ also satisfies the periodicity condition: $F(x) = F(x+l)$. For determining $F(x)$ we obtain a system of linear algebraic equations. Equating the determinant of this system to zero, we find the dispersion relations between the frequency ω and wave number μ. The Bloch-Floquet method is described in detail in the books by Brillouin [98], Bedford and Drumheller [60] and in survey papers by Karpov and Stolyarov [232] and Shul'ga [408, 409].

For one-dimensional periodic materials (for example, layered composites), it is often possible to obtain exact dispersion equations [97, 197, 408]. For two-dimensional and three-dimensional periodic media (fibrous and granular composites), approximate dispersion relations can be found by representing $F(x)$ as Fourier expansions [262–266, 411–414], Rayleigh multipole expansion for the wave potential [193, 326, 367, 368, 380, 472] or their modifications [223, 284, 378].

Elastic waves of deformations which propagate in the microinhomogeneous composites are typically associated with various effects of nonlinearity, dispersion and dissipation.

Nonlinearity yields energy localization, generates higher harmonics and implies energy transfer from low to high frequencies spectrum part. Let in the initial time a wave exhibits the sinusoidal form (dashed curve in Fig. 1.1a), where u is displacement, v is phase velocity and x, t stand for the spatial and time co-ordinate, respectively. The sinusoid undergoes deformation in time (solid curve) and velocity of its points increases with increase of distance measured with respect to the axis x.

Curvature of the wave front continuously increases which might lead to its breaking. The mentioned behavior is well studied in hydrodynamics, where the breaking time is usually recognized as a transition from regular to chaotic motion regime. In solid bodies the mentioned phenomenon is not observed because action of nonlinearities are compensated by an action of dispersion and dissipation phenomena.

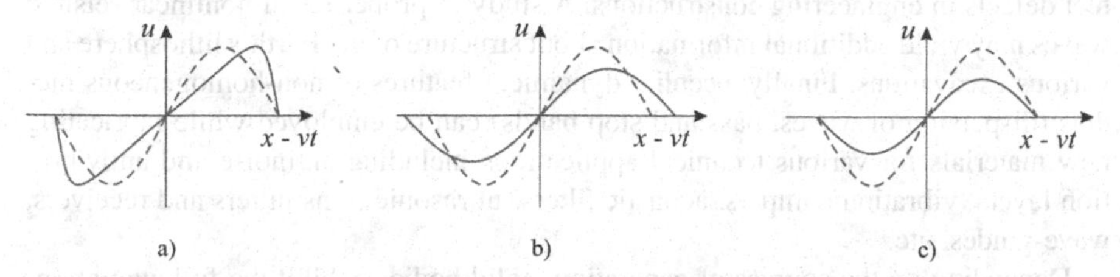

Figure 1.1 Influence of various effects on the wave evolution: (a) - nonlinearity; (b) - dispersion; (c) dissipation

Dispersion implies energy scattering. Curvature of the wave front decreases, and consequently, influence of the nonlinearity is less exhibited (Fig. 1.1b). Scattering is coupled with transition of the kinetic energy of motion into heat energy, causing decrease of the wave amplitude (Fig. 1.1c) and nonlinear effects are less exhibited. It should be noticed that more deeper study of influence of scattering is not considered in frame of this book. We assume that in the case of ideally elastic deformation and with neglection of the internal friction, the scattering effects can be omitted. The latter statement of the problem allows to investigate more rigorously impact of the nonlinearity and dispersion on the studied processes, and reveal their particular role in propagation of the waves of deformations in composites.

If both dynamics features, i.e. nonlinearity and dispersion appear simultaneously, they compensate their mutual influence. In this case one may observe propagation of stationary periodic nonlinear waves. Increase of nonlinearity yields occurrence of the localized waves of bell-type shape (solitons or kinks). The latter waves accumulate large amount of energy and they can propagate on large distances while preserving their forms and velocities. J.S. Russel (1834) (see [456]) is among the pioneers who observed the occurrence of a soliton on the fluid surface. Beginning from that time, the localized nonlinear waves have been detected and measured many times. They include shock waves in air and in fluid (Tsunami), vortices, seismic waves in lithosphere, nerve impulses as well as signals in the optowires. Nonlinear waves of deformation play an important role in mechanics of composite materials [112, 153, 158, 306, 369, 392, 398]. Generation of localized waves is associated with increase of their amplitude. The latter may generate occurrence of plastic zones and development of micro-cracks. It means that nonlinear dynamic effects strongly affect the strength and durability of structures.

It should be emphasized that nonlinear elastic moduli of solid bodies are highly sensitive to the change of the microstructure [469]. Therefore, registration and monitoring of characteristics of nonlinear waves may allow for indirect detection of small changes in the internal material structure that cannot be realized with the traditional methods of control. This opens a possibility of development of new more accurate methods of acoustic diagnostics and health monitoring as well as of diagnostics and control.

The ability of forming and propagation of localized nonlinear waves under initial perturbation with conservation of the form and velocity can be employed to detect defects in engineering constructions. A study of properties of nonlinear seismic waves may yield additional information about structure of the Earth's lithosphere and various excavations. Finally, peculiar dynamical features of non-homogeneous medias (dispersion of waves, pass and stop bands) can be employed while fabricating new materials for various technical applications including antinoise and antivibration layers, vibration dampers, acoustic filters, ultrasonic transmitters and receivers, wave-guides, etc.

Depending on the sources of generation, solid bodies exhibit the following nonlinearities:

1. *Geometric*. It is based on the nonlinear dependencies between deformations and gradients of displacements. Geometric nonlinearity can be governed either by the Cauchy-Green tensor or the Almanzi deformation tensor. In the Cauchy-Green tensor the differentiation is carried out along co-ordinates in the reference configuration (Eulerian description), whereas in the Almanzi tensor it refers to the post-deformation configurations (Lagrangian description). The difference between two mentioned tensors disappears only in the case of linear approximation. It points out for importance of the right configuration estimation while formulating a nonlinear problem.
2. *Physical*. It is based on nonlinear dependencies between stresses and deformations, which are originated from anharmonic atoms interplays. On contrary to the geometric nonlinearity, the mentioned effect cannot be modeled by an exact formula but it is rather based on various hypotheses with regard to deformation of an elastic body. In a general case, physical nonlinearity depends on the way of description of the internal energy. The most popular and recognized in the scientific community are models of weak nonlinearity being based on presentation of the energy deformation density into a truncated power series small deformation or in the deformation tensor. In the literature the mentioned series is referred to the elastic Murnaghan potential.
3. *Structural*. It is caused by non-homogeneity of the internal structure of a given material. It plays a crucial role if the composite consists of components exhibiting high compression properties: voids, cracks, dislocations on the crystals boundaries, etc. Those non-homogeneities yield local concentration of deformations and consequently structural nonlinearity dominates in the dynamical behavior. The material behavior exhibits strong nonlinear properties [469]. In particular, the highly structural nonlinearity characterizes of the rock-forming minerals [101, 390]. In composite materials an effect of structural nonlinearity can be induced by non-ideal contact between a matrix and inclusions.

Nonlinear dynamics properties of composites with pre-stressed states were considered in the works [347, 348, 351].

Dispersion effects can be classified as geometric and structural. The geometric dispersion is mainly featured by finite body size and does not depend on the microstructure. The geometric dispersion is exhibited through the transversal waves occurred in beams and plates (implied by bending stress) as well as through the longitudinal waves in rods (due to reflection of a wave from the size surfaces), etc. The structural dispersion, in contrary to the geometric one, is generated by nonhomogeneity of internal structure of the composite materials. Multiple reflections and refractions of a signal on the boundaries of the components implies dispersion of the wave field. Dispersion causes energy delocalization and annihilates action of nonlinearity.

Nonlinear elastic waves exhibited in solid bodies have been investigated by numerous researchers [152, 306, 369, 398]. However, the main attention was focused on homogeneous materials and constructions, where the effect of dispersion was implied by geometric factors. Complete and detail modeling of a 3D nonlinear medium does not depend to easy tasks. This is why the 3D problems have been reduced to simplified cases of 1D problems [158] or 2D problems [398] that allowed to find analytical solutions qualitatively describing the studied nonlinear phenomena.

Experimental studies of nonlinear deformations in solid bodies were carried out in [141,142,201,398]. In particular, the nonlinear elastic waves of the constant forms were not only experimentally observed but they were well predicted by the earlier obtained theoretical results. Many fundamental results of this research are summarized in the monographs [153, 158, 306, 369, 398].

The governing relations regarding various types of nonlinear waves in composite materials (plane and space, with squared and cubic nonlinearities) are reported in [112, 392].

The so far described methods, from a mathematical point of view, yield occurrence of additional gradient terms appeared in the input wave equations with an account of dispersion effect. It should be noticed that the latter approach contains phenomenological aspect, since the values of coefficients often are not defined. In some cases they can be estimated experimentally (for instance, detailed data regarding properties of rock-forming minerals are given in monograph [101]). However, for majority of the real materials the exact values of the phenomenological parameters remain unknown.

Another way owned on estimate influence of the microstructural effects is based on application of the asymptotic method of homogenization [55, 70, 224]. Let us introduce a perturbation parameter $\eta = l/L$, where L stands for the characteristic size of the problem on macro level (for instance, length of the entire body) and l is the characteristic size of the microstructure (e.g., size of the unit cell). Instead of the input co-ordinate x we use "slow" $x = x$ and "fast" $y = \eta^{-1}x$ variables, whereas the differential operator: $\frac{\partial}{\partial x} = \frac{\partial}{\partial x} + \eta^{-1}\partial/\partial y$. The field of displacements u in a periodically-nonhomogeneous composite material can be approximated by two-scale asymptotic expansion with regard to powers of η in the following form

$$u = u_0(x) + \eta u_1(x,y) + \eta^2 u_2(x,y) + \cdots, \tag{1.1}$$

where the first term u_0 does not depend on fast variables and represents the averaged solution part which "slowly" changes on the microlevel, and subsequent terms $u_i, i = 1, 2, 3 ...$, carry out correctors of the order of η^i and stand for the "fast" oscillatory solutions on the microlevel. Owing to the space periodicity of the investigated material, the terms u_i also satisfy the periodicity condition, i.e. $u_i(x,y) = u_i(x, y+L)$.

Splitting of the input boundary value problem with regard to the powers of the parameter η, one gets a recurrent series of boundary value problem where each of them can be considered in intervals of an isolated periodic cell. Solving the cell problem yields the terms u_i. Next, carrying out integration along the fast co-ordinates, yields the averaged equation governing the wave propagation on the macrolevel. The coefficients of the latter equations are estimated on a basis of data containing physical properties of components and their geometric form. Therefore, the method of homogenization yields explicit information about the internal structure of the composite.

Influence of the non-ideal contact effects among the layers was investigated in [239]. It was shown that the delamination of the layers composite implied decomposition of the nonlinear wave into a sequence of solitons of various amplitude and velocity. The experimental validation of the mentioned effect was reported in [140]. In the case of waves propagated perpendicular to the layers location, the effect of structural dispersion dominates, and it can be analyzed using the method of higher-order homogenization [171, 334].

The differential equations governing propagation of waves in the composite materials are of the Boussinesq type with fourth spatial derivative. The Boussinesq equation has been obtained while modeling nonlinear surface waves [88]. It presents a balance between nonlinearity and dispersion interplay and hence it can be found in all problems associated with those effects. KdV model exhibits the similar properties [246]. Since the latter one usually is presented in the form of the third-order equation, it can be obtained from the Boussinesq equation by carrying out its approximation factorization [258,401,456]. In the second half of the XX century, equations of the Boussinesq and KdV types were widely employed for modeling of dynamic properties of various nonlinear systems including fluid, plasma, discrete chains of particles and solid bodies. The solution in the form of localized nonlinear waves (soliton) were found and studied. Zabusky and Kruskal [468] are among the pioneers who investigated propagation and mutual interaction of the solitons in the nonlinear dispersive media. More information about mathematical models being based on the Boussinesq equations and their application in physics and techniques, as well as about methods of solutions of applied problems are reported in [99].

Zaharov [470] showed that the Boussinesq equations can be integrated by the method of inversed scattering theory. However, finding a solution to the corresponding spectral problem does not belong to easy tasks and exact analytical solutions can be obtained only for certain initial conditions [77, 139]. In the case of plane stationary waves propagating without change of their form, the Boussinesq equation can be reduced to the equation governing dynamics of the anharmonic oscillator, and its solution can be expressed through elliptic functions.

From the practical point of view, investigation of non-stationary dynamical processes play an important role. Analyzing evolution of the initial impulses of various amplitude and form, it is possible to define for which conditions and how fast the localized nonlinear waves may occur in a material. Integration of the Boussinesq equation with arbitrary initial conditions can be always solved in a numerical way. In particular, the pseudo-spectral method has been widely used where integration in time has been carried out by the difference schemes (for instance, Runge-Kutta methods) whereas approximation in space has been realized via expansion into series based on basic functions (for instance, Fourier series) [175, 393]. The mentioned approach is effective while solving dissipative and nonlinear equations having higher-order derivatives [69, 243, 395, 396]. Generation of solitons from perturbation of initial state was considered in [381, 395, 396]. Investigation of evolution of separated impulses was carried out in [243, 372]. It was illustrated that depending on the initial impulse form of deformation and depending on its sign (compression or extension), the different scenarios can be realized: formation of a separated soliton, generation of a few solitons of different amplitude and width, as well as divergence of the initial impulse without creation of a stable wave.

In general, the numerical methods are universal and allow to describe dynamical processes of more general forms than those based on analytical solutions. However, it should be mentioned that in many cases it is difficult to properly interpret the obtained results of the numerical modeling. Moreover, estimation of the area of applicability of the obtained results are impossible without analytical analysis of the studied problem. Therefore, the most effective approach is based on matching numerical and analytical methods. Analytical solutions, found in the special case of stationary waves, can be also used for testing numerical schemes of integration of the governing equations. Besides, the analytical analysis yields important information regarding qualitative properties of solutions like about character of the elastic waves, compression waves, extension waves, on dependence between velocity, width and amplitude of the localized non-linear waves, etc.

1.2 SMALL "TUTORIAL"

Many studies in the theory of composite materials are based on the exploitation of the classical continuum model assuming that the original heterogeneous medium can be simulated by a homogeneous one with certain homogenized (so called effective) mechanical properties. Such approach comes naturally from the hypothesis that the size l of heterogeneities is supposed to be essentially smaller than the macroscopic size L of the whole sample of the material so that in the first approximation one may set $l/L = 0$. However, this limit is never reached for the most of practical problems and in real composite materials microstructural scale effects may result in specific nonlocal phenomena which cannot be predicted in the scope of the homogenized medium theory.

Scale effects can be systematically analyzed by means of the asymptotic homogenization method. According to this approach, physical fields in the composite are represented by multiple scale asymptotic expansions in powers of the small parameter $\eta = l/L$, which characterizes the rate of heterogeneity of the structure. This leads to a decomposition of the solution into macro and micro components which can be evaluated from a recurrent sequence of so called cell boundary value problems. Theoretical foundations of the method were developed in [55,70] and a number of recent applications were presented in the work [224].

Most of the authors restricted to the evaluation only the first-order homogenized coefficients. Higher-order terms in static problems were considered by Gambin and Kröner [181], Boutin [90], and Cherednichenko and Smyshlyaev [121]. They showed that the heterogeneity of the medium results in the induction of an infinite series of displacement fields with successively lower amplitudes. It also leads to nonlocal effects on macro level: instead of homogenized equilibrium equations of continuum mechanics we obtain new equilibrium equations that involve higher-order spatial derivatives and thus represent the influence of the microstructural heterogeneity on the macroscopic material's behaviour. From the quantitative point of view, the degree of this influence is determined by the magnitude of the parameter η.

In dynamic problems the role of the scale effects is even more significant. When the wavelength of a travelling wave signal becomes comparable to the characteristic size of heterogeneities, successive reflections and refractions of the waves at components' interfaces in composite materials lead to such specific phenomena as polarization, dispersion, and attenuation.

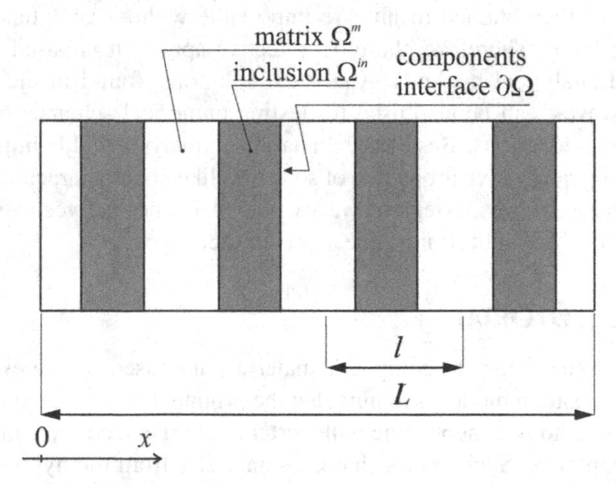

Figure 1.2 Composite structure under consideration.

We consider a uniaxial transverse deformation of a laminated composite material shown at Fig. 1.2. The static equilibrium equation is

$$E^a \frac{\partial^2 u^a}{\partial x^2} = -X, \tag{1.2}$$

where E^a are the elastic moduli of the components, u^a are the displacements in direction x, X is the density of volume forces. Here and in the sequel variables indexed by "*in*" correspond to the inclusions domain Ω^{in}, indexed by "*m*" - to the matrix domain Ω^m, whereas the index "*a*" takes both of these references: "*a*" = "*m*", "*in*".

To avoid problems associated with the behavior of solutions at infinity, we introduce a regularizing parameters λ^a as follows:

$$E^a \frac{\partial^2 u^a}{\partial x^2} = -X + \lambda^a u^a. \tag{1.3}$$

At the components' interface $\partial\Omega$, we assume the perfect bonding conditions implying the equalities of the stresses

$$E^{in} \frac{\partial u^{in}}{\partial x} = E^m \frac{\partial u^m}{\partial x}\bigg|_{\partial\Omega} \tag{1.4}$$

and of the displacements

$$u^{in} = u^m\big|_{\partial\Omega}. \tag{1.5}$$

The presence of two distinct spatial scales gives a possibility to introduce a small parameter

$$\eta = l/L, \tag{1.6}$$

characterizing the rate of heterogeneity of the composite structure. Here the microscopic size l corresponds to the length of a periodically repeatable unit cell, while the macroscopic size L can be associated with the minimal period of Fourier expansion of an external load (in the static case) or with the minimal wavelength of a travelling signal (in the dynamic case). In order to separate macro and micro components of the solution let us introduce so called *slow* x and *fast* y co-ordinate variables

$$x = x, \qquad y = x\eta^{-1} \tag{1.7}$$

and let us search for the displacements in the form of the following expansion

$$u^a = u_0(x) + \eta u_1^a(x,y) + \eta^2 u_2^a(x,y) + \cdots \tag{1.8}$$

The first term u_0 represents the homogeneous part of the solution that changes slowly within the whole sample of the material and does not depend on the fast coordinate y ($\partial u_0/\partial y = 0$). The next terms u_i^a, $i = 1,2,3,...$ describe local variations of the displacement field on the scale of heterogeneities. The micro periodicity of the medium induces the same periodicity for u_i^a with respect to y, i.e. we have

$$u_i^a(x,y) = u_i^a(x,y+L). \tag{1.9}$$

The spatial derivatives takes the following explicit form

$$\frac{\partial}{\partial x} = \frac{\partial}{\partial x} + \eta^{-1}\frac{\partial}{\partial y}, \quad \frac{\partial^2}{\partial x^2} = \frac{\partial^2}{\partial x^2} + 2\eta^{-1}\frac{\partial^2}{\partial x \partial y} + \eta^{-2}\frac{\partial^2}{\partial y^2}. \tag{1.10}$$

Now, we substitute expressions (1.7), (1.8), (1.10) into the input boundary value problem (1.3)–(1.5) and split it with respect to η. Equation (1.3) yields
at η^{-2}:

$$E^a \frac{\partial^2 u_0}{\partial y^2} = 0, \tag{1.11}$$

at η^{-1}:

$$E^a \left(2\frac{\partial^2 u_0}{\partial x \partial y} + \frac{\partial^2 u_1^a}{\partial y^2} \right) = 0, \tag{1.12}$$

at η^0:

$$E^a \left(\frac{\partial^2 u_0}{\partial x^2} + 2\frac{\partial^2 u_1^a}{\partial x \partial y} + \frac{\partial^2 u_2^a}{\partial y^2} \right) = -X + \lambda^a u_0, \tag{1.13}$$

at η^1:

$$E^a \left(\frac{\partial^2 u_1}{\partial x^2} + 2\frac{\partial^2 u_2^a}{\partial x \partial y} + \frac{\partial^2 u_3^a}{\partial y^2} \right) = \lambda^a u_1, \tag{1.14}$$

at η^2:

$$E^a \left(\frac{\partial^2 u_2}{\partial x^2} + 2\frac{\partial^2 u_3^a}{\partial x \partial y} + \frac{\partial^2 u_4^a}{\partial y^2} \right) = \lambda^a u_2, \tag{1.15}$$

and so on.

Equation (1.4) yields
at η^{-1}:

$$E^{in} \frac{\partial u_0}{\partial y} = E^m \frac{\partial u_0}{\partial y} \bigg|_{\partial\Omega}, \tag{1.16}$$

at η^0:

$$E^{in} \left(\frac{\partial u_0}{\partial x} + \frac{\partial u_1^{in}}{\partial y} \right) = E^m \left(\frac{\partial u_0}{\partial x} + \frac{\partial u_1^m}{\partial y} \right) \bigg|_{\partial\Omega}, \tag{1.17}$$

at η^1:

$$E^{in} \left(\frac{\partial u_1^{in}}{\partial x} + \frac{\partial u_2^{in}}{\partial y} \right) = E^m \left(\frac{\partial u_1^m}{\partial x} + \frac{\partial u_2^m}{\partial y} \right) \bigg|_{\partial\Omega}, \tag{1.18}$$

at η^2:

$$E^{in} \left(\frac{\partial u_2^{in}}{\partial x} + \frac{\partial u_3^{in}}{\partial y} \right) = E^m \left(\frac{\partial u_2^m}{\partial x} + \frac{\partial u_3^m}{\partial y} \right) \bigg|_{\partial\Omega}, \tag{1.19}$$

at η^3:

$$E^{in} \left(\frac{\partial u_3^{in}}{\partial x} + \frac{\partial u_4^{in}}{\partial y} \right) = E^m \left(\frac{\partial u_3^m}{\partial x} + \frac{\partial u_4^m}{\partial y} \right) \bigg|_{\partial\Omega}, \tag{1.20}$$

and so on.

Equation (1.5) implies

$$u_i^{in} = u_i^m \big|_{\partial\Omega}. \tag{1.21}$$

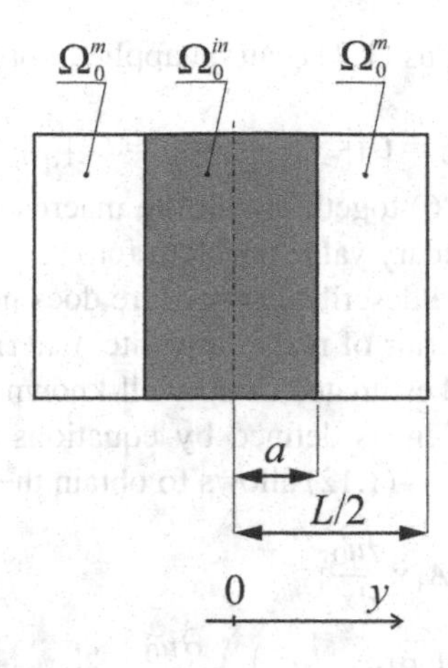

Figure 1.3 Periodically repeatable unit cell.

Equations (1.11) and (1.16) are satisfied trivially since $\partial u_0/\partial y = 0$. All next equations of the system (1.11)–(1.21) define a recurrent sequence of cell boundary value problems; due to the periodicity of u_i^a (1.9) the cell problems can be considered within only one periodically repeatable unit cell of the composite structure (Fig. 1.3). Solution of the cell problems allows to evaluate the terms u_i^a and thus to determine the displacement and the stress fields on micro level. Moreover, application of the homogenizing operator

$$\left(\int_{\Omega_0^{in}} (\cdot)dy + \int_{\Omega_0^m} (\cdot)dy\right) L^{-1} \tag{1.22}$$

over the unit cell domain $\Omega_0 = \Omega_0^{in} + \Omega_0^m$. to equations (1.13), (1.14), (1.15), and so on gives a possibility to derive macroscopic equilibrium equations of different orders. While calculating integrals in the homogenizing operator one should take into account that equations (1.16)–(1.20) imply

$$E^{in} \int\limits_{\Omega_0^{in}} \frac{\partial^2 u_i^{in}}{\partial x \partial y} + \frac{\partial^2 u_{i+1}^{in}}{\partial y^2} dy + E^m \int\limits_{\Omega_0^m} \frac{\partial^2 u_i^m}{\partial x \partial y} + \frac{\partial^2 u_{i+1}^m}{\partial y^2} dy = 0. \tag{1.23}$$

Boundary conditions imposed on u_i^a should be such that the global displacement field (1.8) meets certain macroscopic conditions at $x = 0$ and $x = L$, for example

$$u^m = U_1|_{x=0}, \qquad u^m = U_2|_{x=L}. \tag{1.24}$$

From the other hand, the periodicity equations (1.9) may be replaced by zero boundary conditions at the center and at the outer boundary of the unit cell:

$$u_i^{in} = 0\big|_{y=0}, \qquad u_i^m = 0\big|_{y=\pm\frac{L}{2}}. \tag{1.25}$$

The macroscopic conditions (1.24) can be applied not to the homogeneous component u_0:

$$u_0 = U_1|_{x=0}, \qquad u_0 = U_2|_{x=L}. \tag{1.26}$$

Boundary conditions (1.26) together with the macroscopic equilibrium equation define the macroscopic boundary value problem for u_0.

It should be noted that the described procedure does not predict edge effects appearing nearby outer boundaries of real composite materials. Usually this state can be treated as quasi-static and estimated using well-known methods [19].

The first-order cell problem is defined by equations (1.12), (1.17), (1.22) and (1.25). Integration of equations (1.12) allows to obtain the exact analytical solution:

$$\begin{aligned}
u_1^{in} &= A_1 y \frac{\partial u_0}{\partial x}, \\
u_1^{m} &= \left(B_1 y + \frac{1}{2} C_1 L\right) \frac{\partial u_0}{\partial x} \quad \text{at} \quad y > 0, \\
u_1^{m} &= \left(B_1 y - \frac{1}{2} C_1 L\right) \frac{\partial u_0}{\partial x} \quad \text{at} \quad y < 0,
\end{aligned} \tag{1.27}$$

where $A_1 = -\frac{(1-c)(E^{in}-E^m)}{(1-c)E^{in}+cE^m}$, $B_1 = \frac{c(E^{in}-E^m)}{(1-c)E^{in}+cE^m}$, $C_1 = -B_1$; c is the inclusions' volume fraction, $c = 2a/L$.

Terms containing u_2^a are eliminated using condition (1.23). As the result, we derive the macroscopic equilibrium equation of the order $O(\eta^0)$:

$$\langle E \rangle_0 \frac{\partial^2 u_0}{\partial x^2} = -X + \langle \lambda \rangle u_0, \tag{1.28}$$

where

$$\langle E \rangle_0 = \frac{E^{in} E^m}{(1-c) E^{in} + c E^m}, \qquad \langle \lambda \rangle = (1-c)\lambda^m + c\lambda^{in} \tag{1.29}$$

are the $O(\eta^0)$ order effective modulus.

Expression (1.29) coincides with a well-known formula for the effective elastic modulus of the laminated composite material obtained by simple arithmetical averaging of the components' compliances:

$$\langle E \rangle_0 = \left(\frac{c}{E^{in}} + \frac{1-c}{E^m} \right)^{-1}. \tag{1.30}$$

This is so-called Reuss averaging, and it is known that it gives lower bound for real effective parameter.

Solution of the cell problem (1.13), (1.18), (1.21), (1.25) is:

at $y > 0$:

$$\begin{aligned}
u_2^{in} &= \left[-\frac{1}{2} \left(1 + 2A_1 - \frac{\lambda^{in}}{\langle \lambda \rangle} \frac{\langle E \rangle_0}{E^{in}} \right) y^2 + \frac{1}{4} A_2 L y \right] \frac{\partial^2 u_0}{\partial x^2}, \\
u_2^{m} &= \left[-\frac{1}{2} \left(1 + 2B_1 - \frac{\lambda^{m}}{\langle \lambda \rangle} \frac{\langle E \rangle_0}{E^{m}} \right) y^2 + \frac{1}{4} B_2 L y + \frac{1}{8} C_2 L^2 \right] \frac{\partial^2 u_0}{\partial x^2},
\end{aligned} \tag{1.31}$$

at $y<0$:

$$u_2^{in}=\left[-\frac{1}{2}\left(1+2A_1-\frac{\lambda^{in}}{\langle\lambda\rangle}\frac{\langle E\rangle_0}{E^{in}}\right)y^2-\frac{1}{4}A_2Ly\right]\frac{d^2u_0}{dx^2},$$

$$u_2^{m}=\left[-\frac{1}{2}\left(1+2B_1-\frac{\lambda^{m}}{\langle\lambda\rangle}\frac{\langle E\rangle_0}{E^{m}}\right)y^2-\frac{1}{4}B_2Ly+\frac{1}{8}C_2L^2\right]\frac{\partial^2u_0}{\partial x^2},$$

where

$$A_2=-\frac{c(1-c)E^m\left(E^{in}\lambda^{in}-E^m\lambda^m\right)}{[(1-c)\lambda^m+c\lambda^{in}][(1-c)E^{in}+cE^m]^2}, \tag{1.32}$$

$$B_2=c\frac{(1-c)\left[(1+2c)\lambda^{in}-2c\lambda^m\right](E^{in})^2}{[(1-c)\lambda^m+c\lambda^{in}][(1-c)E^{in}+cE^m]^2}-$$
$$c\frac{\left\{\left[(1-3c+4c^2)\lambda^m-4c^2\lambda^{in}\right]E^{in}E^m+2c\left[(1-c)\lambda^m+c\lambda^{in}\right](E^m)^2\right\}}{[(1-c)\lambda^m+c\lambda^{in}][(1-c)E^{in}+cE^m]^2},$$

$$C_2=-c^2\frac{(1-c)\left(\lambda^{in}-\lambda^m\right)(E^{in})^2}{[(1-c)\lambda^m+c\lambda^{in}][(1-c)E^{in}+cE^m]^2}-$$
$$c^2\frac{\left[(1-2c)\lambda^m+2c\lambda^{in}\right]E^{in}E^m-\left[(1-c)\lambda^m+c\lambda^{in}\right](E^m)^2}{[(1-c)\lambda^m+c\lambda^{in}][(1-c)E^{in}+cE^m]^2}.$$

Substituting expressions (1.31) into equation (1.14) and performing the homogenization we obtain that all integrals in the homogenizing operator vanish. Therefore, in the problem under consideration, the displacement components u_2^a do not contribute to the macroscopic equation.

Solution of the cell problem (1.14), (1.19), (1.21), (1.25) is:

at $y>0$:

$$u_3^{in}=\left\{\frac{1}{6}\left[2+3A_1+(A_1-2)\frac{\lambda^{in}}{\langle\lambda\rangle}\frac{\langle E\rangle_0}{E^{in}}\right]y^3-\frac{1}{4}A_2Ly^2+\frac{1}{24}A_3L^2y\right\}\frac{\partial^3u_0}{\partial x^3},$$

$$u_3^{m}=\left\{\frac{1}{6}\left[2+3B_1+(B_1-2)\frac{\lambda^{m}}{\langle\lambda\rangle}\frac{\langle E\rangle_0}{E^{m}}\right]y^3-\frac{1}{4}\left(B_2+C_1-C_1\frac{\lambda^{m}}{\langle\lambda\rangle}\frac{\langle E\rangle_0}{E^{m}}\right)Ly^2+\right.$$
$$\left.\frac{1}{24}B_3L^2y+\frac{1}{48}C_3L^3\right\}\frac{\partial^3u_0}{\partial x^3}, \tag{1.33}$$

at $y<0$:

$$u_3^{in}=\left\{\frac{1}{6}\left[2+3A_1+(A_1-2)\frac{\lambda^{in}}{\langle\lambda\rangle}\frac{\langle E\rangle_0}{E^{in}}\right]y^3+\frac{1}{4}A_2Ly^2+\frac{1}{24}A_3L^2y\right\}\frac{\partial^3u_0}{\partial x^3},$$

$$u_3^{m}=\left\{\frac{1}{6}\left[2+3B_1+(B_1-2)\frac{\lambda^{m}}{\langle\lambda\rangle}\frac{\langle E\rangle_0}{E^{m}}\right]y^3+\frac{1}{4}\left(B_2+C_1-C_1\frac{\lambda^{m}}{\langle\lambda\rangle}\frac{\langle E\rangle_0}{E^{m}}\right)Ly^2+\right.$$
$$\left.\frac{1}{24}B_3L^2y-\frac{1}{48}C_3L^3\right\}\frac{\partial^3u_0}{\partial x^3},$$

where

$$A_3 = \frac{c(1-c)E^m\left(E^{in}\lambda^{in} - E^m\lambda^m\right)\left[(1-c)E^{in} - cE^m\right]}{\left[(1-c)\lambda^m + c\lambda^{in}\right]\left[(1-c)E^{in} + cE^m\right]^3}, \tag{1.34}$$

$$B_3 = c\frac{(1-c)\left[3c^2\lambda^m + (1-c)(1+3c)\lambda^{in}\right]\left(E^{in}\right)^3}{\left[(1-c)\lambda^m + c\lambda^{in}\right]\left[(1-c)E^{in} + cE^m\right]^3} -$$

$$c\frac{\left[\left(1+c+4c^2-9c^3\right)\lambda^m + c(1-c)(2+9c)\lambda^{in}\right]\left(E^{in}\right)^2E^m}{\left[(1-c)\lambda^m + c\lambda^{in}\right]\left[(1-c)E^{in} + cE^m\right]^3} -$$

$$c^2\frac{\left[\left(1-2c+9c^2\right)\lambda^m - 9c^2\lambda^{in}\right]E^{in}(E^m)^2 + c\left[(1-c)\lambda^m + c\lambda^{in}\right](E^m)^3}{\left[(1-c)\lambda^m + c\lambda^{in}\right]\left[(1-c)E^{in} + cE^m\right]^3},$$

$$C_3 = -c^2\frac{(1-c)\left[c\lambda^m + (1-c)\lambda^{in}\right]\left(E^{in}\right)^3}{\left[(1-c)\lambda^m + c\lambda^{in}\right]\left[(1-c)E^{in} + cE^m\right]^3} -$$

$$c^2\frac{\left[\left(1+c-3c^2\right)\lambda^m + (1-c)(1-3c)\lambda^{in}\right]\left(E^{in}\right)^2E^m}{\left[(1-c)\lambda^m + c\lambda^{in}\right]\left[(1-c)E^{in} + cE^m\right]^3} +$$

$$c^3\frac{\left[(1-c)\lambda^m + c\lambda^{in}\right](E^m)^3}{\left[(1-c)\lambda^m + c\lambda^{in}\right]\left[(1-c)E^{in} + cE^m\right]^3}.$$

Applying the homogenizing operator to equations (1.13), (1.15) and combining the results we come to the macroscopic equilibrium equation of the order $O(\eta^2)$:

$$\langle E\rangle_0\frac{\partial^2 u_0}{\partial x^2} + \eta^2L^2\langle E\rangle_2\frac{\partial^4 u_0}{\partial x^4} = -X + \langle\lambda\rangle u_0, \tag{1.35}$$

where

$$\langle E\rangle_2 = \frac{1}{48}\frac{c^2(1-c)^2E^{in}E^m\left(E^{in}\lambda^{in} - E^m\lambda^m\right)^2}{\left[(1-c)\rho^m + c\rho^{in}\right]^2\left[(1-c)E^{in} + cE^m\right]^3} \tag{1.36}$$

can be treated as the $O(\eta^2)$ order effective modulus.

If we suppose $\lambda^{in} = \lambda^m$, then

$$\langle E\rangle_2 = \frac{1}{48}\frac{c^2(1-c)^2E^{in}E^m\left(E^{in} - E^m\right)^2}{\left[(1-c)E^{in} + cE^m\right]^3}. \tag{1.37}$$

The second term on the left-hand side of equation (1.35) represents the influence of the microstructural scale effect on the macroscopic material's behavior. It can be easily seen that the scale effect disappears in the case of a homogeneous material ($c = 0, c = 1$ or $E^{in} = E^m$). Analysis of the ratio of the second-to-first constitutive terms in equation (1.35) allows to estimate the range of applicability of the homogenization procedure.

Now, let us consider a dynamic problem for the laminated composite material governed by the following PDEs:

$$E^a\frac{\partial^2 u^a}{\partial x^2} = \rho^a\frac{\partial^2 u^a}{\partial t^2}, \tag{1.38}$$

where ρ^a are the mass densities of the components.

Conditions of the perfect bonding between the matrix and inclusions are introduced by equations (1.5) and (1.6).

Following the algorithm presented above we introduce the small parameter η (1.7), *slow* x and *fast* y co-ordinates (1.7), and search the displacements as the asymptotic expansion (1.8). Splitting of equation (1.38) with respect to η yields:

at η^{-2}:

$$E^a \frac{\partial^2 u_0}{\partial y^2} = 0, \tag{1.39}$$

at η^{-1}:

$$E^a \left(2\frac{\partial^2 u_0}{\partial x \partial y} + \frac{\partial^2 u_1^a}{\partial y^2} \right) = 0, \tag{1.40}$$

at η^{0}:

$$E^a \left(\frac{\partial^2 u_0}{\partial x^2} + 2\frac{\partial^2 u_1^a}{\partial x \partial y} + \frac{\partial^2 u_2^a}{\partial y^2} \right) = \rho^a \frac{\partial^2 u_0}{\partial t^2}, \tag{1.41}$$

at η^{1}:

$$E^a \left(\frac{\partial^2 u_1^a}{\partial x^2} + 2\frac{\partial^2 u_2^a}{\partial x \partial y} + \frac{\partial^2 u_3^a}{\partial y^2} \right) = \rho^a \frac{\partial^2 u_1^a}{\partial t^2}, \tag{1.42}$$

at η^{2}:

$$E^a \left(\frac{\partial^2 u_2^a}{\partial x^2} + 2\frac{\partial^2 u_3^a}{\partial x \partial y} + \frac{\partial^2 u_4^a}{\partial y^2} \right) = \rho^a \frac{\partial^2 u_2^a}{\partial t^2} \tag{1.43}$$

and so on.

The set of equations (1.16)–(1.20) as well as equations (1.21)–(1.26) remain the same as in the static case. In the dynamic problem the macroscopic boundary conditions (1.26) should be accompanied by certain initial conditions, for example

$$u_0 = U_3|_{t=0}, \qquad \frac{\partial u_0}{\partial t} = U_4 \bigg|_{t=0}. \tag{1.44}$$

Equations (1.16) and (1.39) are satisfied trivially since $\partial u_0/\partial y = 0$. The first-order cell problem is defined by equations (1.17), (1.21), (1.25), (1.40) and thus coincides with the first-order cell problem of the static case. The solution for u_1^a is given by expressions (1.27). Substituting u_1^a into equation (1.41) and applying the homogenizing operator

$$\left(\int_{\Omega_0^{in}} (\cdot) dy + \int_{\Omega_0^{m}} (\cdot) dy \right) L^{-1}, \tag{1.45}$$

we derive the macroscopic equation of motion of the $O(\eta^0)$ order:

$$\langle E \rangle_0 \frac{\partial^2 u_0}{\partial x^2} = \langle \rho \rangle \frac{\partial^2 u_0}{\partial t^2}, \tag{1.46}$$

where $\langle E\rangle_0$ is the $O(\eta^0)$ order effective modulus determined by formula (1.29), $\langle\rho\rangle$ is the homogenized mass density of the composite material, $\langle\rho\rangle = (1-c)\rho^m + c\rho^{in}$.

Solution of the cell problem (1.18), (1.21), (1.25), (1.41) is:

at $y > 0$:

$$
\begin{aligned}
u_2^{in} &= \left[-\frac{1}{2}\left(1+2A_1-\frac{\rho^{in}}{\langle\rho\rangle}\frac{\langle E\rangle_0}{E^{in}}\right)y^2+\frac{1}{4}A_2Ly\right]\frac{\partial^2 u_0}{\partial x^2},\\
u_2^{m} &= \left[-\frac{1}{2}\left(1+2B_1-\frac{\rho^{m}}{\langle\rho\rangle}\frac{\langle E\rangle_0}{E^{m}}\right)y^2+\frac{1}{4}B_2Ly+\frac{1}{8}C_2L^2\right]\frac{\partial^2 u_0}{\partial x^2},
\end{aligned}
\tag{1.47}
$$

at $y < 0$:

$$
\begin{aligned}
u_2^{in} &= \left[-\frac{1}{2}\left(1+2A_1-\frac{\rho^{in}}{\langle\rho\rangle}\frac{\langle E\rangle_0}{E^{in}}\right)y^2-\frac{1}{4}A_2Ly\right]\frac{d^2 u_0}{dx^2},\\
u_2^{m} &= \left[-\frac{1}{2}\left(1+2B_1-\frac{\rho^{m}}{\langle\rho\rangle}\frac{\langle E\rangle_0}{E^{m}}\right)y^2-\frac{1}{4}B_2Ly+\frac{1}{8}C_2L^2\right]\frac{\partial^2 u_0}{\partial x^2},
\end{aligned}
$$

where

$$
A_2 = -\frac{c\,(1-c)\,E^m\left(E^{in}\rho^{in}-E^m\rho^m\right)}{\left[(1-c)\,\rho^m+c\rho^{in}\right]\left[(1-c)\,E^{in}+cE^m\right]^2},
\tag{1.48}
$$

$$
\begin{aligned}
B_2 &= c\frac{(1-c)\left[(1+2c)\,\rho^{in}-2c\rho^m\right]\left(E^{in}\right)^2}{\left[(1-c)\,\rho^m+c\rho^{in}\right]\left[(1-c)\,E^{in}+cE^m\right]^2}-\\
&\quad c\frac{\left[\left(1-3c+4c^2\right)\rho^m-4c^2\rho^{in}\right]E^{in}E^m+2c\left[(1-c)\,\rho^m+c\rho^{in}\right]\left(E^m\right)^2}{\left[(1-c)\,\rho^m+c\rho^{in}\right]\left[(1-c)\,E^{in}+cE^m\right]^2},
\end{aligned}
$$

$$
\begin{aligned}
C_2 &= -c^2\frac{(1-c)\left(\rho^{in}-\rho^m\right)\left(E^{in}\right)^2}{\left[(1-c)\,\rho^m+c\rho^{in}\right]\left[(1-c)\,E^{in}+cE^m\right]^2}-\\
&\quad c^2\frac{\left[(1-2c)\,\rho^m+2c\rho^{in}\right]E^{in}E^m-\left[(1-c)\,\rho^m+c\rho^{in}\right]\left(E^m\right)^2}{\left[(1-c)\,\rho^m+c\rho^{in}\right]\left[(1-c)\,E^{in}+cE^m\right]^2}.
\end{aligned}
$$

Substituting expressions (1.47) into equation (1.42) and performing the homogenization we obtain that all integrals in the homogenizing operator vanish. Therefore, in the problem under consideration, the displacement components u_2^a do not contribute to the macroscopic equation. However, Boutin and Auriault [92] have shown that such contribution do take place in dynamic problems for anisotropic composite materials, where the $O(\eta^1)$ order macroscopic equation describes the phenomenon of polarization.

Solution of the cell problem (1.19), (1.21), (1.25), (1.42) is:

at $y > 0$:

$$u_3^{in} = \left\{ \frac{1}{6} \left[2 + 3A_1 + (A_1 - 2) \frac{\rho^{in}}{\langle \rho \rangle} \frac{\langle E \rangle_0}{E^{in}} \right] y^3 - \frac{1}{4} A_2 L y^2 + \frac{1}{24} A_3 L^2 y \right\} \frac{\partial^3 u_0}{\partial x^3},$$

$$u_3^{m} = \left\{ \frac{1}{6} \left[2 + 3B_1 + (B_1 - 2) \frac{\rho^{m}}{\langle \rho \rangle} \frac{\langle E \rangle_0}{E^{m}} \right] y^3 - \frac{1}{4} \left(B_2 + C_1 - C_1 \frac{\rho^{m}}{\langle \rho \rangle} \frac{\langle E \rangle_0}{E^{m}} \right) L y^2 + \right.$$

$$\left. \frac{1}{24} B_3 L^2 y + \frac{1}{48} C_3 L^3 \right\} \frac{\partial^3 u_0}{\partial x^3}, \tag{1.49}$$

at $y < 0$:

$$u_3^{in} = \left\{ \frac{1}{6} \left[2 + 3A_1 + (A_1 - 2) \frac{\rho^{in}}{\langle \rho \rangle} \frac{\langle E \rangle_0}{E^{in}} \right] y^3 + \frac{1}{4} A_2 L y^2 + \frac{1}{24} A_3 L^2 y \right\} \frac{\partial^3 u_0}{\partial x^3},$$

$$u_3^{m} = \left\{ \frac{1}{6} \left[2 + 3B_1 + (B_1 - 2) \frac{\rho^{m}}{\langle \rho \rangle} \frac{\langle E \rangle_0}{E^{m}} \right] y^3 + \frac{1}{4} \left(B_2 + C_1 - C_1 \frac{\rho^{m}}{\langle \rho \rangle} \frac{\langle E \rangle_0}{E^{m}} \right) L y^2 + \right.$$

$$\left. \frac{1}{24} B_3 L^2 y - \frac{1}{48} C_3 L^3 \right\} \frac{\partial^3 u_0}{\partial x^3},$$

where

$$A_3 = \frac{c(1-c) E^m \left(E^{in} \rho^{in} - E^m \rho^m \right) \left[(1-c) E^{in} - c E^m \right]}{\left[(1-c) \rho^m + c \rho^{in} \right] \left[(1-c) E^{in} + c E^m \right]^3}, \tag{1.50}$$

$$B_3 = c \frac{(1-c) \left[3c^2 \rho^m + (1-c)(1+3c) \rho^{in} \right] \left(E^{in} \right)^3}{\left[(1-c) \rho^m + c \rho^{in} \right] \left[(1-c) E^{in} + c E^m \right]^3} +$$

$$c \frac{\left[(1 + c + 4c^2 - 9c^3) \rho^m + c(1-c)(2+9c) \rho^{in} \right] \left(E^{in} \right)^2 E^m}{\left[(1-c) \rho^m + c \rho^{in} \right] \left[(1-c) E^{in} + c E^m \right]^3}$$

$$c^2 \frac{\left[(1 - 2c + 9c^2) \rho^m - 9c^2 \rho^{in} \right] E^{in} \left(E^m \right)^2}{\left[(1-c) \rho^m + c \rho^{in} \right] \left[(1-c) E^{in} + c E^m \right]^3}$$

$$c^3 \frac{\left[(1-c) \rho^m + c \rho^{in} \right] \left(E^m \right)^3}{\left[(1-c) \rho^m + c \rho^{in} \right] \left[(1-c) E^{in} + c E^m \right]^3},$$

$$C_3 = -c^2 \frac{(1-c) \left[c \rho^m + (1-c) \rho^{in} \right] \left(E^{in} \right)^3}{\left[(1-c) \rho^m + c \rho^{in} \right] \left[(1-c) E^{in} + c E^m \right]^3} +$$

$$c^2 \frac{\left[(1 + c - 3c^2) \rho^m + (1-c)(1-3c) \rho^{in} \right] \left(E^{in} \right)^2 E^m}{\left[(1-c) \rho^m + c \rho^{in} \right] \left[(1-c) E^{in} + c E^m \right]^3}$$

$$c^2 \frac{\left[(1 - c - 3c^2) \rho^m + 3c^2 \rho^{in} \right] E^{in} \left(E^m \right)^2}{\left[(1-c) \rho^m + c \rho^{in} \right] \left[(1-c) E^{in} + c E^m \right]^3} +$$

$$c^3 \frac{\left[(1-c) \rho^m + c \rho^{in} \right] \left(E^m \right)^3}{\left[(1-c) \rho^m + c \rho^{in} \right] \left[(1-c) E^{in} + c E^m \right]^3}.$$

Applying the homogenizing operator to equations (1.41), (1.43) and combining the results we come to the macroscopic equation of motion of the $O(\eta^2)$ order:

$$\langle E\rangle_0 \frac{\partial^2 u_0}{\partial x^2} + \eta^2 L^2 \langle E\rangle_2 \frac{\partial^4 u_0}{\partial x^4} = \langle \rho\rangle \frac{\partial^2 u_0}{\partial t^2}, \tag{1.51}$$

where

$$\langle E\rangle_2 = \frac{1}{48} \frac{c^2(1-c)^2 E^{in} E^m \left(E^{in}\rho^{in} - E^m\rho^m\right)^2}{[(1-c)\rho^m + c\rho^{in}]^2 [(1-c)E^{in} + cE^m]^3} \tag{1.52}$$

is the $O(\eta^2)$ order effective modulus in the dynamic case.

The second term in equation (1.51) represents the influence of the microstructural scale effect, which leads to the phenomenon of dispersion on macro level. If the impedances of the components are identical ($E^{in}\rho^{in} = E^m\rho^m$) then $\langle E\rangle_2$ vanishes and the macroscopic equation (1.51) turns to the non-dispersive form (1.46). From the physical point of view, this means that there are no reflections of the waves at the components' interfaces.

Assuming $\rho^{in} = \rho^m$ yields (1.37). Expression (1.52) also coincides with those obtained in [120, 170, 171, 399]. Also, as shown in section 2.3, the first terms of the expansion of the solution of the exact dispersion equation (which can be obtained for the problem under consideration) in a series in a small parameter η, coincide with those obtained with the asymptotic homogenization method.

In [399], a solution of equation (1.52) was compared with a numerical solution of the original problem. Gaussian initial disturbance was chosen. The discrepancy between the numerical solution of exact and the homogenized problems becomes sufficiently large only at very large times.

Surprisingly, some mathematicians call the representation of the form (1.51) "criminal algorithm", because it mixed different powers of small parameter in one higher-order homogenized equations [10, 269, 399]. At the same time, it is noted that such equations have a higher degree of approximation [10].

Solution (1.52) shows that in this case the homogenization approach gives an upper bound estimation for the eigenvalues. Since it stands for a constructive estimate, it is interesting to trace what is known about a priori estimations. In the general case, the problem of constructing such estimations can be formulated as follows. For many problems, the eigenvalues are the minima of some quadratic functionals. Moreover, often the initial formulation of the problem exhibits a symmetry, while the variable coefficients of quadratic forms are asymmetric. A study of the behavior of the minima of such functionals with a change in the degree of asymmetry of the coefficient was carried out in [207, 473, 475]. In [475], it was shown that for a wide range of problems, the natural vibration frequencies do not decrease when the density is symmetrized. In [207, 473] this result was generalized to symmetrization of the stiffnesses of the original system.

1.3 ANALYTICAL AND NUMERICAL SOLUTIONS IN THE THEORY OF COMPOSITE MATERIALS

"The speed of computation as a limiting factor in simulating physical systems has largely been replaced by the difficulty of extracting useful information from large data sets".

J. Guckenheimer [191]

We live in the era of Big Data and AI, ANNs and powerful commercial codes. Large datasets are becoming increasingly available and important in science and technology. However, our ability to understand the ongoing processes, as before, is based on simple models [35]. Like 50 years ago, "The purpose of computing is insight, not numbers" [200]. In our opinion, there is no question of "asymptotic or computer (numerical) modeling". Both of these scientific directions are equally important [14, 15].

As shown in [13], without a clear understanding of the asymptotic nature of any model of applied mathematics, it is impossible to correctly determine its place in the hierarchy of describing various aspects of the phenomenon. In addition, the more powerful numerical algorithms become, the more the role of a successful zero approximation increases. All these algorithms require a suitable initial seed values to start the approximation. For essentially nonlinear problems, it is important to know in the vicinity of which we construct the solution. Otherwise, the most advanced numerical algorithm will be unstable, the solution may diverge, or we will get a parasitic solution. Asymptotic solutions often allow finding such suitable initial seed values.

"When we bundle exacting algorithms into libraries and wrap them into packages to facilitate easy use, we create de facto standards that make it easy to ignore numerical analysis. We regard the existing base as static and invest in the development of problem-solving environments and high-level languages. This is needed, but we also need to maintain our investment in continuing research on algorithms themselves" [192]. An increase of the number of program packages developed on the basis of unknown for users' principles results in an increase of chance of omitting the existence of their limitations and principal systematic errors.

From this point of view, asymptotic methods are not only a useful analysis tool, but also provide important and useful benchmark solutions.

A possibility of getting information has significantly increased in recent years. This implies high quality tools for its safe keeping and a transformation in a manner to be understandable by a human being. The last requirement is related to: a) construction of low dimensional models; b) aggregation of high dimensional information; c) extraction of the most important singularities in a system's behaviour (for instance, the bifurcation points), and so on. The most suitable tools to realize the mentioned requirements are related to analytical, and in particular, to asymptotic methods.

Discrete numerical solution refers to applications of the finite elements and difference methods. These methods are powerful and their application is reasonable when the geometry and the physical parameters are fixed. Many experts perceive a pristine computational block (package) as an exact formula: just substitute data and get the result! However, a sackful of numbers is not as useful as an analytical formula. Pure numerical procedures can fail as a rule for the critical parameters and analytical matching with asymptotic solutions can be useful even for the numerical computations [457]. Moreover, numerical packages sometimes are presented as a remedy from all deceases. It is worth noting again that numerical solutions are useful if we are interested in a fixed geometry and fixed set of parameters for engineering purposes. Analytical formula are useful to specialists developing codes for composites, especially for optimal design. We are talking about the creation of highly specialized codes, which enable to solve a narrow class of problems with exceptionally high speed. "It should be emphasized that in problem of design optimization the requirements of accuracy are not very high. A key role plays the ability of the model to predict how the system reacts on the change of the design parameters. This combination of requirements opens a road to renaissance of approximate analytical and semi-analytical models, which in the recent decades have been practically replaced by "universal codes" [169].

Asymptotic approaches allow us to define really important parameters of the system. The important parameters in a boundary value problem are those that, when slightly perturbed, yield large variations in the solutions. In other words, asymptotic methods make it possible to evaluate the sensitivity of the system. There is no need to recall that in real problems the parameters of composites are known with a certain (often not very high) degree of accuracy. This causes the popularity of various kinds of assessments in engineering practice. In addition, the fuzzy object oriented (robust in some sense) model is useful to the engineer. Multiparameter models rarely have this quality. "You can make your model more complex and more faithful to reality, or you can make it simpler and easier to handle. Only the most naive scientist believes that the perfect model is the one that represents reality" [186]. It should also be remembered that for the construction of multiparameter models it is necessary to have very detailed information on the state of the system. Obtaining such information for an engineer is often very difficult for a number of objective reasons, or it requires a lot of time and money. It should be emphasized that the most natural way of constructing sufficiently accurate low-parametric models is based on the use of the asymptotic methods.

1.4 SOME GENERAL RESULTS OF THE HOMOGENIZATION THEORY

For the past years' homogenization methods have proven to be powerful techniques for the study of heterogeneous media. Homogenization approach can be used for local and integral functionals [338]. For analysis of ODEs and PDEs the classical tools include multiple-scale expansions [70] or Bakhvalov's ansatz [55, 345] usually used.

An approach based on Fourier analysis was proposed in [128]. This method works in the following way. First, original operator is transformed into an equivalent operator in the Fourier space. The standard Fourier series is used to expand the coefficients of the operator and a Fourier transform is used to decompose the integrals. Next, the

Fourier transforms of the integrals are expanded using a suitable two-scale expansion, and the homogenized problem is finally derived by neglecting higher-order terms in the above expansions when moving to the limit as the period tends to zero.

Another use of Fourier analysis for homogenization of ODEs and PDEs was proposed in [34]. This variant of homogenization is strongly based on Vishik amd Lyusternik asymptotic results for ODEs and PDEs with rapidly changing coefficients and boundary conditions [451]. We also mention papers on homogenization using nonsmooth transformations [365, 366].

G- and Γ-variational convergences theories [132, 338] are used to identify the asymptotic limit for the integral functionals.

In our book we deal with local operators, i.e. PDEs and ODEs with rapidly oscillating or periodically discontinuous coefficients and we use homogenization method, based on multiple-scale expansions.

Mathematical homogenization method gives possibility to obtain many general results useful for problems considered in our book [48–52, 56, 149, 150].

An important issue is the variational formulation of the problems in the theory of composites. In [48], the following two problems are solved. (1) It is shown that for a certain interpretation of the the Euler equation for a stationary point of a periodic functional in the approximation of any order of accuracy with respect to a small microinhomogeneity parameter, the Euler equation corresponds to a certain "averaged" functional. (2) Let the periodic medium be such that for some processes there exists an energy integral in it. In this case, the homogenized equation also has an energy integral.

An important question is whether the occurrence of dissipation in a composite composed of components without dissipation is possible. For composites of a periodic structure, dissipation does not occur [48]. In random composites viscosity, and hence dissipation, occurs even if it is absent in all components [50].

Homogenization problems for mixtures are mathematically rigorously considered in [51, 56]. The averaged equations that describe the propagation of small long-wave disturbances in a strongly inhomogeneous mixture of weakly viscous liquids and gases are substantiated. Such equations were previously constructed and justified for mixtures of solids. Under some additional restrictions on the determining parameters of the medium, the averaged equations are justified for mixtures of a random structure, and under these conditions the averaged equations turn out to be the same as for the periodic case.

The equations of dynamics of composites of periodic or random structures composed of weakly compressible elastic components were considered in [49]. Under certain restrictions on the determining parameters (the ratio of the linear scale of inhomogeneities to the linear scale of disturbances and the ratio of the Lamé coefficients), the averaged equations are justified. These equations are written explicitly and describe, in particular, the effect of reducing the speed of sound in composites composed of substantially heterogeneous materials.

Let us dwell on the curious "paradox" described in [52]. The following cases are considered. In the first case, homogenization is used for media with voids. In the second case, homogenization procedure is employed for media with soft inclusions, and then in the homogenized expressions the characteristics of inclusions were assumed to be zero. The corresponding homogenized systems do not coincide.

2 Waves in Layered Composites: Linear Problems

One-dimensional propagation of linear waves in nonhomogeneous media is studied. Higher-order homogenization method is employed for spatially infinite domain. Then, the Bloch-Floquet method is used and the dispersion equation is derived analytically. A few of limiting cases are investigated. Two numerical example are given and the results are presented graphically in the form of dispersion curves and the wave attenuation coefficients for the composites of steel-aluminium and CE plastic-steel.

2.1 ONE-DIMENSIONAL (1D) DYNAMIC PROBLEM

Processes of propagation of waves in homogeneous and nonhomogeneous medium are qualitatively different. In composite materials the local reflection and refraction on the boundaries of the components separation implies the waves dispersion on the microlevel. The latter phenomenon can be described with a help of the homogenization method of a higher order, and hence reliable solution of the problem can be obtained for the long wave analysis. If the wave length is decreased and becomes compared with a size of internal structure, the pass and stop bands appear, and hence the composite material plays the role of a wave filter. The mentioned phenomenon can be described with a help of the piecewise homogeneous models, in particular, by employment of the Bloch-Floquet method [82, 176].

We consider the problem of elastic waves propagation in the layered composite consisting of the components $\Omega^{(1)}$ and $\Omega^{(2)}$ (Fig. 2.1). Let a longitudinal wave is propagated in the direction perpendicular to the layers plane and let the stresses depend on co-ordinate x. The governing equations take the following form

$$E^{(a)}\frac{\partial^2 u^{(a)}}{\partial x^2} = \rho^{(a)}\frac{\partial^2 u^{(a)}}{\partial t^2}. \tag{2.1}$$

Ideal contact on the boundaries of the components $\partial\Omega$ is assumed, i.e. we take

$$\left\{u^{(1)} = u^{(2)}\right\}\Big|_{\partial\Omega}, \tag{2.2}$$

$$\left\{E^{(1)}\frac{\partial u^{(1)}}{\partial x} = E^{(2)}\frac{\partial u^{(2)}}{\partial x}\right\}\Bigg|_{\partial\Omega}. \tag{2.3}$$

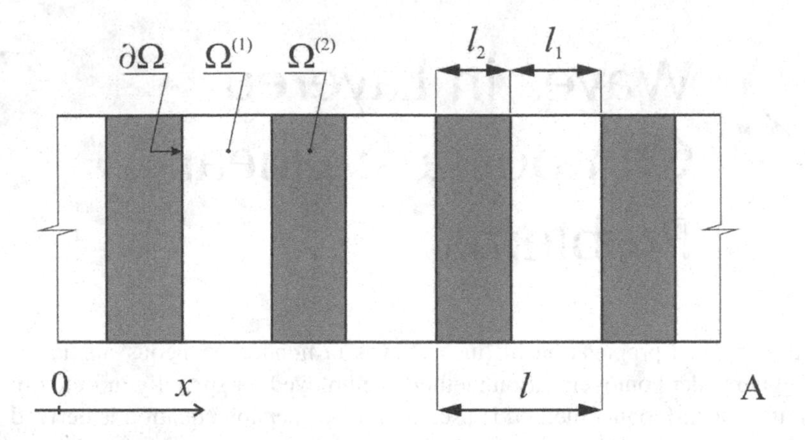

Figure 2.1 Layered composite.

The 1D problem (2.1)–(2.3) allows to find an exact dispersive equation. Therefore, it can be considered as an benchmark example for investigation of the area of applicability of the higher-order homogenization method. The practical advantage of the homogenization approach is displayed in the cases when the exact dispersion equations are not known (2D and 3D problems, nonlinear composites, etc.).

2.2 HIGHER-ORDER HOMOGENIZATION METHOD

We begin with the introduction of a small parameter $\eta = l/L$, where l stands for the periodicity cell size, and L describes the wave length. Let us introduce "slow" x and "fast" $y = \eta^{-1}x$ variables, then

$$\partial/\partial x = \partial/\partial x + \eta^{-1}\partial/\partial y. \tag{2.4}$$

The being searched solution is presented in the form of the following series

$$u^{(a)} = u_0(x) + \eta u_1^{(a)}(x,y) + \eta^2 u_2^{(a)}(x,y) + \cdots, \tag{2.5}$$

where $u_i^{(a)}(x,y) = u_i^{(a)}(x,y+L)$, $i = 1,2,3,\ldots$.

Assuming that the components properties have the same order: $E^{(2)}/E^{(1)} = O(\eta^0)$, $\rho^{(2)}/\rho^{(1)} = O(\eta^0)$, we carry out the splitting of the boundary value problem (2.1)–(2.3) with regard to η. In result, one gets a sequence of the boundary value problems on a cell including the microscopic equations of motion:

$$E^{(a)}\left(\frac{\partial^2 u_{i-2}^{(a)}}{\partial x^2} + 2\frac{\partial^2 u_{i-1}^{(a)}}{\partial x \partial y} + \frac{\partial^2 u_i^{(a)}}{\partial y^2}\right) = \rho^{(a)}\frac{\partial^2 u_{i-2}^{(a)}}{\partial t^2}, \tag{2.6}$$

where $u_{-1}^{(a)} = 0$, as well as the microscopic conditions of an ideal contact

$$\left\{u_i^{(1)} = u_i^{(2)}\right\}\Big|_{\partial\Omega}, \tag{2.7}$$

$$\left\{ E^{(1)} \left(\frac{\partial u_{i-1}^{(1)}}{\partial x} + \frac{\partial u_i^{(1)}}{\partial y} \right) = E^{(2)} \left(\frac{\partial u_{i-1}^{(2)}}{\partial x} + \frac{\partial u_i^{(2)}}{\partial y} \right) \right\} \Bigg|_{\partial\Omega} . \tag{2.8}$$

In the case of the space-infinite composite, the periodicity conditions for $u_i^{(a)}$ can be replaced by zeroth boundary conditions in the cell center and on the cell boundary (Fig. 2.2)

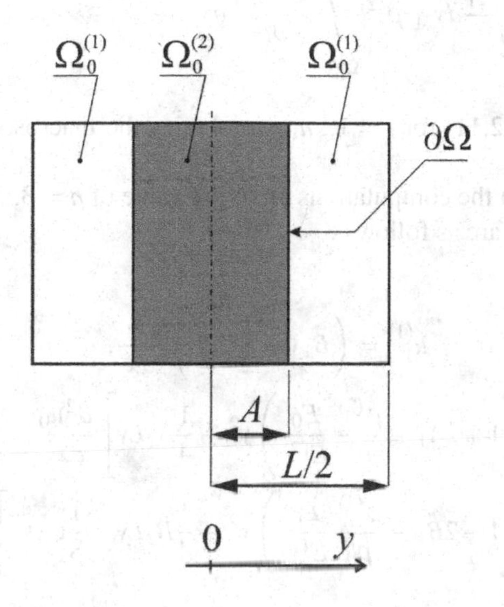

Figure 2.2 Cell of periodicity (with respect to co-ordinate y).

$$\left\{ u_i^{(2)} = 0 \right\} \Big|_{y=0}, \qquad \left\{ u_i^{(1)} = 0 \right\} \Big|_{y=\pm\frac{L}{2}} . \tag{2.9}$$

Algorithm of the homogenization procedure is as follows. A solution to i-th boundary value problems (2.6)–(2.9) allows to determine the term $u_i^{(a)}$. Knowing $u_i^{(a)}$, we employ to the $(i+1)$-th equation (2.6) the averaging operator $l^{-1} \int_{\Omega_0} (\cdot) dy$ along the cell area $\Omega_0 = \Omega_0^{(1)} + \Omega_0^{(2)}$.

The terms $u_{i+1}^{(a)}$ are cancelled based on boundary condition (2.8), which with a help of periodicity $u_{i+1}^{(a)}$ in y gives

$$E^{(1)} \int\limits_{\Omega_0^{(1)}} \frac{\partial^2 u_i^{(1)}}{\partial x \partial y} + \frac{\partial^2 u_{i+1}^{(1)}}{\partial y^2} dy + E^{(2)} \int\limits_{\Omega_0^{(2)}} \frac{\partial^2 u_i^{(2)}}{\partial x \partial y} + \frac{\partial^2 u_{i+1}^{(2)}}{\partial y^2} dy = 0. \tag{2.10}$$

In result, the following homogenized equation of order η^{i-1} is obtained

$$
\begin{gathered}
E^{(1)} \int\limits_{\Omega_0^{(1)}} \frac{\partial^2 u_{i-1}^{(1)}}{\partial x^2} + \frac{\partial^2 u_i^{(1)}}{\partial x \partial y} dy + E^{(2)} \int\limits_{\Omega_0^{(2)}} \frac{\partial^2 u_{i-1}^{(2)}}{\partial x^2} + \frac{\partial^2 u_i^{(2)}}{\partial x \partial y} dy = \\
\rho^{(1)} \int\limits_{\Omega_0^{(1)}} \frac{\partial^2 u_{i-1}^{(1)}}{\partial t^2} dy + \rho^{(2)} \int\limits_{\Omega_0^{(2)}} \frac{\partial^2 u_{i-1}^{(2)}}{\partial t^2} dy.
\end{gathered}
\tag{2.11}
$$

Unite equations (2.11) for $i = 1...n$, one obtains the macroscopic wave equation of order η^n.

Here we carry out the computations up to the value of $n = 3$. The exact analytical solutions on the cell are as follows

for $y > 0$

$$
\begin{aligned}
u_1^{(2)} &= A_1 y \frac{\partial u_0}{\partial x}, \qquad u_1^{(1)} = \left(B_1 y + \frac{1}{2} C_1 L \right) \frac{\partial u_0}{\partial x}, \\
u_2^{(2)} &= \left[-\frac{1}{2} \left(1 + 2A_1 - \frac{\rho^{(2)}}{\rho_0} \frac{E_0}{E^{(2)}} \right) y^2 + \frac{1}{4} A_2 L y \right] \frac{\partial^2 u_0}{\partial x^2}, \\
u_2^{(1)} &= \left[-\frac{1}{2} \left(1 + 2B_1 - \frac{\rho^{(1)}}{\rho_0} \frac{E_0}{E^{(1)}} \right) y^2 + \frac{1}{4} B_2 L y + \frac{1}{8} C_2 L^2 \right] \frac{\partial^2 u_0}{\partial x^2},
\end{aligned}
\tag{2.12}
$$

$$
\begin{aligned}
u_3^{(2)} &= \left\{ \frac{1}{6} \left[2 + 3A_1 + (A_1 - 2) \frac{\rho^{(2)}}{\rho_0} \frac{E_0}{E^{(2)}} \right] y^3 - \frac{1}{4} A_2 L y^2 + \frac{1}{24} A_3 L^2 y \right\} \frac{\partial^3 u_0}{\partial x^3}, \\
u_3^{(1)} &= \left\{ \frac{1}{6} \left[2 + 3B_1 + (B_1 - 2) \frac{\rho^{(1)}}{\rho_0} \frac{E_0}{E^{(1)}} \right] y^3 - \right. \\
&\left. \frac{1}{4} \left(B_2 + C_1 - C_1 \frac{\rho^{(1)}}{\rho_0} \frac{E_0}{E^{(1)}} \right) L y^2 + \frac{1}{24} B_3 L^2 y + \frac{1}{48} C_3 L^3 \right\} \frac{\partial^3 u_0}{\partial x^3},
\end{aligned}
$$

for $y < 0$

$$
\begin{aligned}
u_1^{(2)} &= A_1 y \frac{\partial u_0}{\partial x}, \qquad u_1^{(1)} = \left(B_1 y - \frac{1}{2} C_1 L \right) \frac{\partial u_0}{\partial x}, \\
u_2^{(2)} &= \left[-\frac{1}{2} \left(1 + 2A_1 - \frac{\rho^{(2)}}{\rho_0} \frac{E_0}{E^{(2)}} \right) y^2 - \frac{1}{4} A_2 L y \right] \frac{d^2 u_0}{dx^2}, \\
u_2^{(1)} &= \left[-\frac{1}{2} \left(1 + 2B_1 - \frac{\rho^{(1)}}{\rho_0} \frac{E_0}{E^{(1)}} \right) y^2 - \frac{1}{4} B_2 L y + \frac{1}{8} C_2 L^2 \right] \frac{\partial^2 u_0}{\partial x^2},
\end{aligned}
\tag{2.13}
$$

$$u_3^{(2)} = \left\{ \frac{1}{6} \left[2 + 3A_1 + (A_1 - 2) \frac{\rho^{(2)}}{\rho_0} \frac{E_0}{E^{(2)}} \right] y^3 + \frac{1}{4} A_2 L y^2 + \frac{1}{24} A_3 L^2 y \right\} \frac{\partial^3 u_0}{\partial x^3},$$

$$u_3^{(1)} = \left\{ \frac{1}{6} \left[2 + 3B_1 + (B_1 - 2) \frac{\rho^{(1)}}{\rho_0} \frac{E_0}{E^{(1)}} \right] y^3 + \right.$$

$$\left. \frac{1}{4} \left(B_2 + C_1 - C_1 \frac{\rho^{(1)}}{\rho_0} \frac{E_0}{E^{(1)}} \right) L y^2 + \frac{1}{24} B_3 L^2 y - \frac{1}{48} C_3 L^3 \right\} \frac{\partial^3 u_0}{\partial x^3}.$$

The coefficients A_i, B_i, C_i follow

$$A_1 = -\left(1 - c^{(2)}\right)\left(E^{(2)} - E^{(1)}\right)/D, \qquad B_1 = c^{(2)}\left(E^{(2)} - E^{(1)}\right)/D,$$

$$C_1 = -B_1, \qquad D = \left(1 - c^{(2)}\right) E^{(2)} + c^{(2)} E^{(1)},$$

$$A_2 = -\left[\left(1 - c^{(2)}\right)\left\{\left[\left(1 - c^{(2)}\right)\rho^{(1)} + 2c^{(2)}\rho^{(2)}\right] E_0 + \rho_0 E^{(1)}\right\} E^{(2)} + \right.$$

$$\left.\left(c^{(2)}\right)^2 \rho^{(2)} E^{(1)} E_0\right] / \left(\rho_0 E^{(2)} D\right),$$

$$B_2 = \left(c^{(2)} E^{(1)} \left[c^{(2)}\left(\rho^{(2)} - 2\rho^{(1)}\right) E_0 - 2\rho_0 E^{(1)}\right] + \right.$$

$$\left. E^{(2)} \left\{\left(1 + 2c^{(2)}\right)\rho_0 E^{(1)} - \left[1 - \left(c^{(2)}\right)^2\right]\rho^{(1)} E_0\right\}\right) / \left(\rho_0 E^{(1)} D\right),$$

$$C_2 = c^{(2)} \left[\rho_0 \left(E^{(1)}\right)^2 + \left(1 - c^{(2)}\right)\rho^{(1)} E^{(2)} E_0 - E^{(1)} \left\{\rho_0 E^{(2)} + \right.\right.$$

$$\left.\left.\left[c^{(2)}\rho^{(2)} + \rho^{(1)}\left(1 - 2c^{(2)}\right)\right] E_0\right\} / \left(\rho_0 E^{(1)} D\right),$$

$$A_3 = -\left\{\left(c^{(2)}\right)^3 \rho^{(2)} \left(E^{(1)}\right)^2 E_0 + \left(1 - c^{(2)}\right)\left(E^{(2)}\right)^2 \left[\rho^{(1)} E_0 \left(1 - c^{(2)}\right)^2 - \right.\right.$$

$$\left. \rho_0 E^{(1)}\right] + \left\{3E_0\left(1 - c^{(2)}\right)\left[c^{(2)}\rho^{(2)} + \left(1 - c^{(2)}\right)\rho^{(1)}\right] - \rho_0 E^{(1)}\right\} \times$$

$$\left. c^{(2)} E^{(1)} E^{(2)}\right\} / \left(\rho_0 E^{(2)} D^2\right),$$

$$B_3 = \left(\left\{\left(c^{(2)}\right)^2 E_0 \left[5c^{(2)}\rho^{(2)} + 3\rho^{(1)}\left(1 - 4c^{(2)}\right)\right] - 2c\rho_0 E^{(2)}\left(1 - 3c^{(2)}\right)\right\} \times \right.$$

$$\left(E^{(1)}\right)^2 + E^{(1)} E^{(2)} \left\{\rho_0 E^{(2)}\left(1 - c^{(2)}\right)\left(1 + 3c^{(2)}\right) + 3\left(c^{(2)}\right)^2 E_0 \times \right.$$

$$\left.\left[\rho^{(2)}\left(1 - c^{(2)}\right) - \rho^{(1)}\left(4 - 5c^{(2)}\right)\right]\right\} - 3\left(c^{(2)}\right)^2 \rho_0 \left(E^{(1)}\right)^3 +$$

$$\left.\rho^{(1)}\left(E^{(2)}\right)^2 E_0 \left(1 - c^{(2)}\right)\left[1 + c^{(2)} - 5\left(c^{(2)}\right)^2\right]\right) / \left(\rho_0 E^{(1)} D^2\right),$$

(2.14)

$$C_3 = c^{(2)} \left[\left(E^{(1)}\right)^2 \left\{ \rho_0 E^{(2)} \left(1 - 2c^{(2)}\right) - c^{(2)} E_0 \left[2c^{(2)} \rho^{(2)} + 3\rho^{(1)} \left(1 - 2c^{(2)}\right) \right] \right\} - \right.$$
$$E^{(1)} E^{(2)} \left\{ \rho_0 E^{(2)} \left(1 - c^{(2)}\right) + \rho^{(1)} E_0 \left[1 - 6c^{(2)} + 6\left(c^{(2)}\right)^2 \right] \right\} +$$
$$\left. c^{(2)} \rho_0 \left(E^{(1)}\right)^3 + \rho^{(1)} \left(E^{(2)}\right)^2 E_0 \left(1 - c^{(2)}\right) \left(1 - 2c^{(2)}\right) \right] / \left(\rho_0 E^{(1)} D^2\right).$$

In the reported formulas

$$E_0 = \left[c^{(2)}/E^{(2)} + \left(1 - c^{(2)}\right)/E^{(1)} \right]^{-1} \tag{2.15}$$

stands for the effective elasticity modulus of the layered composite in the quasi-homogeneous cases (for $\eta = 0$),

$$\rho_0 = \left(1 - c^{(2)}\right) \rho^{(1)} + c^{(2)} \rho^{(2)} \tag{2.16}$$

is the averaged density, and $c^{(2)} = l_2/l = 2A/L$ is the volume fraction of the component $\Omega^{(2)}$, $A = l_2/(2\eta)$.

Macroscopic wave equation of order η^2 takes the following form

$$E_0 \frac{\partial^2 u_0}{\partial x^2} + \eta^2 L^2 E_2 \frac{\partial^4 u_0}{\partial x^4} + O(\eta^4) = \rho_0 \frac{\partial^2 u_0}{\partial t^2}, \tag{2.17}$$

where the coefficient

$$E_2 = \frac{1}{12} \left(c^{(2)}\right)^2 \left(1 - c^{(2)}\right)^2 \frac{E_0^3 \left(E^{(2)} \rho^{(2)} - E^{(1)} \rho^{(1)}\right)^2}{\rho_0^2 \left(E^{(1)} E^{(2)}\right)^2} \tag{2.18}$$

can be interpreted as the effective elastic modulus of order η^2.

The second term on the left hand side of equation (2.17) exhibits the dispersion effect of the wave scattering on the microinhomogeneties of the composite. The coefficient E_2 is equal to zero and the dispersion vanishes in the case of (a) homogeneous material ($c^{(2)} = 0$, $c^{(2)} = 1$) and (b) under equality of the acoustic impedance of the components ($E^{(2)} \rho^{(2)} = E^{(1)} \rho^{(1)}$), where local reflections on the boundary of separation $\partial\Omega$ are absent.

Consider a harmonic wave

$$u_0 = U \exp(i\mu x) \exp(i\omega t) \tag{2.19}$$

with the amplitude U, frequency ω and the wave number $\mu = 2\pi/L$.

Substituting (2.19) into (2.17), the following dispersion relation is obtained

$$\omega^2 = \omega_0^2 \left[1 - 4\pi^2 \frac{E_2}{E_0} \eta^2 + O(\eta^4) \right], \tag{2.20}$$

where $\eta = l/L = \mu l/(2\pi)$, $\omega_0 = \mu v_0$ is frequency and $v_0 = \sqrt{E_0/\rho_0}$ stands for velocity in the quasi-homogeneous case.

Phase v_p and group v_g velocities are governed by the following equations, respectively

$$v_p^2 = \left(\frac{\omega}{\mu}\right)^2 = v_0^2\left[1 - 4\pi^2\eta^2\frac{E_2}{E_0} + O(\eta^4)\right], \tag{2.21}$$

$$v_g^2 = \left(\frac{d\omega}{d\mu}\right)^2 = v_0^2\frac{(1 - 8\pi^2\eta^2E_2/E_0)^2}{1 - 4\pi^2\eta^2E_2/E_0} + O(\eta^4). \tag{2.22}$$

Asymptotic solution (2.20)–(2.22) describes the long-wave approximation. Decreasing the wave length L (and increasing the frequency ω) implies decrease of the group velocity v_g. The condition $v_g = 0$ defines a bound of the first stop band. The upper frequency limit ω_{max} can be found based on equations (2.20), (2.22)

$$\omega_{max}^2 = \frac{\langle E\rangle_0^2}{4l^2\langle E\rangle_2\langle\rho\rangle} + O(\eta^4). \tag{2.23}$$

It should be noticed that the given homogenization procedure is developed for the spatially infinite domain. It allowed for defining the dispersive relations as the self-characteristic of the composite material independently on the kind of the boundary conditions on the macrolevel. If one considers composite structure of the finite size then the homogenized solution must be supplemented by the macroscopic boundary conditions. The formal transition of the results obtained for the infinite domain into the case of a composite of finite size can produce errors in the boundary conditions on the macro level.

2.3 THE BLOCH-FLOQUET METHOD AND EXACT DISPERSION EQUATION

A solution suitable for short waves can be found using the Bloch-Floquet approach [82, 176]. Owing to the given method, let us write the ansatz governing propagation of the harmonic wave in the piece-wise medium

$$u^{(a)} = F^{(a)}(x)\exp(i\mu x)\exp(i\omega t), \tag{2.24}$$

where $F^{(a)}(x)$ is periodic function taking into account of the material microstructure, $F^{(a)}(x) = F^{(a)}(x+l)$.

Periodicity of $F^{(a)}(x)$ implies

$$u^{(a)}(x+l) = u^{(a)}(x)\exp(i\mu l). \tag{2.25}$$

Substituting ansatz (2.24) into the wave equation (2.1) yields

$$F^{(a)}(x) = F_1^{(a)}\exp(i\left[\mu^{(a)} - \mu\right]x) + F_2^{(a)}\exp(-i\left[\mu^{(a)} + \mu\right]x), \tag{2.26}$$

where: $\mu^{(a)} = \omega\sqrt{\rho^{(a)}/E^{(a)}}$ stands for the wave complex number.

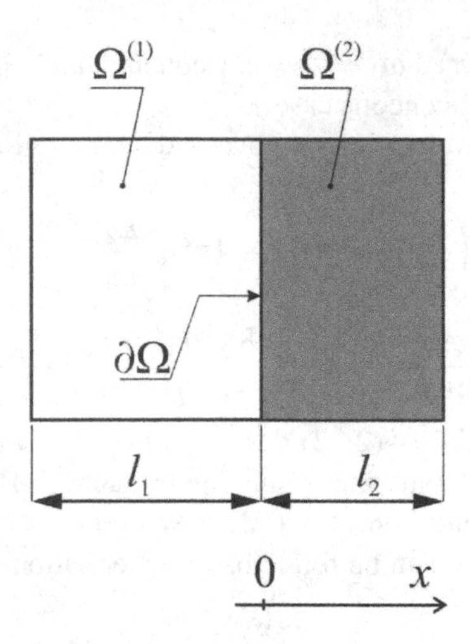

Figure 2.3 Periodicity cell (with respect to co-ordinate x).

In order to define the coefficients $F_1^{(a)}$, $F_2^{(a)}$, we need to employ the conditions (2.2), (2.3). Let us consider the cell of periodicity shown in Fig. 2.3. On the boundary $x = 0$ we have

$$\left\{u^{(1)} = u^{(2)}\right\}\Big|_{x=0}, \qquad \left\{E^{(1)}\frac{\partial u^{(1)}}{\partial x} = E^{(2)}\frac{\partial u^{(2)}}{\partial x}\right\}\Bigg|_{x=0}. \tag{2.27}$$

Displacements and stresses on the boundaries $x = -l_1$ can be linked with a help of equation (2.25) in the following way

$$\begin{aligned} u^{(2)}\Big|_{x=l_2} &= u^{(1)}\Big|_{x=-l_1} \exp(i\mu l), \\ E^{(2)}\frac{\partial u^{(2)}}{\partial x}\Bigg|_{x=l_2} &= E^{(1)}\frac{\partial u^{(1)}}{\partial x}\Bigg|_{x=-l_1} \exp(i\mu l). \end{aligned} \tag{2.28}$$

Substituting ansatz (2.24), (2.26) into boundary conditions (2.27), (2.28), the following four linear algebraic equations are obtained

$$\begin{aligned} &F_1^{(1)} + F_2^{(1)} - F_1^{(2)} - F_2^{(2)} = 0, \\ &F_1^{(1)} z_1 - F_2^{(1)} z_1 - F_1^{(2)} z_2 + F_2^{(2)} z_2 = 0, \\ &F_1^{(1)} \exp\left[-i(\mu^{(1)} - \mu)l_1\right] + F_2^{(1)} \exp\left[i(\mu^{(1)} + \mu)l_1\right] - \\ &F_1^{(2)} \exp\left[i(\mu^{(2)} - \mu)l_2\right] - F_2^{(2)} \exp\left[-i(\mu^{(2)} + \mu)l_2\right] = 0, \\ &F_1^{(1)} z_1 \exp\left[-i(\mu^{(1)} - \mu)l_1\right] - F_2^{(1)} z_1 \exp\left[i(\mu^{(1)} + \mu)l_1\right] - \\ &F_1^{(2)} z_2 \exp\left[i(\mu^{(2)} - \mu)l_2\right] + F_2^{(2)} z_2 \exp\left[-i(\mu^{(2)} + \mu)l_2\right] = 0, \end{aligned} \tag{2.29}$$

where $z^{(a)} = \sqrt{E^{(a)}\rho^{(a)}}$ stands for the components impedance.

System (2.29) has non-trivial solution only if the determinant of the matrix composed of its coefficients is equal to zero. The given condition allows to write the dispersion equation in the explicit analytical form

$$\cos(\mu l) = \cos(\mu^{(1)}l_1)\cos(\mu^{(2)}l_2) - \frac{1}{2}\left(\frac{z^{(1)}}{z^{(2)}} + \frac{z^{(2)}}{z^{(1)}}\right)\sin(\mu^{(1)}l_1)\sin(\mu^{(2)}l_2). \tag{2.30}$$

Exact equation (2.30) allows for investigation of numerous simplified cases. The limiting transitions $l_2 \to 0, \rho^{(2)}l_2 \to m, \rho^{(1)} \to 0, E^{(1)}/l_1 \to C$ yields the known dispersive relations for the oscillating chain of masses m, coupled by springs of stiffness C

$$\cos(\mu l) = 1 - \frac{m\omega^2}{2C}. \tag{2.31}$$

Employing $l_2 \to 0, \rho^{(2)}l_2 \to m, \rho^{(1)} = \rho, E^{(1)} = E$, yields the dispersive equation of the rod with the attached masses longitudinal vibrations

$$\cos(\mu l) = \cos\left(\omega l\sqrt{\frac{\rho}{E}}\right) - \frac{m\omega}{2\sqrt{E\rho}}\sin\left(\omega l\sqrt{\frac{\rho}{E}}\right). \tag{2.32}$$

In the case of long waves one may consider the frequency ω in the η series, where $\eta = l/L = \mu l/(2\pi) \to 0$. The coefficients of the latter series are computed from equation (2.30), and first two of them coincide with the first two coefficients of the asymptotic series (2.20), obtained with the help of homogenization method.

Let us consider the material to be made up of layers of equal travel time. Such a material is known as the Goupillaud medium in geophysics [189]. In this case

$$T^{(1)} = T^{(2)} = T, \tag{2.33}$$

where $T^{(a)} = \sqrt{\rho^{(a)}/E^{(a)}}\,l_a$.

Condition (2.33) provided that (2.30) is solved exactly

$$\omega = \pm\frac{1}{T}\arccos\frac{\sqrt{2z^{(1)}z^{(2)}\cos(\mu l) + \left(z^{(1)}\right)^2 + \left(z^{(2)}\right)^2}}{z^{(1)} + z^{(2)}}. \tag{2.34}$$

In the case when the components of the composite have the high contrast parameters, $z^{(2)} << z^{(1)}$, one can introduce a small parameter $\varepsilon_1^2 = z^{(2)}/z^{(1)} << 1$ and recast equation (2.30) in the following form

$$\varepsilon_1^2\cos rd = \varepsilon_1^2\cos\Omega\cos(\Omega a) - \frac{1}{2}\left(1 + \varepsilon_1^4\right)\sin\Omega\sin(\Omega a), \tag{2.35}$$

where $\Omega = \mu^{(1)}l_1$, $a = (\mu^{(2)}/\mu^{(1)})(l_2/l_1)$.

Possible simplifications of this equation depend on the value of a.

If $a \sim 1$, i.e. $l_2/l_1 \sim \varepsilon_1^2$ (the lengths of the components of the composite differ significantly), then the solution of equation (2.35) can be sought in the form of an expansion

$$\Omega = \varepsilon_1\Omega_0 + \varepsilon_1^2\Omega_1 + \cdots \qquad (2.36)$$

Substituting the expansion (2.36) into equation (2.35), one obtains in a first approximation

$$\cos rd = 1 - \frac{1}{2\varepsilon_1^2}\sin(\varepsilon_1\Omega_0)\sin(\varepsilon_1\Omega_0 a). \qquad (2.37)$$

By expanding the right-hand side of expression (2.37) into a series with preserving terms of the second and fourth orders, it is possible to approximate the original system by lattice with alternating masses with fairly high accuracy [326] (lattice approximation [109])

$$\Omega_0^4 - \frac{6}{\varepsilon_1^2(1+a^2)}\Omega_0^2 + \frac{24}{\varepsilon_1^2 a(1+a^2)}\sin^2\frac{rd}{2} = 0. \qquad (2.38)$$

Equation (2.24) can be splitted with respect to real μ_R and imaginary μ_I part of the wave number $\mu = \mu_R + i\mu_I$:

$$u^{(a)} = F^{(a)}(x)\exp(-\mu_I x)\exp(i\mu_R x)\exp(i\omega t). \qquad (2.39)$$

The case of $\mu_I = 0$ corresponds to pass band, whereas $\mu_I > 0$ corresponds to stop band. The parameter μ_I is wave attenuation parameter. As it follows from equation (2.30), boundaries of stop band and pass band are defined by the conditions $\cos(\mu l) = \pm 1$. The corresponding waves length $L = 2l/n$, $n = 1,2,3,\ldots$

General form of solutions are shown in Figs 2.4, 2.5, where stop bands are marked in grey colour. Observe that equation (2.30) has infinitely many roots. A choice of unique (physical meaning) solutions to the mentioned equation stands for the one of principal problems of theory of waves propagation in the periodic structures.

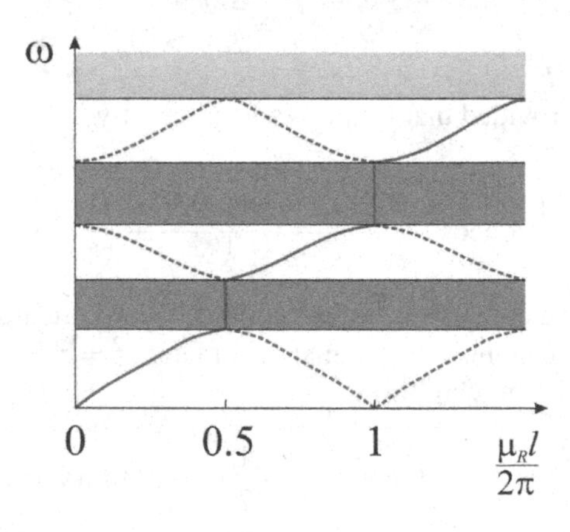

Figure 2.4 Dispersion curves.

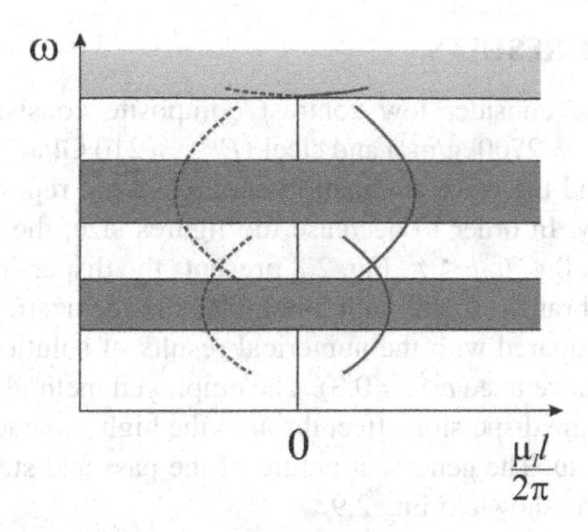

Figure 2.5 Wave attenuation coefficient.

For waves described dynamics of cristal lattices [241] often the so called multi-mode regimes are considered, when one deals with occurrence of a few harmonics simultaneously which refer to the distinct branches of dispersion curves. Interaction of modes with the normal ($d\omega/d\mu > 0$) and anomalous ($d\omega/d\mu < 0$) dispersion has been investigated in reference [272]. In mechanics of materials the physically important is one mode regime where for a given harmonic (2.24) there is preserved unique dependence between ω and μ [60,272,408,409]. The corresponding branches of the spectrum are denoted by solid curves in Fig. 2.4, whereas the remaining branches are marked by dashed curves. The branches choice has been validated through an asymptotic transition from solution of the piece-wise homogenous medium to the solution of the corresponding homogenous medium: equality of the components impedance ($z^{(2)}/z^{(1)} \to 1$) implies decrease of the stop bands and solid branches tend to coincide with a straight line ($\omega \to \omega_0 = \mu v_0$).

The coefficient μ_I is associated with symmetric pair of positive and negative roots (Fig. 2.5). Positive values μ_I (solid curves) describe the attenuation effect. Negative values μ_I (dashed curves) should correspond to the exponential increase of a signal which has no physical interpretation in the case of the considered conservative system.

Solution behavior can be described in the following way. For low frequencies ω the dispersion curve is close to the straight line, $\omega \approx \omega_0$, phase velocity $v_p = \omega/\mu$ and group velocity $v_p = d\omega/d\mu$ are close to each other and do not depend on ω, $v_p \approx v_g \approx v_0$. The so considered case is referred to as quasi-homogenous. Increase of frequency, angle of inclination of the dispersive curve is decreased. Both phase and group velocities are decreased though the group velocity decreases more fast $v_p > v_g$. On the boundary of the stop band the group velocity tends to zero $v_g = 0$, and the composite exhibits a standing wave. In the successive pass bands the group velocity changes from zero on the boundaries up to maximum values in the center.

2.4 NUMERICAL RESULTS

As the first case we consider low contrast composite consisting of aluminium ($E^{(1)} = 70$ GPa, $\rho^{(1)} = 2700$kg/m^3) and steel ($E^{(2)} = 210$ GPa, $\rho^{(2)} = 7800$kg/m^3). Dispersion curves and the wave attenuation coefficient are reported in Fig. 2.6 and Fig. 2.7, respectively. In order to decrease the figures size, the spectrum branches are shown in interval $0 \leq \mu_R l \leq \pi$. Fig. 2.8 presents the dispersion curve in the first pass band (acoustic branch) found with a help of homogenization method (formula (2.20)), which is compared with the numerical results of solution to the dispersion equation (2.30) (we have fixed $c^{(2)} = 0.3$). The employed method of homogenization allows to account of the dispersion effect through the high accuracy only for low values of the frequency ω. The general structure of the pass and stop bands versus the volume fraction $c^{(2)}$ is shown in Fig. 2.9.

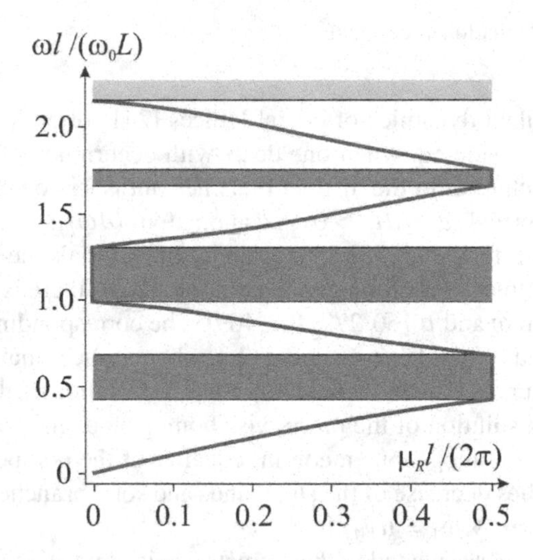

Figure 2.6 Dispersion curves of composite "steel-aluminium".

As the second example we consider the high-contrast composite composed of carbon-epoxide (CE) plastic ($E^{(1)} = 8.96$ GPa, $\rho^{(1)} = 1600$ kg/m^3) and steel ($E^{(2)} = 210$ GPa, $\rho^{(2)} = 7800$ kg/m^3). Dispersion curves are reported in Fig. 2.10, whereas the attenuation coefficient is shown in Fig. 2.11. The results obtained for the acoustic branch using the homogenization method coincide well with the exact solution (Fig. 2.12). The computations have been carried out for fixed $c^{(2)} = 0.5$. Structure of the pass and stop bands is shown in Fig. 2.13.

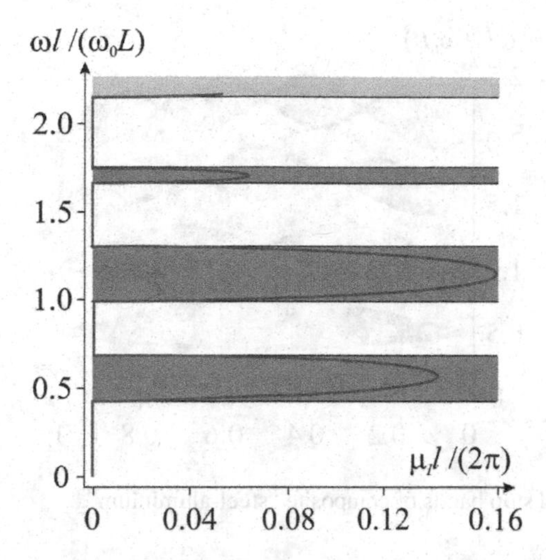

Figure 2.7 Wave attenuation coefficient of composite "steel-aluminium".

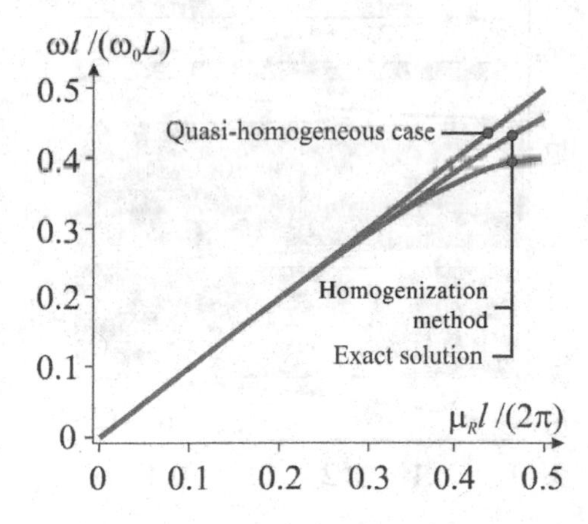

Figure 2.8 Acoustic branched of composite "steel-aluminium".

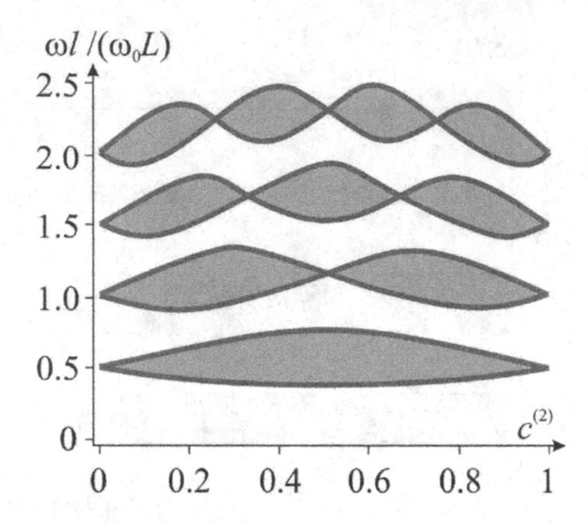

Figure 2.9 Pass and stop bands of composite "steel-aluminium".

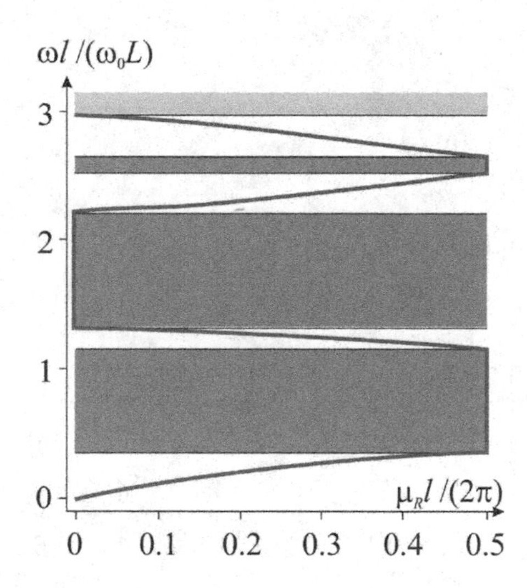

Figure 2.10 Dispersion curves of composite "CE plastic-steel".

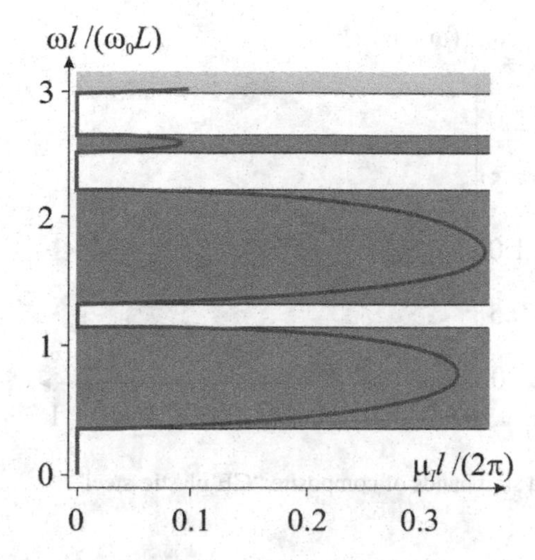

Figure 2.11 Wave attenuation coefficient of composite "CE plastic-steel".

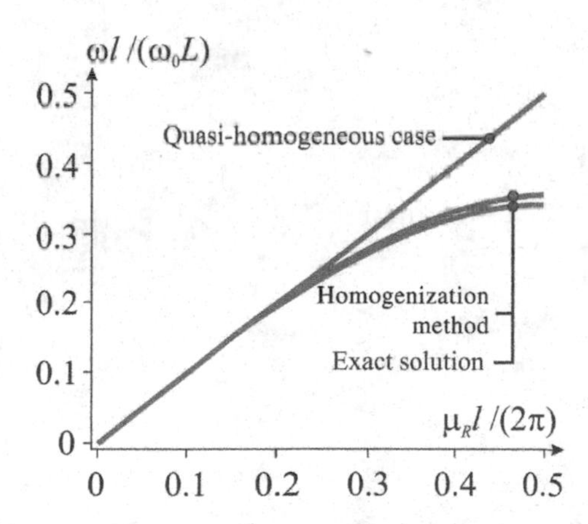

Figure 2.12 Acoustic branch of composite "CE plastic-steel".

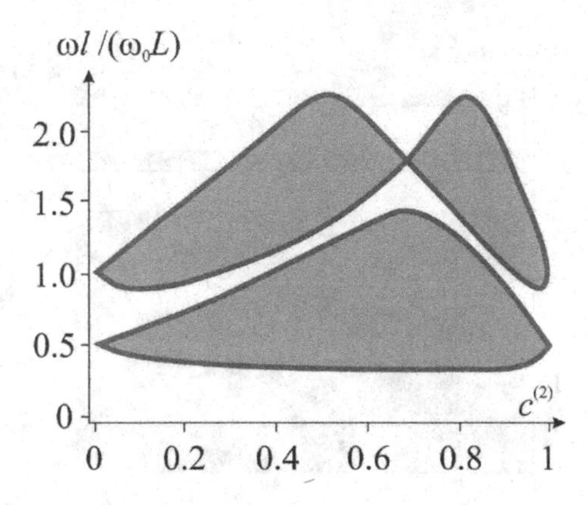

Figure 2.13 Pass and stop bands of composite "CE plastic-steel".

3 Waves in Fiber Composites: Linear Problems

Two-dimensional linear elastic waves propagation in the fiber composite material are investigated. Higher-order homogenization method is employed. The obtained solutions realibility is numerically validated. Effective moduli in the case of strong fibers interaction are obtained. Then the Bloch-Floquet method is used as well as the Fourier series approach to get a solution to the studied problem. Numerical results allowed to estimate dispersive curves and acoustic branches of the composite "nickel-aluminium" and "carbon-epoxide plastic". Moreover, shear wave dispersion in cylindrically structured cancellous viscoelastic bones is analyzed.

3.1 TWO-DIMENSIONAL (2D) DYNAMIC PROBLEM

If displacements and stresses in composite material depend on two space coordinates x_1, x_2, then the problem of propagation of elastic waves is splitted into two independent problems with regard to plane and antiplane deformation. In what follows we consider the antiplane problem which deals with a shear wave propagating in plane x_1x_2 in the fiber composite with the cylindrical inclusions (see Fig. 3.1).

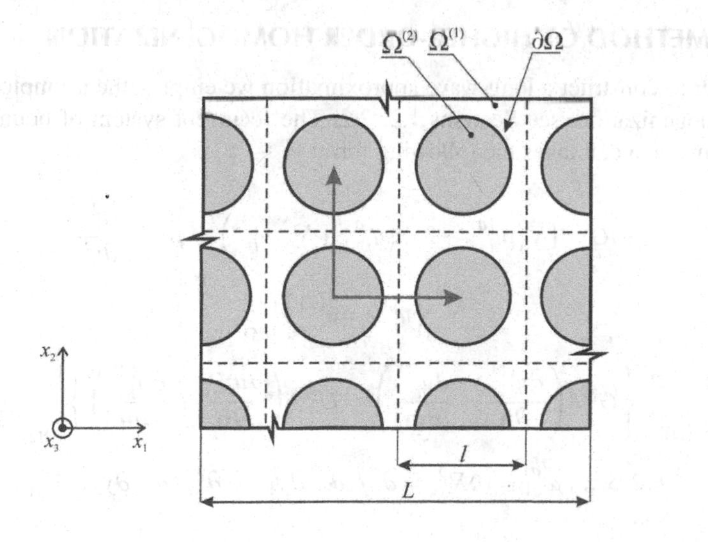

Figure 3.1 Fiber composite material.

The governing wave equation takes the following form

$$\nabla_x \cdot (G \nabla_x u) = \rho \frac{\partial^2 u}{\partial t^2}, \tag{3.1}$$

where u is the displacement in x_3 direction and $\nabla_x = \mathbf{e}_1 \partial/\partial x_1 + \mathbf{e}_2 \partial/\partial x_2$.

Since we have the nonhomogeneous medium, the shear modulus G and density ρ are discontinuous functions of the co-ordinates

$$G(\mathbf{x}) = G^{(a)}, \quad \rho(\mathbf{x}) = \rho^{(a)}(\mathbf{x}) \quad \text{for} \quad \mathbf{x} \in \Omega^{(a)}, \quad a = 1,2. \tag{3.2}$$

Equation (3.1) can be presented in equivalent form

$$G^{(a)} \nabla^2_{xx} u^{(a)} = \rho^{(a)} \frac{\partial^2 u^{(a)}}{\partial t^2}, \tag{3.3}$$

where $\nabla^2_{xx} = \partial^2/\partial x_1^2 + \partial/\partial x_2^2$.

On the boundary $\partial\Omega$, separating the composite component, we account of the following ideal contact conditions

$$\left\{u^{(1)} = u^{(2)}\right\}\Big|_{\partial\Omega}, \quad \left\{G^{(1)} \frac{\partial u^{(1)}}{\partial \mathbf{n}} = G^{(2)} \frac{\partial u^{(2)}}{\partial \mathbf{n}}\right\}\Big|_{\partial\Omega}. \tag{3.4}$$

It should be emphasized that the boundary value problem (3.1)–(3.4) allows for various physical interpretations. Beside considered here the example of elastic shear wave, considered problems describe processes of propagation of the electromagnetic waves in composites with dielectric inclusions.

3.2 METHOD OF HIGHER-ORDER HOMOGENIZATION

In order to construct a long-wave approximation we employ the asymptotic method of homogenization, see Sections 1.2, 2.2. The recurrent system of boundary value problems on a cell takes the following form:

$$G^{(a)} \left(\nabla^2_{xx} u^{(a)}_{i-2} + 2\nabla^2_{xy} u^{(a)}_{i-1} + \nabla^2_{yy} u^{(a)}_i\right) = \rho^{(a)} \frac{\partial^2 u^{(a)}_{i-2}}{\partial t^2}, \tag{3.5}$$

$$\left\{u^{(1)}_i = u^{(2)}_i\right\}\Big|_{\partial\Omega}, \tag{3.6}$$

$$\left\{G^{(1)} \left(\frac{\partial u^{(1)}_{i-1}}{\partial \mathbf{n}} + \frac{\partial u^{(1)}_i}{\partial \mathbf{m}}\right) = G^{(2)} \left(\frac{\partial u^{(2)}_{i-1}}{\partial \mathbf{n}} + \frac{\partial u^{(2)}_i}{\partial \mathbf{m}}\right)\right\}\Big|_{\partial\Omega}, \tag{3.7}$$

where $i = 1,2,3,\ldots$, $u^{(a)}_{-1} = 0$, $\nabla^2_{xy} = \partial^2/(\partial x_1 \partial y_1) + \partial^2/(\partial x_2 \partial y_2)$, $\nabla^2_{yy} = \partial^2/\partial y_1^2 + \partial/\partial y_2^2$.

In the case of spatially infinite composite we change the periodicity conditions by zeroth boundary conditions in the cell center and its external boundary $\partial\Omega_0$

$$\left\{u_i^{(2)}=0\right\}\Big|_{\substack{x=0\\y=0}}, \qquad \left\{u_i^{(1)}=0\right\}\Big|_{\partial\Omega_0}. \tag{3.8}$$

After defining the $u_i^{(a)}$ term from the i-th boundary value problem (3.5)–(3.8) we employ the averaging operator

$$S_0^{-1}\iint_{\Omega_0}(\cdot)dS, \tag{3.9}$$

where $dS = dy_1 dy_2$.

The terms $u_{i+1}^{(a)}$ are removed with the help of Gauss-Ostrogradsky's theorem after integration with regard to the cell area

$$G^{(1)}\iint\limits_{\Omega_0^{(1)}}\left(\nabla_{xy}^2 u_i^{(1)}+\nabla_{yy}^2 u_{i+1}^{(1)}\right)dS+G^{(2)}\iint\limits_{\Omega_0^{(2)}}\left(\nabla_{xy}^2 u_i^{(2)}+\nabla_{yy}^2 u_{i+1}^{(2)}\right)dS=0. \tag{3.10}$$

The homogenized equation of motion of η^{i-1} order has the following form

$$\begin{gathered}G^{(1)}\iint\limits_{\Omega_0^{(1)}}\left(\nabla_{xx}^2 u_{i-1}^{(1)}+\nabla_{xy}^2 u_i^{(1)}\right)dS+G^{(2)}\iint\limits_{\Omega_0^{(2)}}\left(\nabla_{xx}^2 u_{i-1}^{(2)}+\nabla_{xy}^2 u_i^{(2)}\right)dS=\\ \rho^{(1)}\iint\limits_{\Omega_0^{(1)}}\frac{\partial^2 u_{i-1}^{(1)}}{\partial t^2}dS+\rho^{(2)}\iint\limits_{\Omega_0^{(2)}}\frac{\partial^2 u_{i-1}^{(2)}}{\partial t^2}dS.\end{gathered} \tag{3.11}$$

Adding the equations (3.11) for $i = 1,...,n$, one gets the macroscopic wave equation of order η^n. In order to solve boundary value problems on cell (3.5)–(3.8), one may employ various approaches. As a rule, the largest computational problems appear in the case of strong interaction of the inclusions when significant oscillations on the microlevel are exhibited. Let us consider the asymptotic solution using non-dimensional width of the clearance between the neighboring fibers $\nu = 2(1-A/L)$ (Fig. 3.2). Let $\nu << 1$ and $G^{(2)}/G^{(1)} >> 1$. In zeroth approximation $O(\nu^0)$ for the zones $d\Omega_1$, $d\Omega_2$ the following estimations hold

$$\begin{aligned}\frac{\partial^2 u_i^{(1)}}{\partial y_1^2} &>> \frac{\partial^2 u_i^{(1)}}{\partial y_2^2} \quad \text{for} \quad \mathbf{y}\in d\Omega_1,\\ \frac{\partial^2 u_i^{(1)}}{\partial y_1^2} &<< \frac{\partial^2 u_i^{(1)}}{\partial y_2^2} \quad \text{for} \quad \mathbf{y}\in d\Omega_2.\end{aligned} \tag{3.12}$$

Physical meaning of the estimations (3.9) relies on observation that for zone $d\Omega_1$ in the case of narrow clearance between fibers the change of the local displacements in direction y_1 is essentially larger than in direction y_2. Therefore, the term

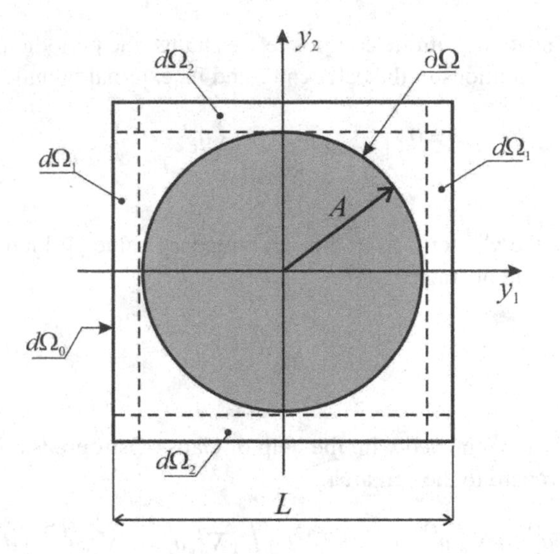

Figure 3.2 Cell of periodicity.

$\partial^2 u_i^{(1)}/\partial y_2^2$ can be neglected while comparing it with $\partial^2 u_i^{(1)}/\partial y_1^2$. Analogously, in the case of zone $d\Omega_2$ one may omit the term $\partial^2 u_i^{(1)}/\partial y_1^2$ in comparison with $\partial^2 u_i^{(1)}/\partial y_2^2$. The so far used simplification is well known in fluid mechanics where it is named lubrication theory [123].

Relations (3.12) allow to write equation (3.5) in the following approximate form

$$G^{(1)}\left(\frac{\partial^2 u_{i-2}^{(1)}}{\partial x_s^2}+2\frac{\partial^2 u_{i-1}^{(1)}}{\partial x_s \partial y_s}+\frac{\partial^2 u_i^{(1)}}{\partial y_s^2}\right)=\rho^{(1)}\frac{\partial^2 u_{i-2}^{(1)}}{\partial t^2}, \qquad s=1,2,$$
$$G^{(2)}\left(\nabla_{xx}^2 u_{i-2}^{(2)}+2\nabla_{xy}^2 u_{i-1}^{(2)}+\nabla_{yy}^2 u_i^{(2)}\right)=\rho^{(2)}\frac{\partial^2 u_{i-2}^{(2)}}{\partial t^2}. \tag{3.13}$$

Solving the boundary value problems (3.6)–(3.8), (3.13) for $i=1,2,3$, we get for $y_s>0$

$$u_1^{(2)}=A_{1,s}y_s\frac{\partial u_0}{\partial x_s}, \qquad u_1^{(1)}=\left(B_{1,s}y_s+\frac{1}{2}C_{1,s}L\right)\frac{\partial u_0}{\partial x_s},$$

$$u_2^{(2)}=\left[-\frac{1}{2}\left(1+2A_{1,s}-\frac{\rho^{(2)}}{\rho_0}\frac{G_0}{G^{(2)}}\right)y_s^2+\frac{1}{4}A_{2,s}Ly_s\right]\frac{\partial^2 u_0}{\partial x_s^2},$$

$$u_2^{(1)}=\left[-\frac{1}{2}\left(1+2B_{1,s}-\frac{\rho^{(1)}}{\rho_0}\frac{G_0}{G^{(1)}}\right)y_s^2+\frac{1}{4}B_{2,s}Ly_s+\frac{1}{8}C_{2,s}L^2\right]\frac{\partial^2 u_0}{\partial x_s^2},$$

$$
\begin{aligned}
u_3^{(2)} &= \left\{ \frac{1}{6}\left[2 + 3A_{1,s} + (A_{1,s} - 2)\frac{\rho^{(2)}}{\rho_0}\frac{G_0}{G^{(2)}} \right] y_s^3 - \right. \\
&\left. \frac{1}{4}A_{2,s}Ly_s^2 + \frac{1}{24}A_{3,s}L^2 y_s \right\} \frac{\partial^3 u_0}{\partial x_s^3}, \\
u_3^{(1)} &= \left\{ \frac{1}{6}\left[2 + 3B_{1,s} + (B_{1,s} - 2)\frac{\rho^{(1)}}{\rho_0}\frac{G_0}{G^{(1)}} \right] y_s^3 - \right. \\
&\frac{1}{4}\left(B_{2,s} + C_{1,s} - C_{1,s}\frac{\rho^{(1)}}{\rho_0}\frac{G_0}{G^{(1)}} \right) Ly_s^2 + \\
&\left. \frac{1}{24}B_{3,s}L^2 y_s + \frac{1}{48}C_{3,s}L^3 \right\} \frac{\partial^3 u_0}{\partial x_s^3},
\end{aligned} \tag{3.14}
$$

and for $y_s < 0$

$$
\begin{aligned}
u_1^{(2)} &= A_{1,s} y_s \frac{\partial u_0}{\partial x_s}, \qquad u_1^{(1)} = \left(B_{1,s} y_s - \frac{1}{2}C_{1,s} L \right) \frac{\partial u_0}{\partial x_s}, \\
u_2^{(2)} &= \left[-\frac{1}{2}\left(1 + 2A_{1,s} - \frac{\rho^{(2)}}{\rho_0}\frac{G_0}{G^{(2)}} \right) y_s^2 - \frac{1}{4}A_{2,s}Ly_s \right] \frac{\partial^2 u_0}{\partial x_s^2}, \\
u_2^{(1)} &= \left[-\frac{1}{2}\left(1 + 2B_{1,s} - \frac{\rho^{(1)}}{\rho_0}\frac{G_0}{G^{(1)}} \right) y_s^2 - \frac{1}{4}B_{2,s}Ly_s + \frac{1}{8}C_{2,s}L^2 \right] \frac{\partial^2 u_0}{\partial x_s^2}, \\
u_3^{(2)} &= \left\{ \frac{1}{6}\left[2 + 3A_{1,s} + (A_{1,s} - 2)\frac{\rho^{(2)}}{\rho_0}\frac{G_0}{G^{(2)}} \right] y_s^3 + \right. \\
&\left. \frac{1}{4}A_{2,s}Ly_s^2 + \frac{1}{24}A_{3,s}L^2 y_s \right\} \frac{\partial^3 u_0}{\partial x_s^3}, \\
u_3^{(1)} &= \left\{ \frac{1}{6}\left[2 + 3B_{1,s} + (B_{1,s} - 2)\frac{\rho^{(1)}}{\rho_0}\frac{G_0}{G^{(1)}} \right] y_s^3 + \right. \\
&\left. \frac{1}{4}\left(B_{2,s} + C_{1,s} - C_{1,s}\frac{\rho^{(1)}}{\rho_0}\frac{G_0}{G^{(1)}} \right) Ly_s^2 + \frac{1}{24}B_{3,s}L^2 y_s - \frac{1}{48}C_{3,s}L^3 \right\} \frac{\partial^3 u_0}{\partial x_s^3}.
\end{aligned} \tag{3.15}
$$

Here G_0 stands for the effective shear modulus in the quasi-homogeneous case (it is yielded by equation (3.11) for $i = 1$); $\rho_0 = \left(1 - c^{(2)}\right)\rho^{(1)} + c^{(2)}\rho^{(2)}$ stands for the averaged density, whereas $c^{(2)} = \pi A^2/L^2$ stands for the volume fraction of fibers, $0 \le c^{(2)} \le c_{\max}^{(2)}$, $c_{\max}^{(2)} = \pi/4 = 0.7853...$

The coefficients $A_{i,s}$, $B_{i,s}$, $C_{i,s}$ follow

$$
\begin{aligned}
&A_{1,s} = -(1-\chi_s)\left(G^{(2)}-G^{(1)}\right)/D_s, \quad B_{1,s} = \chi_s\left(G^{(2)}-G^{(1)}\right)/D_s\,,\\
&C_{1,s} = -B_{1,s}, \quad D_s = (1-\chi_s)\,G^{(2)} + \chi_s G^{(1)},\\
&A_{2,s} = -\left[(1-\chi_s)\left\{\left[(1-\chi_s)\,\rho^{(1)} + 2\chi_s\rho^{(2)}\right]G_0 + \rho_0 G^{(1)}\right\}G^{(2)} + \right.\\
&\left.\chi_s^2\rho^{(2)}G^{(1)}G_0\right]/\left(\rho_0 G^{(2)}D_s\right),\\
&B_{2,s} = \left\{\chi_s G^{(1)}\left[\chi_s\left(\rho^{(2)} - 2\rho^{(1)}\right)G_0 - 2\rho_0 G^{(1)}\right] + \right.\\
&\left.G^{(2)}\left[(1+2\chi_s)\,\rho_0 G^{(1)} - (1-\chi_s^2)\,\rho^{(1)}G_0\right]\right\}/\left(\rho_0 G^{(1)}D_s\right),\\
&C_{2,s} = \chi_s\left[\rho_0\left(G^{(1)}\right)^2 + (1-\chi_s)\,\rho^{(1)}G^{(2)}G_0 - G^{(1)}\left\{\rho_0 G^{(2)} + \right.\right.\\
&\left.\left.\left[\chi_s\rho^{(2)} + \rho^{(1)}(1-2\chi_s)\right]G_0\right\}/\left(\rho_0 G^{(1)}D_s\right)\right.,\\
&A_{3,s} = -\left\{\chi_s^3\rho^{(2)}\left(G^{(1)}\right)^2 G_0 + (1-\chi_s)\left(G^{(2)}\right)^2\left[\rho^{(1)}G_0(1-\chi_s)^2 - \right.\right.\\
&\left.\rho_0 G^{(1)}\right] + \left\{3G_0(1-\chi_s)\left[\chi_s\rho^{(2)} + (1-\chi_s)\rho^{(1)}\right] - \rho_0 G^{(1)}\right\}\times\\
&\left.\chi_s G^{(1)}G^{(2)}\right\}/\left(\rho_0 G^{(2)}D_s^2\right),\\
&B_{3,s} = \left[\left\{\chi_s^2 G_0\left[5\chi_s\rho^{(2)} + 3\rho^{(1)}(1-4\chi_s)\right] - 2\chi_s\rho_0 G^{(2)}(1-3\chi_s)\right\}\times\right.\\
&\left(G^{(1)}\right)^2 + G^{(1)}G^{(2)}\left\{\rho_0 G^{(2)}(1-\chi_s)(1+3\chi_s) + 3\chi_s^2 G_0\times\right.\\
&\left.\left.\left[\rho^{(2)}(1-\chi_s) - \rho^{(1)}(4-5\chi_s)\right]\right\} - 3\chi_s^2\rho_0\left(G^{(1)}\right)^3 + \right.\\
&\left.\rho^{(1)}\left(G^{(2)}\right)^2 G_0(1-\chi_s)(1+\chi_s-5\chi_s^2)\right]/\left(\rho_0 G^{(1)}D_s^2\right),\\
&C_{3,s} = \chi_s\left[\left(G^{(1)}\right)^2\left\{\rho_0 G^{(2)}(1-2\chi_s) - \chi_s G_0\left[2\chi_s\rho^{(2)} + 3\rho^{(1)}(1-2\chi_s)\right]\right\} - \right.\\
&G^{(1)}G^{(2)}\left[\rho_0 G^{(2)}(1-\chi_s) + \rho^{(1)}G_0(1-6\chi_s+6\chi_s^2)\right] + \\
&\left.\chi_s\rho_0\left(G^{(1)}\right)^3 + \rho^{(1)}\left(G^{(2)}\right)^2 G_0(1-\chi_s)(1-2\chi_s)\right]/\left(\rho_0 G^{(1)}D_s^2\right),
\end{aligned}
\tag{3.16}
$$

where $\chi_1 = \sqrt{\left(c^{(2)}/c^{(2)}_{\max}\right)(1-y_2^2/A^2)}$, $\chi_2 = \sqrt{\left(c^{(2)}/c^{(2)}_{\max}\right)(1-y_1^2/A^2)}$.

The macroscopic wave equation of order η^2 takes the following form

$$
G_0\nabla^2_{xx}u_0 + \eta^2 L^2 G_2\nabla^4_1 u_0 + O(\eta^4) = \rho_0\frac{\partial^2 u_0}{\partial t^2}, \tag{3.17}
$$

where $\nabla_1^4 = \partial^4/\partial x_1^4 + \partial/\partial x_2^4$, and the coefficient G_2 presents the effective modulus of order η^2 (it is computed from equation (3.11) for $i = 3$).

In the order to check reliability of the obtained solution, we consider numerical results for the nondimensional effective shear modulus $\lambda_0 = G_0/G^{(1)}$. Fig. 3.3 presents λ_0 in the case of strong interaction of the neighboring fibers ($\lambda^{(2)} = G^{(2)}/G^{(1)} \to \infty$, $c^{(2)} \to c^{(2)}_{\max}$). Owing to the mathematical analogy between the problems of conductivity and longitudinal shear, the obtained solution (solid curves) are compared with those obtained through the asymptotic formula (dashed curves) which have been obtained for two highly conducted cylinders approaching a contact [358].

The estimations (3.12) are valid only in the case of a high contrast composite material with large volume fraction of inclusions. Yet, the obtained solution exhibits high accuracy for arbitrary values of the parameters $1 \le \lambda^{(2)} < \infty$ and $0 \le c^{(2)} \le c^{(2)}_{\max}$. This statement is illustrated in Fig. 3.4, where the results obtained for λ_0 (solid curves) are compared with the results reported in [358] (circles).

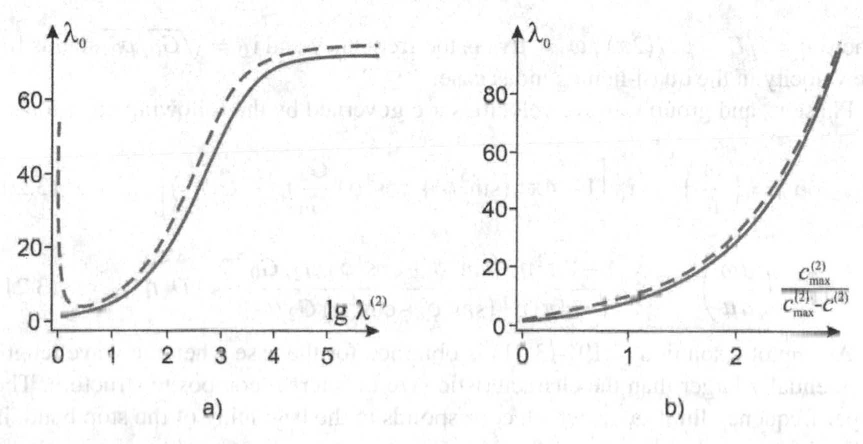

Figure 3.3 Effective modulus in the case of strong interaction of the neighboring fibers; a) $\lambda^{(2)} \to \infty$, $c^{(2)} = 0.784$, b) $\lambda^{(2)} = \infty$, $c^{(2)} \to c^{(2)}_{\max}$.

Consider the following harmonic wave

$$u_0 = U \exp(i\mu \cdot \mathbf{x}) \exp(i\omega t), \tag{3.18}$$

where U is the amplitude, ω stands for the frequency, $\mu = \mu_1 \mathbf{e}_1 + \mu_2 \mathbf{e}_2$ is wave vector, $\mu \equiv |\mu| = 2\pi/L$ stands for the wave number.

Projections of the vector μ on the axis of co-ordinates are as follows: $\mu_1 = \mu \cos\phi$, $\mu_2 = \mu \sin\phi$, where ϕ stands for the angle between the axis x_1 and direction of the wave propagation, $\phi = (\mathbf{e}_1, \mu)$.

Substituting of ansatz (3.18) into equation (3.17), the following dispersion relation is obtained

$$\omega^2 = \omega_0^2 \left[1 - 4\pi^2 \left(\sin^4\varphi + \cos^4\varphi \right) \frac{G_2}{G_0} \eta^2 + O(\eta^4) \right], \tag{3.19}$$

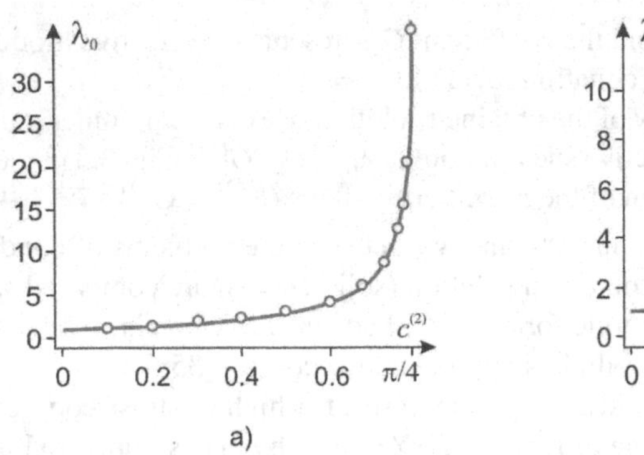

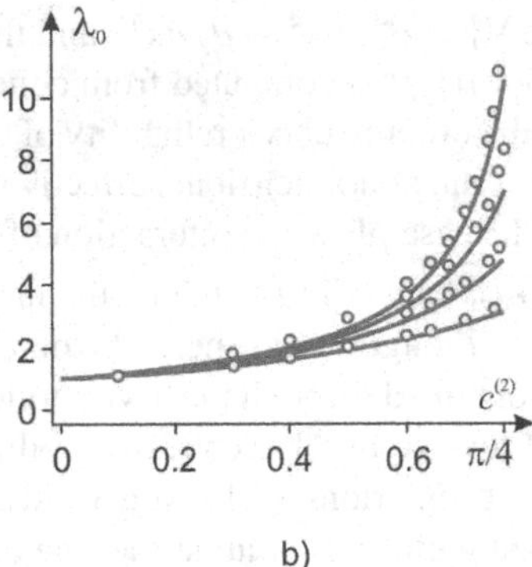

Figure 3.4 Effective modulus for different values of the fibers stiffness; a) $\lambda^{(2)} = \infty$, b) $\lambda^{(2)} = 5, 10, 20, 50$.

where $\eta = l/L = \mu l/(2\pi)$, $\omega_0 = \mu v_0$ is the frequency and $v_0 = \sqrt{G_0/\rho_0}$ stands for the velocity in the quasi-homogenous case.

Phase v_p and group v_g wave velocities are governed by the following equations:

$$v_p^2 = \left(\frac{\omega}{\mu}\right)^2 = v_0^2\left[1 - 4\pi^2\left(\sin^4\phi + \cos^4\phi\right)\frac{G_2}{G_0}\eta^2 + O(\eta^4)\right], \tag{3.20}$$

$$v_g^2 = \left(\frac{d\omega}{d\mu}\right)^2 = v_0^2\frac{\left[1 - 8\pi^2\eta^2\left(\sin^4\phi + \cos^4\phi\right)G_2/G_0\right]^2}{1 - 4\pi^2\eta^2\left(\sin^4\phi + \cos^4\phi\right)G_2/G_0} + O(\eta^4). \tag{3.21}$$

Asymptotic solution (3.19)–(3.21) is obtained for the case when the wave length is essentially larger than the characteristic size of internal composite structure. The upper frequency limit $\omega_{\max}$, which corresponds to the beginning of the stop band, is defined from the condition $v_g = 0$:

$$\omega_{\max}^2 = \frac{G_0^2}{4l^2 G_2 \rho_0\left(\sin^4\phi + \cos^4\phi\right)} + O(\eta^4). \tag{3.22}$$

In the quasi-homogeneous case the considered fiber composite is transversally-isotropic; the solution for G_0, v_0, ω_0 does not depend on the direction of wave propagation in plane x_1x_2. The higher-order approximations allow to detect the anisotropic effects of the problem. Namely, beginning with the terms of order η^2, the relations (3.19)–(3.21) depend on direction ϕ of the wave vector.

3.3 THE BLOCH-FLOQUET METHOD AND SOLUTION BASED ON FOURIER SERIES

Owing to the Bloch-Floquet method, a solution to the wave equation (3.1) is searched in the following form

$$u = F(\mathbf{x})\exp(i\mu\cdot\mathbf{x})\exp(i\omega t), \tag{3.23}$$

where $F(\mathbf{x})$ is periodic functions of co-ordinates, $F(\mathbf{x}) = F(\mathbf{x}+\mathbf{l}_p)$, $\mathbf{l}_p = p_1\mathbf{l}_1 + p_2\mathbf{l}_2$, $p_1, p_2 = 0, \pm 1, \pm 2, \ldots$, $\mathbf{l}_1$, $\mathbf{l}_2$ are translations vectors of the square lattice.

We present the function $F(\mathbf{x})$ and properties of components $G(\mathbf{x})$, $\rho(\mathbf{x})$ in the form of Fourier series

$$
\begin{aligned}
F(\mathbf{x}) &= \sum_{n_1=-\infty}^{\infty} \sum_{n_2=-\infty}^{\infty} A_{n_1 n_2} \exp\left[i\frac{2\pi}{l}(n_1 x_1 + n_2 x_2)\right], \\
G(\mathbf{x}) &= \sum_{n_1=-\infty}^{\infty} \sum_{n_2=-\infty}^{\infty} B_{n_1 n_2} \exp\left[i\frac{2\pi}{l}(n_1 x_1 + n_2 x_2)\right], \\
\rho(\mathbf{x}) &= \sum_{n_1=-\infty}^{\infty} \sum_{n_2=-\infty}^{\infty} C_{n_1 n_2} \exp\left[i\frac{2\pi}{l}(n_1 x_1 + n_2 x_2)\right],
\end{aligned}
\tag{3.24}
$$

where the coefficients $B_{n_1 n_2}$, $C_{n_1 n_2}$ are defined in the following way

$$
\begin{aligned}
B_{n_1 n_2} &= \frac{1}{S_0} \iint_{\Omega_0} G(\mathbf{x}) \exp\left[-i\frac{2\pi}{l}(n_1 x_1 + n_2 x_2)\right] dS, \\
C_{n_1 n_2} &= \frac{1}{S_0} \iint_{\Omega_0} \rho(\mathbf{x}) \exp\left[-i\frac{2\pi}{l}(n_1 x_1 + n_2 x_2)\right] dS,
\end{aligned}
\tag{3.25}
$$

and the operator $\iint_{\Omega_0} (\cdot) dS$ denotes integration along the cell area Ω_0, $dS = dx_1 dx_2$, $S_0 = l^2$ stands for the cell area.

Substituting relations (3.23), (3.24) into equation (3.1) and comparing the coefficients standing by the terms $\exp\left[i2\pi l^{-1}(j_1 x_1 + j_2 x_2)\right]$, $j_1, j_2 = 0, \pm 1, \pm 2, \ldots$, we obtain the infinite system of linear algebraic equations for unknowns $A_{n_1 n_2}$

$$
\begin{aligned}
&\sum_{n_1=-\infty}^{\infty} \sum_{n_2=-\infty}^{\infty} A_{n_1 n_2} \left\{ B_{\substack{j_1-n_1,\\ j_2-n_2}} \left[\left(\frac{2\pi}{l} n_1 + \mu_1\right)\left(\frac{2\pi}{l} j_1 + \mu_1\right) + \right.\right. \\
&\left.\left. \left(\frac{2\pi}{l} n_2 + \mu_2\right)\left(\frac{2\pi}{l} j_2 + \mu_2\right) \right] - C_{\substack{j_1-n_1,\\ j_2-n_2}} \omega^2 \right\} = 0.
\end{aligned}
\tag{3.26}
$$

The condition for existence of solution to the system (3.26) is defined by equating to zero the determinant of the matrix composed of their coefficients. It gives the dispersive relation for ω and μ. It should be emphasized that the given method does not employ boundary conditions (3.4) in the explicit form. Condition of ideal contact between fibers and matrix are "hidden" in equation (3.1) and in development (3.24) for $F(\mathbf{x})$, where the fields of displacements and stresses are implicitly treated as continuous.

In the case of the numerical studies reported below, the dispersive relations are computed on a basis of truncation of system (3.26) assuming $-j_{\max} \leq j_1, j_2 \leq j_{\max}$. The number of employed equations is equal to $(2j_{\max}+1)^2$. The introduced truncation has physical interpretation: we neglect high frequencies.

In order to illustrate the effect of wave filtering, let us split the problem (3.23) into real μ_R and imaginary part μ_I of the vector $\mu = \mu_R + i\mu_I$

$$
u = F(\mathbf{x}) \exp(-\mu_I \cdot \mathbf{x}) \exp(i\mu_R \cdot \mathbf{x}) \exp(i\omega t). \tag{3.27}
$$

The imaginary part $\mu_I \equiv |\mu_I|$ of the wave number presents the attenuation coefficient. The values $\mu_I = 0$ correspond to pass band, whereas $\mu_I \neq 0$ correspond to stop band. Boundaries of the pass and stop bands are defined through condition $\mu l = \pi n / \sqrt{\sin^4\phi + \cos^4\phi}$, $n = 1, 2, 3, \ldots$. The corresponding length of waves are defined by the formula $L = (2l/n)\sqrt{\sin^4\phi + \cos^4\phi}$.

3.4 NUMERICAL RESULTS

In order to compare solutions obtained with a help of the Fourier series with result obtained by other authors, we consider the non-homogeneous material composed of the matrix with properties $G^{(1)} = 1$, $\rho^{(1)} = 1$ and with voids, $G^{(2)} = 0$, $\rho^{(2)} = 0$, $A/L = 0.4$, $c^{(2)} \approx 0.503$.

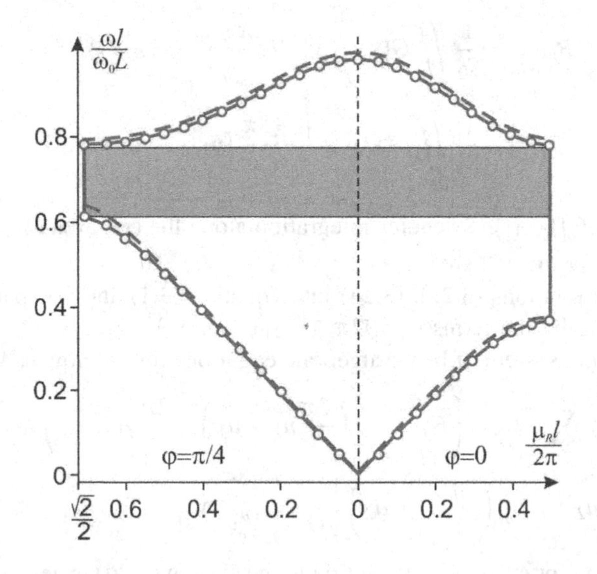

Figure 3.5 Dispersive curves of composite with voids.

Dispersive curves are shown in Fig. 3.5. The dashed lines correspond to computations for $j_{\max} = 1$, whereas the solid lines correspond to $j_{\max} = 2$; circles refer to the results reported in work [376] and obtained by the Rayleigh method. The dispersive diagram is composed of two parts separated by the vertical dashed line. The first part corresponds to the orthogonal ($\phi = 0$), and left part corresponds to the diagonal ($\phi = \pi/4$) direction of the wave propagation. In the quasi-homogenous case ($\omega \to 0$) the solution is isotropic and does not depend on the angle ϕ. However, with increase of the frequency ω the composite exhibits anisotropic properties. Brown color refers to the full stop band, where the signal propagation is not possible in any of directions. Increase of $j_{\max}$ increases accuracy of the numerical results. As an example of the low-contrast composite we consider material with aluminium matrix ($G^{(1)} = 27.9$

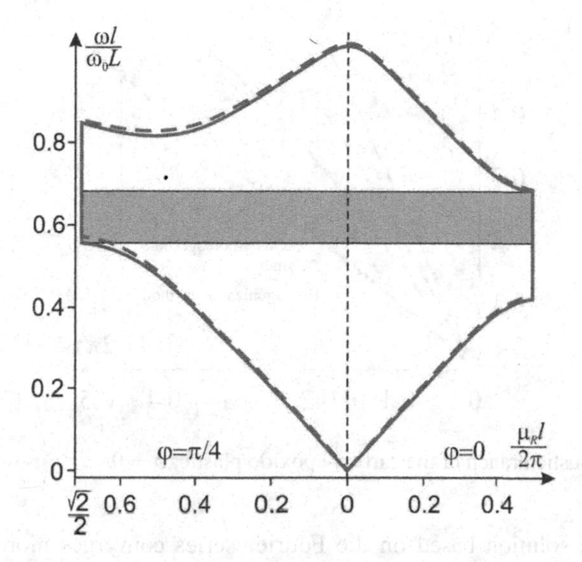

Figure 3.6 Dispersive curves of the composite "nickel-aluminium".

GPa, $\rho^{(1)} = 2700$ kg/m^3) and nickel fibers ($G^{(2)} = 75.4$ GPa, $\rho^{(2)} = 8936$ kg/m^3, $c^{(2)} = 0.35$). The related dispersive curves are shown in Fig. 3.6.

The results obtained for $j_{\max} = 1$ (dashed curves) and for $j_{\max} = 2$ (solid curves) are close to each other which confirm the fast convergence of the solution. As it follows from computation of the acoustic branch of spectrum (Fig. 3.7), the homogenization method allows to account of the dispersion effect, though the good accuracy is achieved only on the interval of low frequencies.

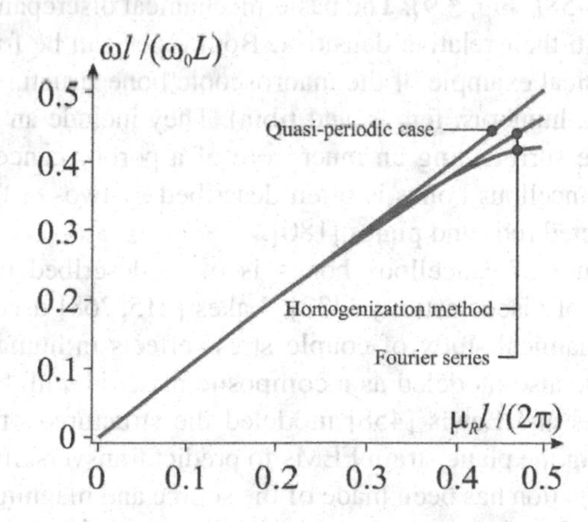

Figure 3.7 Acoustic branch of the composite "nickel-aluminium", $\phi = 0$.

In the case of high-contrast composite (epoxy matrix with $G^{(1)} = 1.53$ GPa, $\rho^{(1)} = 1250$ kg/m^3 and carbon fibers with $G^{(2)} = 86$ GPa, $\rho^{(2)} = 1800$ kg/m^3,

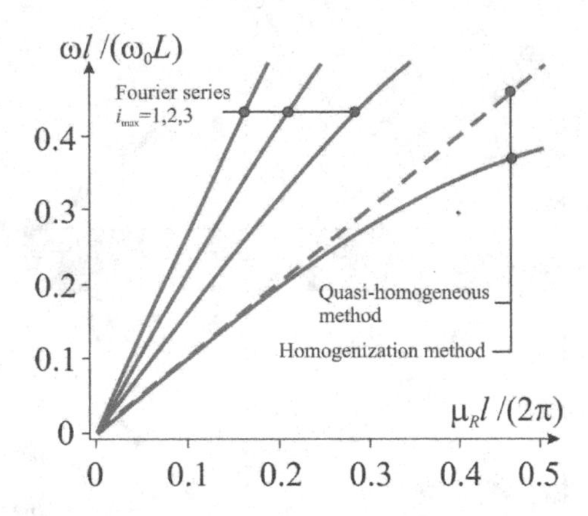

Figure 3.8 Acoustic branch of the carbon-epoxide plastic, $\phi = 0$.

$c^{(2)} = 0.5$), the solution based on the Fourier series converges more slowly (Fig. 3.8). The homogenization method yields qualitatively correct results up to the first stop band (for $\mu_R l = \pi/\sqrt{\sin^4\phi + \cos^4\phi}$ we get $v_g \approx 0$).

3.5 SHEAR WAVES DISPERSION IN CYLINDRICALLY STRUCTURED CANCELLOUS VISCOELASTIC BONES

Animal and human bones are heterogeneous materials with a complicated hierarchical structure. Bone tissues occur in the two main forms: as a dense solid (cortical or compact bone) and as a porous media filled by a viscous marrow (trabecular or cancellous bone) [219,458] (Fig. 3.9). The basic mechanical discrepancy between these two types consists in their relative densities. Both types can be found in the bones of the body. A classical example of the macroscopic bone structure can be given by the long bones (e.g., humerus, femur, and tibia). They include an outer shell of the dense cortical tissue surrounding an inner core of a porous cancellous tissue. The microstructure of cancellous bones is often described by two- or three-dimensional mesh of interconnected rods and plates [180].

The microstructure of cancellous bones is often described by two- or three-dimensional theory of viscoelasticity [432]. Lakes [215, 268] used nonlocal theory of elasticity for dynamical study of couple stress effects in human compact bone. Cortical bone can be also modeled as a composite material with hierarchical structure [185]. Williams and Lewis [458] modeled the structured of a 2D section of trabecular bone using the plane strain FEMs to predict transversally isotropic elastic constant. An investigation has been made of the source and magnitude of anisotropic material properties of cancellous bone in the proximal epiphysis of the human tibia. Results are reported for stiffness measurements made in three orthogonal directions on 21 cubes of cancellous bone before testing to failure along one of the three principal axes. The structure is approximately transversely isotropic. Strength and

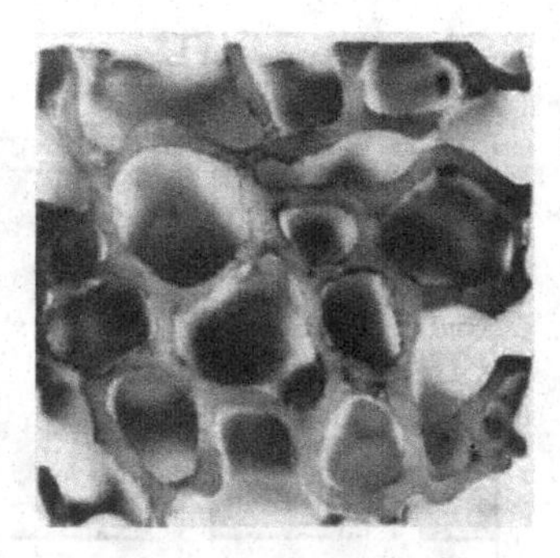

Figure 3.9 Scanning micrograph of a cylindrically structured cancellous bone [458] [by courtesy of the ASME].

stiffness are linear with area fraction for loading along the isotropic axis. Strength is proportional to stiffness for all directions. A FEM is proposed, based on experimental observations, which enables one to predict the elastic constants of cylindrically structured cancellous bone in the tibia from morphological measurements in the transverse plane.

In order to take into account the influence of bone marrow, Kasra and Grynpas [233] performed suitable numerical calculation using FEM. Many authors applied the homogenization theory to investigation of mechanical behaviour of the structures modeling trabecular bones [210]. Parnell and Grimal [352] investigate the influence of mesoscale porosity on the induced anisotropy of the material using asymptotic homogenization and new solution of the cell problem.

Despite the obvious simplicity, such idealized models may provide a satisfactory agreement between theoretical predictions and experimental results for the mechanical properties of real bones. Here we deal with a two-dimensional model of cylindrically structured cancellous bones.

A challenging problem consists in the detection of the bone structure using non-invasive measurements. The inverse homogenization approach ("dehomogenization procedure") [122] can help to derive information about the microgeometry of the bone tissue from the values of effective moduli. Another possibility is to extract this information from a dynamic response of the bone, measuring velocity and attenuation of ultrasonic waves of different frequencies.

Acoustic waves propagating through cancellous bones undergo dispersion and damping. There are two different physical effects influencing on the dynamic properties of the bone: (i) transmission of mechanical energy to heat due to the viscosity of the marrow (viscoelastic damping and dispersion) [215] and (ii) successive reflections and refractions of local waves at the trabecula-marrow interfaces (Bloch dispersion). From the theoretical standpoint, both effects are realized simultaneously. However, their intensities are very frequency dependent. For many real materials the effects of viscoelastic damping and Bloch dispersion are observed in a rather distant frequency ranges. In such a case they can be analyzed separately.

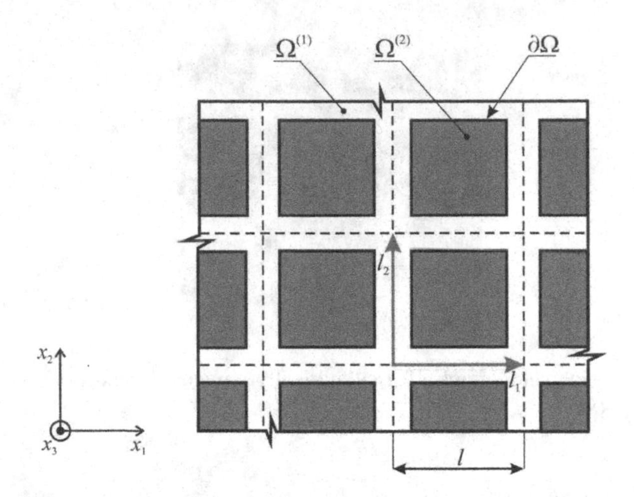

Figure 3.10 Cancellous structure under consideration

We study transverse antiplane shear waves propagating in the (x_1,x_2) plane through a regular cancellous structure consisting of a spatially infinite elastic matrix (trabeculae) $\Omega^{(1)}$ and viscous inclusions (marrow) $\Omega^{(2)}$ (Fig. 3.10). The governing two-dimensional wave equation is:

$$\nabla_x \cdot (G\nabla_x w) = \rho \frac{\partial^2 u}{\partial t^2}, \tag{3.28}$$

where G is the complex shear modulus, ρ is the mass density, u is the longitudinal displacement (in the x_3 direction), $\nabla_x = \mathbf{e}_1\partial/\partial x_1 + \mathbf{e}_2\partial/\partial x_2$, $\mathbf{e}_1$, $\mathbf{e}_2$ are the unit Cartesian vectors.

Due to the heterogeneity of the medium the physical properties G and ρ are represented by piece-wise continuous functions of co-ordinates:

$$\rho G(\mathbf{x}) = G^{(a)}(\mathbf{x}), \quad (\mathbf{x}) = \rho^{(a)}(\mathbf{x}) \quad for \quad \mathbf{x} \in \Omega^{(a)}, \quad \mathbf{x} = x_1\mathbf{e}_1 + x_2\mathbf{e}_2. \tag{3.29}$$

Here and in the sequel the superscript (a) denotes different components of the structure, $a = 1,2$. In Eq. (3.29), $G^{(1)}$ is the real shear modulus of the elastic matrix, $G^{(2)} = i\omega\nu^{(2)}$, where ω is the frequency of a harmonic wave, $\nu^{(2)}$ is the viscosity of the marrow.

Because Eq. (3.28) contains delta functions, below we deal with the generalized solution of this equation [275]. Equation (3.28) can be written in the equivalent form:

$$G^{(a)}\nabla_{xx}^2 u^{(a)} = \rho^{(a)} \frac{\partial^2 u^{(a)}}{\partial t^2}, \tag{3.30}$$

$$\left\{u^{(1)} = u^{(2)}\right\}\Big|_{\partial\Omega}, \quad \left\{G^{(1)}\frac{\partial u^{(1)}}{\partial \mathbf{n}} = G^{(2)}\frac{\partial u^{(2)}}{\partial \mathbf{n}}\right\}\Big|_{\partial\Omega}, \tag{3.31}$$

where $\nabla_{xx}^2 = \partial^2/\partial x_1^2 + \partial/\partial x_2^2$ and $\partial/\partial \mathbf{n}$ is the normal derivative to the contour $\partial\Omega$.

From the physical standpoint, Eq. (3.31) mean the perfect bonding conditions at the trabeculae-marrow interface $\partial\Omega$.

We start with the case, when the wavelength L is essentially larger than the internal size l of the cancellous structure, $l << L$. The original heterogeneous bone can be approximately substituted by a homogeneous one with a certain homogenized effective complex shear modulus G_0. Such an approach neglects local reflections and refractions of the waves on microlevel. The effect of dispersion is caused by the transmission of the mechanical energy of the acoustic wave to heat due to the viscosity of the marrow (so called viscoelastic damping).

Let us study the input boundary value problem (3.30), (3.31) by the asymptotic homogenization method. In order to separate macro- and microscale components of the solution we introduce slow $\mathbf{x}$ and fast $\mathbf{y}$ co-ordinate variables:

$$\mathbf{x} = \mathbf{x}, \qquad \mathbf{y} = \eta^{-1}\mathbf{x}, \tag{3.32}$$

where $\mathbf{y} = y_1\mathbf{e}_1 + y_2\mathbf{e}_2$, $\eta = l/L$ is a small parameter, and search the displacement as an expansion

$$u^{(a)} = u_0(\mathbf{x}) + \eta u_1^{(a)}(\mathbf{x},\mathbf{y}) + \eta^2 u_2^{(a)}(\mathbf{x},\mathbf{y}) + \cdots \tag{3.33}$$

The first term u_0 of expansion (3.33) represents the homogenized part of the solution; it does not depend on the fast co-ordinates ($\partial u_0/\partial y_1 = \partial u_0/\partial y_2 = 0$). The next terms $u_i^{(a)}$, $i = 1,2,3,\ldots$, provide corrections of the orders η^i and describe local variations of the displacements on microlevel.

The differential operators read

$$\nabla_x = \nabla_x + \eta^{-1}\nabla_y, \qquad \nabla_{xx}^2 = \nabla_{xx}^2 + 2\eta^{-1}\nabla_{xy}^2 + \eta^{-2}\nabla_{yy}^2, \tag{3.34}$$

where $\nabla_y = \mathbf{e}_1\partial/\partial y_1 + \mathbf{e}_2\partial/\partial y_2$, $\nabla_{xy}^2 = \partial^2/(\partial x_1\partial y_1) + \partial^2/(\partial x_2\partial y_2)$, $\nabla_{yy}^2 = \partial^2/\partial y_1^2 + \partial/\partial y_2^2$.

Splitting the input problem (3.30)–(3.31) with respect to η leads to a recurrent sequence of cell boundary value problems:

$$G^{(a)}\left(\nabla_{xx}^2 u_{i-2}^{(a)} + 2\nabla_{xy}^2 u_{i-1}^{(a)} + \nabla_{yy}^2 u_i^{(a)}\right) = \rho^{(a)}\frac{\partial^2 u_{i-2}^{(a)}}{\partial t^2}, \tag{3.35}$$

$$\left\{u_i^{(1)} = u_i^{(2)}\right\}\Big|_{\partial\Omega},$$

$$\left\{G^{(1)}\left(\frac{\partial u_{i-1}^{(1)}}{\partial \mathbf{n}} + \frac{\partial u_i^{(1)}}{\partial \mathbf{m}}\right) = G^{(2)}\left(\frac{\partial u_{i-1}^{(2)}}{\partial \mathbf{n}} + \frac{\partial u_i^{(2)}}{\partial \mathbf{m}}\right)\right\}\Bigg|_{\partial\Omega}, \tag{3.36}$$

where $i = 1,2,3,\ldots$, $u_{-1}^{(a)} = 0$, $\partial/\partial\mathbf{m}$ is the normal derivative to the interface $\partial\Omega$ written in fast variables.

For a spatially periodic medium the terms $u_i^{(a)}$ have to satisfy the conditions of periodicity

$$u_i^{(a)}(\mathbf{x},\mathbf{y}) = u_i^{(a)}(\mathbf{x},\mathbf{y}+\mathbf{L}_p) \tag{3.37}$$

and normalization

$$\left\langle u_i^{(a)} \right\rangle = 0, \tag{3.38}$$

where $\mathbf{L}_p = \eta^{-1}\mathbf{l}_p$, $\mathbf{l}_p = p_1\mathbf{l}_1 + p_2\mathbf{l}_2$, $p_1, p_2 = 0, \pm 1, \pm 2, \ldots$, $\mathbf{l}_1, \mathbf{l}_2$ are the fundamental translation vectors of the cancellous structure.

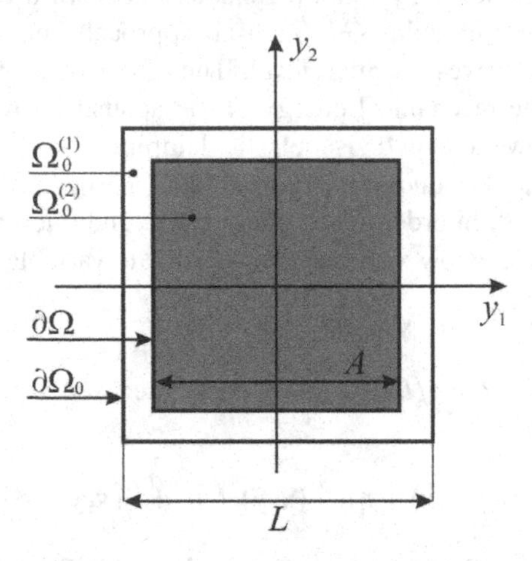

Figure 3.11 Periodically repeated unit cell.

Let us introduce the homogenizing operator over the unit cell domain $\Omega_0 = \Omega_0^{(1)} + \Omega_0^{(2)}$ (Fig. 3.11), $S_0 = L^2$ is the area of the unit cell in fast co-ordinates

$$\langle \cdot \rangle = \frac{1}{S_0} \left(\iint\limits_{\Omega_0^{(1)}} (\cdot) dy_1 dy_2 + \iint\limits_{\Omega_0^{(2)}} (\cdot) dy_1 dy_2 \right). \tag{3.39}$$

The conditions (3.37), (3.38) we replace by the following zero boundary conditions for these functions in the center of cell and along the outer contour $\partial\Omega_0$

$$\left\{ u_i^{(2)} = 0 \right\} \Big|_{x,y=0}, \qquad \left\{ u_i^{(1)} = 0 \right\} \Big|_{\partial\Omega_0}. \tag{3.40}$$

The "zero boundary conditions" approximation appears quite reasonable. For 1D case they are equivalent to the periodicity conditions. For 2D case replacing of the periodicity conditions to the zero boundary conditions leads to the more stiff system and provides an upper bound for the effective properties. Numerical calculations showed that discrepancy between the solutions in both cases is not essential, but using boundary conditions (3.40) sufficiently simplifies of the cell problem.

Due to the periodicity of $u_i^{(a)}$ (3.37), Eqs. (3.35), (3.36) can be considered within only one unit cell. Solution of the cell problem (3.35), (3.36), (3.37) at $i = 1$ determines the term $u_1^{(a)}$. In order to find the effective modulus G_0, the homogenizing operator (3.39) is applied to equation (3.37) at $i = 2$. Terms $u_2^{(a)}$ are eliminated by means of Green-Ostrogradsky's theorem, which together with the boundary conditions (3.36) and the periodicity relation (3.37) implies

$$\left\langle G^{(a)} \left(\nabla^2_{xy} u_i^{(a)} + \nabla^2_{yy} u_{i+1}^{(a)} \right) \right\rangle = 0. \tag{3.41}$$

As the result, the homogenized wave equation of the order η^0 is obtained:

$$\left\langle G^{(a)} \left(\nabla^2_{xx} u_0 + \nabla^2_{xy} u_1^{(a)} \right) \right\rangle = \left\langle \rho^{(a)} \right\rangle \frac{\partial^2 u_0}{\partial t^2}. \tag{3.42}$$

Substituting to (3.42) expressions for $u_1^{(a)}$, evaluated below, we shall come to a macroscopic wave equation

$$G_0 \nabla^2_{xx} u_0 = \rho_0 \frac{\partial^2 u_0}{\partial t^2}, \tag{3.43}$$

where $\rho_0 = \left(1 - c^{(2)}\right) \rho^{(1)} + c^{(2)} \rho^{(2)}$ is the effective mass density, $c^{(2)}$ is the volume fraction of the inclusions, $c^{(2)} = A^2/S_0$, A is the size of the inclusion (Fig. 3.11).

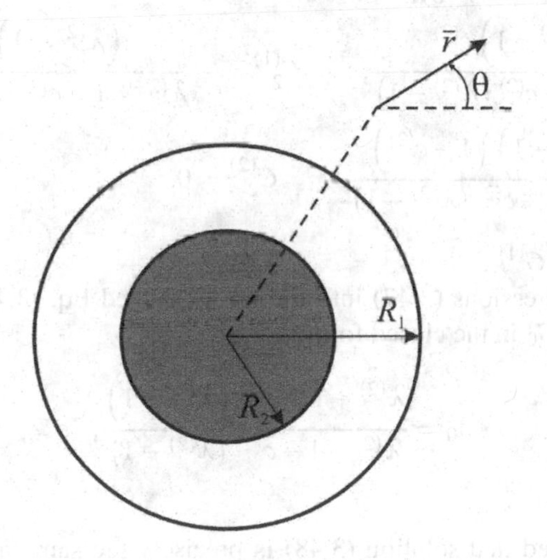

Figure 3.12 Simplification of the unit cell in the case $c^{(2)} << 1$.

The effective modulus G_0 can be derived after evaluation of the integrals in Eq. (3.42).

Below we find approximate solutions of the cell problem (3.35), (3.36), (3.40) and determine the effective shear modulus G_0 using a boundary shape perturbation and a lubrication theory approaches.

If the volume fraction $c^{(2)}$ of the marrow inclusions is relatively small, the square shapes of the domains $\Omega^{(1)}$, $\Omega^{(2)}$ can be approximately substituted by the equal circles of radii R_1, R_2 so that $c^{(2)} = R_2^2/R_1^2$ (Fig. 3.12). This simplification can be considered as the first approximation of the method of the boundary shape perturbation [196].

Let us introduce in the unit cell the polar co-ordinates $r^2 = y_1^2 + y_2^2$, $\tan\theta = y_2/y_1$. Eqs. (3.35), (3.36), (3.40) at $i = 1$ reads

$$\frac{\partial^2 u_1^{(a)}}{\partial r^2} + \frac{1}{r}\frac{\partial u_1^{(a)}}{\partial r} + \frac{1}{r^2}\frac{\partial^2 u_1^{(a)}}{\partial \theta^2} = 0, \tag{3.44}$$

$$\left\{u_1^{(1)} = u_1^{(2)}\right\}\Big|_{r=R_2},$$
$$\left\{G^{(1)}\left(\frac{\partial u_0}{\partial \mathbf{n}} + \frac{\partial u_1^{(1)}}{\partial r}\right) = G^{(2)}\left(\frac{\partial u_0}{\partial \mathbf{n}} + \frac{\partial u_1^{(2)}}{\partial r}\right)\right\}\Bigg|_{r=R_2}, \tag{3.45}$$

$$\left\{u_1^{(2)} = 0\right\}\Big|_{r=0}, \qquad \left\{u_1^{(1)} = 0\right\}\Big|_{r=R_1}, \tag{3.46}$$

where $\partial/\partial\mathbf{n} = \cos\theta\,\partial/\partial x_1 + \sin\theta\,\partial/\partial x_2$.

Solution of the simplified cell problem (3.44)–(3.46) is

$$u_1^{(a)} = \left(C_1^{(a)} r + C_2^{(a)} r^{-1}\right)\frac{\partial u_0}{\partial \mathbf{n}},$$

$$C_1^{(1)} = \frac{\left(\lambda^{(2)} - 1\right)c^{(2)}}{\lambda^{(2)} + 1 - c^{(2)}\left(\lambda^{(2)} - 1\right)}, \qquad C_2^{(1)} = -\frac{\left(\lambda^{(2)} - 1\right)R_2^2}{\lambda^{(2)} + 1 - c^{(2)}\left(\lambda^{(2)} - 1\right)}, \tag{3.47}$$

$$C_1^{(2)} = -\frac{\left(\lambda^{(2)} - 1\right)\left(1 - c^{(2)}\right)}{\lambda^{(2)} + 1 - c^{(2)}\left(\lambda^{(2)} - 1\right)}, \qquad C_2^{(2)} = 0,$$

where $\lambda^{(2)} = G^{(2)}/G^{(1)}$.

Substituting expressions (3.47) into the homogenized Eq. (3.42), we obtain the effective modulus G_0 in the closed form:

$$\lambda_0 = \frac{\lambda^{(2)} + 1 + c^{(2)}\left(\lambda^{(2)} - 1\right)}{\lambda^{(2)} + 1 - c^{(2)}\left(\lambda^{(2)} - 1\right)}, \tag{3.48}$$

where $\lambda_0 = G_0/G^{(1)}$.

It should be noted that solution (3.48) is precisely the same as can be obtained by the composite cylinder assemblage model and by the generalized self-consistent scheme.

In the case of densely-packed marrow inclusions, when the volume fraction $c^{(2)}$ is close to unit, an asymptotic solution of the cell problem can be obtained using as a natural small parameter the non-dimensional width $\delta = A/L$ of the trabecula (Fig. 3.13). Let us suppose $\delta << 1$. Being restricted by the $O(\delta^0)$ approximation, for the matrix strips $d\Omega_1$, $d\Omega_2$, which separate neighbouing inclusions, one can show:

$$\frac{\partial^2 u_1^{(1)}}{\partial y_1^2} >> \frac{\partial^2 u_1^{(1)}}{\partial y_2^2} \quad \text{for} \quad \mathbf{y} \in d\Omega_1, \qquad \frac{\partial^2 u_1^{(1)}}{\partial y_1^2} << \frac{\partial^2 u_1^{(1)}}{\partial y_2^2} \quad \text{for} \quad \mathbf{y} \in d\Omega_2. \tag{3.49}$$

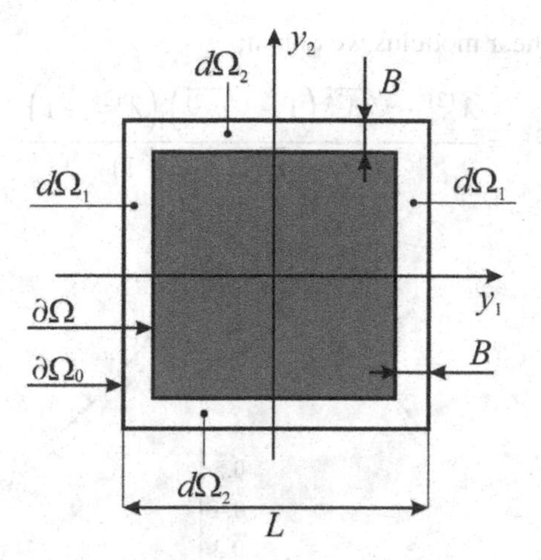

Figure 3.13 Unit cell in the case $c^{(2)} \to 1$.

The physical meaning of estimations (3.49) is that in the narrow strip $d\Omega_1$ the variation of local stresses in the direction y_1 is dominant and, hence, the term $\partial^2 u_1^{(1)}/\partial y_2^2$ can be neglected in comparison with $\partial^2 w_1^{(1)}/\partial y_1^2$. Vice versa, in the strip $d\Omega_2$ the dominant variation of the local stress field takes place in the direction y_2, so the term $\partial^2 w_1^{(1)}/\partial y_1^2$ can be neglected in comparison with $\partial^2 w_1^{(1)}/\partial y_2^2$. Such a simplification is similar to the basic idea of the well-known lubrication theory, which was used in the theory of composites for many years [123].

Following estimations (3.48), in the $O(\delta^0)$ approximation Eq. (3.35) reads

$$G^{(1)}\left(\frac{\partial^2 u_{i-2}^{(1)}}{\partial x_s^2}+2\frac{\partial^2 u_{i-1}^{(1)}}{\partial x_s \partial y_s}+\frac{\partial^2 u_i^{(1)}}{\partial y_s^2}\right)=\rho^{(1)}\frac{\partial^2 u_{i-2}^{(1)}}{\partial t^2},$$
$$G^{(2)}\left(\nabla_{xx}^2 u_{i-2}^{(2)}+2\nabla_{xy}^2 u_{i-1}^{(2)}+\nabla_{yy}^2 u_i^{(2)}\right)=\rho^{(2)}\frac{\partial^2 u_{i-2}^{(2)}}{\partial t^2}. \tag{3.50}$$

Solution of the simplified cell problem (3.36), (3.40), (3.50) at $i=1$ is:

$$u_1^{(2)}=-\frac{\left(1-\sqrt{c^{(2)}}\right)\left(\lambda^{(2)}-1\right)}{\lambda^{(2)}-\sqrt{c^{(2)}}\left(\lambda^{(2)}-1\right)}y_s\frac{\partial u_0}{\partial x_s},$$
$$u_1^{(1)}=-\frac{\sqrt{c^{(2)}}\left(\lambda^{(2)}-1\right)}{\lambda^{(2)}-\sqrt{c^{(2)}}\left(\lambda^{(2)}-1\right)}\left(\frac{L}{2}-y_s\right)\frac{\partial u_0}{\partial x_s}\quad \text{at} \quad y_s>0, \tag{3.51}$$
$$u_1^{(1)}=\frac{\sqrt{c^{(2)}}\left(\lambda^{(2)}-1\right)}{\lambda^{(2)}-\sqrt{c^{(2)}}\left(\lambda^{(2)}-1\right)}\left(\frac{L}{2}+y_s\right)\frac{\partial u_0}{\partial x_s}\quad \text{at} \quad y_s<0,$$

where $s=1,2$.

For the effective shear modulus we obtain:

$$\lambda_0 = \frac{\lambda^{(2)} - \sqrt{c^{(2)}}\left(1 - \sqrt{c^{(2)}}\right)\left(\lambda^{(2)} - 1\right)}{\lambda^{(2)} - \sqrt{c^{(2)}}\left(\lambda^{(2)} - 1\right)}. \tag{3.52}$$

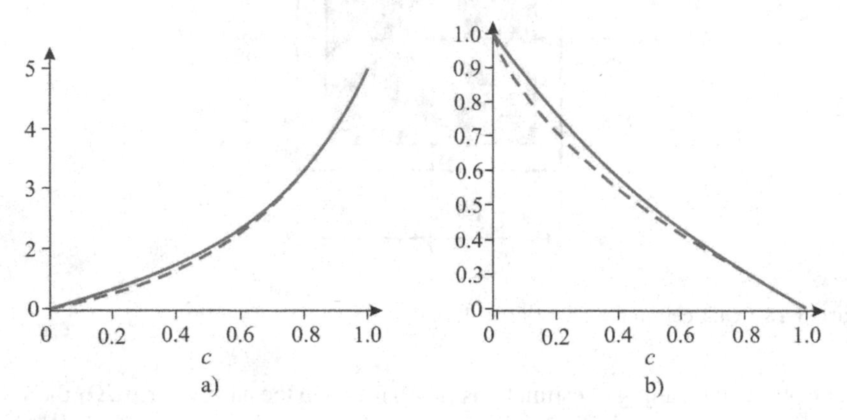

Figure 3.14 Effective modulus in the elastic case. Solids - formula (3.48), dashes - formula (3.52): a) $\lambda^{(2)} = 5$, b) $\lambda^{(2)} = 0.2$.

Numerical results, calculated by formulas (3.48), (3.52), are very close (except the case $\lambda^{(2)} < 1$, $c^{(2)} \to 0$). This is illustrated at Fig. 3.14 for real values of $\lambda^{(2)}$, which correspond to elastic materials. Moreover, in the limit $c^{(2)} \to 1$ the approximate solutions (3.48), (3.52) exhibits the same asymptotic behavior and give identical expansions for λ_0 until the order $O\left[(1-c)^2\right]$:

$$\lambda_0 = \lambda^{(2)} - \frac{1}{2}\left(\lambda^{(2)} - 1\right)\left(\lambda^{(2)} + 1\right)\left(1 - c^{(2)}\right) + O\left[\left(1 - c^{(2)}\right)^2\right] \quad \text{at} \quad c^{(2)} \to 1. \tag{3.53}$$

This fact reveals that for the cancellous structure under consideration expression (3.48), originally obtained for the case $c^{(2)} << 1$, provides a reasonable approximation in the whole region of the inclusions volume fraction $0 \leq c^{(2)} \leq 1$.

Let us consider a harmonic wave

$$u_0 = U \exp(-i\mu \cdot \mathbf{x}) \exp(i\omega t), \tag{3.54}$$

where U is the amplitude, ω is the frequency, $\mu = \mu_1 \mathbf{e}_1 + \mu_2 \mathbf{e}_2$ is the wave vector, the direction of propagation is determined by the angle $\alpha = (\mathbf{e}_1, \mu)$ and $\tan\alpha = \mu_2/\mu_1$.

Separating real μ_R and imaginary μ_I parts of the wave vector $\mu = \mu_R - i\mu_I$, expression (3.54) reads:

$$u_0 = U \exp(-\mu_I \cdot \mathbf{x}) \exp(-i\mu_R \cdot \mathbf{x}) \exp(i\omega t). \tag{3.55}$$

Here $\mu_I = |\mu_I|$ is the attenuation factor and $\mu_R = |\mu_R| = 2\pi/L$ is the wave number.

For the viscoelastic composite medium the effective complex modulus G_0, the attenuation coefficient μ_I and the phase velocity $v_p = \omega/\mu_R$ depend on the frequency of the travelling signal. Substituting expression (3.29) into the macroscopic wave equation (3.43), we obtain:

$$(G_{0,R} + iG_{0,I})(\mu_I + i\mu_R)^2 = -\rho_0\omega^2, \tag{3.56}$$

where $G_{0,R}$, $G_{0,I}$ are, respectively, the real and the imaginary part of $G_0 = G_{0,R} + iG_{0,I}$.

After routine transformations we derive:

$$\mu_I = \mu_R \tan(\varphi_0/2), \qquad v_p^2 = \frac{v_0^2}{\cos(\varphi_0/2)^2}, \tag{3.57}$$

where $\tan(\varphi_0) = G_{0,I}/G_{0,R}$ is the effective loss tangent, $v_0 = \sqrt{|G_0|/\rho_0}$ is the effective velocity in the elastic case.

Adopting for G_0 the solution (3.48), we obtain:

$$\begin{aligned} G_0 &= G^{(1)}\frac{G^{(1)}\left(1-c^{(2)}\right)+i\omega\nu^{(2)}\left(1+c^{(2)}\right)}{G^{(1)}\left(1+c^{(2)}\right)+i\omega\nu^{(2)}\left(1-c^{(2)}\right)}, \\ \tan(\varphi_0) &= \frac{4G^{(1)}c^{(2)}\omega\nu^{(2)}}{\left[1-\left(c^{(2)}\right)^2\right]\left[\left(G^{(1)}\right)^2+\left(\omega\nu^{(2)}\right)^2\right]}. \end{aligned} \tag{3.58}$$

In the numerical examples we accept some rough estimations of the properties of the components following [100, 194, 444]. The shear modulus of the trabeculae is $G^{(1)} = 3.85 \cdot 10^9$ Pa, the viscosity of the marrow is $\nu^{(2)} = 0.15$ Pa · s (at the room temperature of 20°C) and $\nu^{(2)} = 0.05$ Pa · s (at the body temperature of 37°C). The trabeculae volume fraction $c^{(1)} = 1 - c^{(2)}$ can vary from 0.05...0.1 for aged osteoporotic bones to 0.3...0.35 for young normal bones.

Dependencies of the attenuation factor μ_I upon the frequency ω are displayed at Fig. 3.15 (normal bone, $c^{(1)} = 0.3$) and Fig. 3.16 (osteoporotic bone, $c^{(1)} = 0.1$). The dispersion effect vanishes (i) at $\omega \to 0$, when the deformation rate is small and the stiffness of the marrow is negligible, and (ii) at $\omega \to \infty$, when the deformation rate is high, so the marrow acts like a perfectly stiff medium. Decrease of the trabeculae volume fraction $c^{(1)}$ leads to the intensifying of the dispersion: the damping frequency region extends and the attenuation factor μ_I grows. Decrease in temperature (i.e., increase in the marrow viscosity $\nu^{(2)}$) leads to a reduction of the damping frequency. In any case, for physically meaningful values of the bone properties the effect of viscoelastic damping can be observed starting from the frequencies of the order 100 MHz and higher.

It should be noted that in the long-wave limit ($l << L$) the cancellous structure under consideration is transversely-orthotropic. The obtained solution for antiplane shear waves is isotropic in the plane x_1x_2, so the parameters G_0, φ_0 do not depend on the direction of the wave propagation. The effect of anisotropy is predicted in the case of short waves.

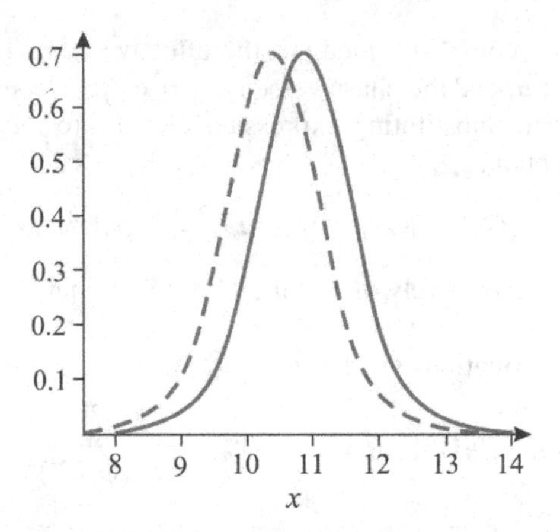

Figure 3.15 Attenuation factor of a normal bone. Solids - $\nu^{(2)} = 0.05$ Pa $\cdot$ s, dashes - $\nu^{(2)} = 0.15$ Pa $\cdot$ s.

When the wavelength L is comparable to the internal size l of the cancellous structure, the effect of dispersion is caused by successive reflections and refractions of local waves at the trabecula-marrow interfaces. Decrease in the wavelength reveals a sequence of pass and stop frequency bands. Thus, a heterogeneous bone can act as a discrete wave filter. If the frequency of the signal falls within a stop band, a stationary wave is excited and neighbouring trabeculae vibrate in alternate directions. On macrolevel the amplitude of the global wave attenuates exponentially, so no propagation is possible.

In order to explore such a case, let us assume that the beginning of the first stop band is essentially lower than the viscoelastic damping frequencies. The marrow is not involved into the shear deformation, so we can set $G^{(2)} = 0$, $\rho^{(2)} = 0$.

Following the Bloch-Floquet theorem (Chapter 2.3), a harmonic wave, propagating through a periodic cancellous structure, is represented in the form:

$$w = F(\mathbf{x}) \exp(i\mu \cdot \mathbf{x}) \exp(i\omega t), \tag{3.59}$$

where $F(\mathbf{x})$ is a spatially periodic function, $F(\mathbf{x}) = F(\mathbf{x} + \mathbf{l}_p)$.

We use the plane-wave expansions method [411] and express the function $F(\mathbf{x})$ and the material properties $G(\mathbf{x})$, $\rho(\mathbf{x})$ as Fourier series:

$$\begin{aligned} F(\mathbf{x}) &= \sum_{k_1=-\infty}^{\infty} \sum_{k_2=-\infty}^{\infty} A_{k_1 k_2} \exp\left[i\frac{2\pi}{l}(k_1 x_1 + k_2 x_2)\right], \\ G(\mathbf{x}) &= \sum_{k_1=-\infty}^{\infty} \sum_{k_2=-\infty}^{\infty} B_{k_1 k_2} \exp\left[i\frac{2\pi}{l}(k_1 x_1 + k_2 x_2)\right], \\ \rho(\mathbf{x}) &= \sum_{k_1=-\infty}^{\infty} \sum_{k_2=-\infty}^{\infty} C_{k_1 k_2} \exp\left[i\frac{2\pi}{l}(k_1 x_1 + k_2 x_2)\right], \end{aligned} \tag{3.60}$$

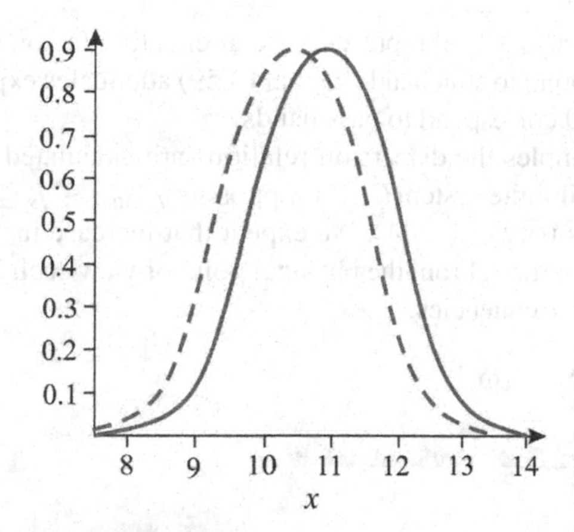

Figure 3.16 Attenuation factor of an osteoporotic bone. Solids - $\nu^{(2)} = 0.05$ Pa · s, dashes - $\nu^{(2)} = 0.15$ Pa · s.

where

$$B_{k_1k_2} = \frac{1}{l^2}\iint\limits_{\Omega_0} G(\mathbf{x})\exp\left[-i\frac{2\pi}{l}(k_1x_1+k_2x_2)\right]dx_1dx_2,$$

$$C_{k_1k_2} = \frac{1}{l^2}\iint\limits_{\Omega_0} \rho(\mathbf{x})\exp\left[-i\frac{2\pi}{l}(k_1x_1+k_2x_2)\right]dx_1dx_2,$$

the operator $\iint_{\Omega_0}(\cdot)dx_1dx_2$ denotes integration over a unit cell Ω_0.

Substituting ansatz (3.59) and expansions (3.60) into the wave equation (3.28) and collecting the terms $\exp\left[i2\pi l^{-1}(j_1x_1+j_2x_2)\right]$, $j_1, j_2 = 0, \pm1, \pm2, ...,$ we come to an infinite system of linear algebraic equations for the unknown coefficients $A_{k_1k_2}$:

$$\sum_{k_1=-\infty}^{\infty}\sum_{k_2=-\infty}^{\infty} A_{k_1k_2}\left\{B_{\substack{j_1-k_1,\\ j_2-k_2}}\left[\left(\frac{2\pi}{l}k_1+\mu_1\right)\left(\frac{2\pi}{l}j_1+\mu_1\right)+\right.\right.$$
$$\left.\left.\left(\frac{2\pi}{l}k_2+\mu_2\right)\left(\frac{2\pi}{l}j_2+\mu_2\right)\right]-C_{\substack{j_1-k_1,\\ j_2-k_2}}\omega^2\right\}=0. \tag{3.61}$$

System (3.61) has a nontrivial solution if and only if the determinant of the matrix of the coefficients is zero. Equating the determinant to zero, we derive a dispersion relation for ω and μ. It should be noted that the plane-wave expansions method does not use explicitly the bonding conditions (3.31), whereas they are "embedded" implicitly into equation (3.28) and expansions (3.60).

To illustrate the appearance of phononic band gaps let us rewrite ansatz (3.59) separating real μ_R and imaginary μ_I parts of the wave vector $\mu = \mu_R + i\mu_I$:

$$w = F(\mathbf{x})\exp(-\mu_I\cdot\mathbf{x})\exp(i\mu_R\cdot\mathbf{x})\exp(i\omega t). \tag{3.62}$$

The imaginary part $\mu_I \equiv |\mu_I|$ represents the attenuation factor. Frequency regions where $\mu_I \neq 0$ correspond to stop bands (signal (3.59) attenuates exponentially), while regions where $\mu_I = 0$ correspond to pass bands.

In numerical examples the dispersion relations are calculated approximately by the truncation of the infinite system (3.61) supposing $-j_{\max} \leq j_s \leq j_{\max}$. The number of the kept equations is $(2j_{\max}+1)^2$. We expect that increase in $j_{\max}$ shall improve the accuracy of the solution. From the physical point of view such a truncation means cutting off the higher frequencies.

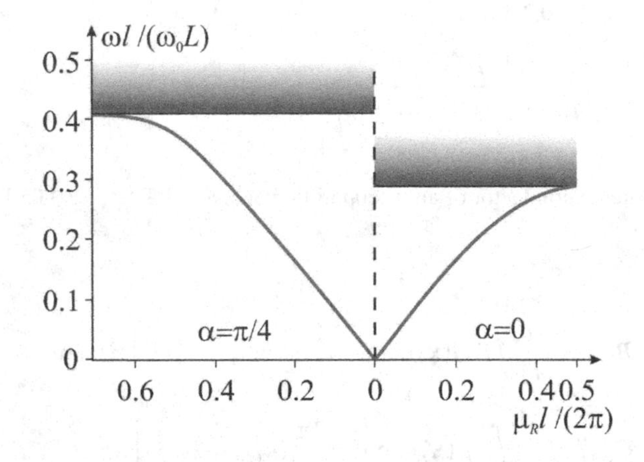

Figure 3.17 Dispersion curves of a normal bone.

Fig. 3.17 displays dispersion curves for a normal bone with the following properties of the trabecular tissue: $G^{(1)} = 3.85 \cdot 10^9$ Pa, $\rho^{(1)} = 1900$ kg/m^3, $c^{(1)} = 0.3$. Calculations are performed at $j_{\max} = 3$. The diagram consists of two (left and right) parts separated by a vertical dash line. The right part displays a solution for the orthogonal direction ($\alpha = 0$) and the left part - for the diagonal direction ($\alpha = \pi/4$) of the wave propagation. The results for the frequency ω are normalized to $\omega_0 = v_0\mu_R = 2\pi v_0/L$. We can observe that in the long-wave case ($\omega \to 0$, $l/L \to 0$) the solution is isotropic. However, with the increase in ω and decrease in L the cancellous structure exhibits an anisotropic behavior.

Shaded areas in Fig. 3.17 indicate the beginning of the first stop bands. Let us estimate the corresponding values ω_s of the frequency. We obtain $\omega_s l/(\omega_0 L) \approx 0.29$ at $\alpha = 0$ and $\omega_s l/(\omega_0 L) \approx 0.41$ at $\alpha = \pi/4$. The typical length of trabeculae is about $l \approx 10^{-3}$m. Taking into account $\omega_0 = 2\pi v_0/L$, $v_0 = \sqrt{G_0/\rho_0}$, we derive: $\omega_s \approx 2.0$ MHz at $\alpha = 0$ and $\omega_s \approx 2.8$ MHz at $\alpha = \pi/4$.

4 Longitudinal Waves in Layered Composites with Account of Physical and Geometrical Nonlinearities

Wave propagation in nonlinear elastic media with microstructure is studied. As an illustrative example, a 1D model of a layered composite material is considered. Geometrical nonlinearity is described by the Cauchy-Green strain tensor. For predicting physical nonlinearity, energy of deformation as a series expansion in powers of the strains is employed. The effective wave equation is derived by the higher-order asymptotic homogenization method. An asymptotic solution of the nonlinear cell problem is obtained employing series expansions in powers of the gradients of displacements. Analytical expressions for the effective moduli are presented. The balance between nonlinearity and dispersion results in formation of stationary nonlinear waves that are described explicitly in terms of elliptic functions. In the case of weak nonlinearity, an asymptotic solution is developed. A number of nonlinear phenomena are detected, such as generation of higher-order modes and localization. Numerical results are presented and practical significance of the nonlinear effects is illustrated and discussed.

4.1 FUNDAMENTAL RELATIONS OF NONLINEAR THEORY OF ELASTICITY

In this section, we consider the elastic waves of deformations in a periodically nonhomogeneous material with geometrical and physical nonlinearity. In the beginning we revisit the fundamental equations of the nonlinear theory of elasticity for the homogeneous isotropic medium. The geometrical nonlinearity can be taken into account by introducing of the exact relations between deformations ε_{ij} and displacements u_i. In the predeformed state those relations are governed by the following Cauchy-Green deformation tensor [290]:

$$\varepsilon_{ij} = \frac{1}{2}\left(\frac{\partial u_i}{\partial x_j} + \frac{\partial u_j}{\partial x_i} + \frac{\partial u_k}{\partial x_j}\frac{\partial u_k}{\partial x_i}\right), \qquad \varepsilon_{ij} = \varepsilon_{ji}, \quad i,j,k = 1,2,3. \tag{4.1}$$

Here and later in this section the summation takes place within all repeated indices. The Piola-Kirchhoff stress tensor is defined through density of the potential energy of deformation W in the following way

$$\sigma_{ij} = \frac{\partial W}{\partial \left(\partial u_i / \partial x_j\right)}. \tag{4.2}$$

The motion equations take the following form

$$\frac{\partial \sigma_{ij}}{\partial x_j} = \rho \frac{\partial^2 u_i}{\partial t^2}, \tag{4.3}$$

where ρ stands for the density.

In order to describe the physical nonlinearity we present density of the internal energy W in the form of series development with regard to the powers of invariants of the deformation tensor [290] of the following form

$$W = \frac{\lambda}{2}\varepsilon_{ii}^2 + \mu\varepsilon_{ij}^2 + \frac{A}{3}\varepsilon_{ij}\varepsilon_{jk}\varepsilon_{ik} + B\varepsilon_{ii}\varepsilon_{ij}\varepsilon_{ji} + \frac{C}{3}\varepsilon_{ii}^3 + O\left(\varepsilon_{ij}^4\right). \tag{4.4}$$

First two terms of series (4.4) correspond to the linear-elastic model, whereas λ, μ are the Lamé coefficients. The successive terms take into account physical nonlinearity, whereas A, B, C are elastic moduli of third order (Landau coefficients). Equation (4.4) describes the elastic Murnaghan potential [328]. It should be emphasized that the Landau coefficients are known [111, 148, 178, 214, 290, 370] (see also Table 4.1) for majority of materials.

Table 4.1

Elastic properties of some materials and rock-forming minerals [370], GPa.

Material	λ	μ	A	B	C
Steel Helca 37	111	82.1	-720	-280	-180
Aluminium D16T	57	27.6	-260	-180	-110
Organic glass	3.9	1.9	-14.4	-7.2	-4
Polystyrol	1.7	0.95	-10	-8	-11
Granite	22	23.6	-14070	-20230	-1150
Limestone	22.7	20.6	-9730	-6435	-1870
Sandstone	1.9	6.3	-17530	-5670	-2230

However, the Murnaghan model holds only for weak physical nonlinearity. The latter increases with increase of the deformation amplitude. The series (4.4) can be used in practical/engineering computations if the relations of three successive terms to the first two does not overcome the value of 10^{-1}. In the case of majority of the design materials, including metals and polymers, the elasticity third-order moduli are negative and the absolute value larger then the elasticity moduli of the second order. Therefore, the area of applicability of the series (4.4) is bounded by the value of maximum possible deformation $\varepsilon \leq 10^{-2}$, whereas in solid bodies the elastic deformations are less than 10^{-3}.

The Murnaghan model might be unsuitable for modeling of the rubber-type materials and elastomers, which allow for account of large elastic deformations (up to the order of 10^0) as well as for the rock-forming minerals which exhibit large values of the third-order elastic moduli (Table 4.1). In the latter cases, other relations for internal energy W are used [337].

4.2 INPUT BOUNDARY VALUE PROBLEMS

We consider a layered composite material composed of two components $\Omega^{(1)}$ and $\Omega^{(2)}$ (Fig. 4.1). Let the plane waves propagate in direction to the layers location; hence, the parameters of the stress-strain state depend only on one spatial co-ordinate x_1, $u_i = u_i(x_1, t)$. Substitution of (4.1), (4.2), (4.4) into equation (4.3) yields the following set of three PDEs

$$E_1^{(n)} \frac{\partial^2 u_1^{(n)}}{\partial x_1^2} + E_2^{(n)} \frac{\partial u_1^{(n)}}{\partial x_1} \frac{\partial^2 u_1^{(n)}}{\partial x_1^2} + \\ E_{20}^{(n)} \left(\frac{\partial u_2^{(n)}}{\partial x_1} \frac{\partial^2 u_2^{(n)}}{\partial x_1^2} + \frac{\partial u_3^{(n)}}{\partial x_1} \frac{\partial^2 u_3^{(n)}}{\partial x_1^2} \right) = \rho^{(n)} \frac{\partial^2 u_1^{(n)}}{\partial t^2}, \tag{4.5}$$

$$\mu^{(n)} \frac{\partial^2 u_2^{(n)}}{\partial x_1^2} + E_{20}^{(n)} \left(\frac{\partial u_1^{(n)}}{\partial x_1} \frac{\partial^2 u_2^{(n)}}{\partial x_1^2} + \frac{\partial u_2^{(n)}}{\partial x_1} \frac{\partial^2 u_1^{(n)}}{\partial x_1^2} \right) = \rho^{(n)} \frac{\partial^2 u_2^{(n)}}{\partial t^2}, \tag{4.6}$$

$$\mu^{(n)} \frac{\partial^2 u_3^{(n)}}{\partial x_1^2} + E_{20}^{(n)} \left(\frac{\partial u_1^{(n)}}{\partial x_1} \frac{\partial^2 u_3^{(n)}}{\partial x_1^2} + \frac{\partial u_3^{(n)}}{\partial x_1} \frac{\partial^2 u_1^{(n)}}{\partial x_1^2} \right) = \rho^{(n)} \frac{\partial^2 u_3^{(n)}}{\partial t^2}. \tag{4.7}$$

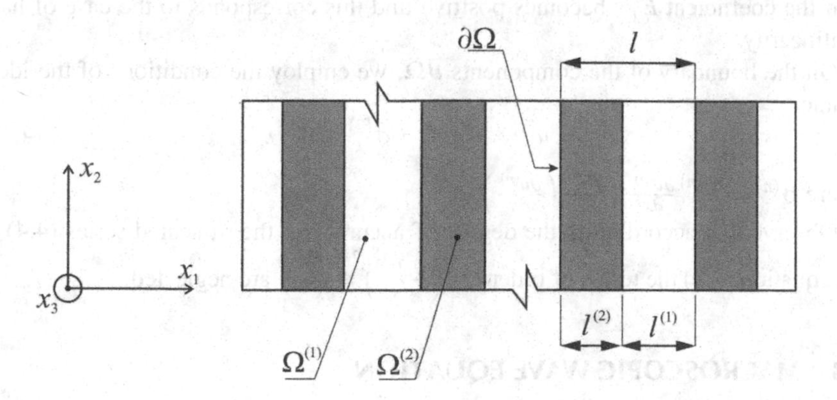

Figure 4.1 Layered composite material.

Here and later, index (n) denotes layers number $n = 1, 2$, and $\rho^{(n)}$ stand for their density.

In the above

$$E_1^{(n)} = \lambda^{(n)} + 2\mu^{(n)}, \qquad E_{20}^{(n)} = \lambda^{(n)} + 2\mu^{(n)} + A^{(n)}/2 + B^{(n)},$$
$$E_2^{(n)} = 3\left(\lambda^{(n)} + 2\mu^{(n)}\right) + 2\left(A^{(n)} + 3B^{(n)} + C^{(n)}\right) E_2^{(n)} =$$
$$3\left(\lambda^{(n)} + 2\mu^{(n)}\right) + 2\left(A^{(n)} + 3B^{(n)} + C^{(n)}\right). \tag{4.8}$$

Equation (4.5) describes the longitudinal extension-compression wave, whereas equations (4.6), (4.7) stand for the two transversal waves. In contrary to the linear case, the longitudinal and transversal waves are coupled through nonlinear terms, and elastic constant $E_{20}^{(n)}$ plays the role of coupling coefficient.

We consider further the longitudinal wave assuming that displacement component $u_2^{(n)}$, $u_3^{(n)}$ are equal zero ($u_2^{(n)} = u_3^{(n)} = 0$) and we introduce the notation $x \equiv x_1$, $u \equiv u_1$, $\sigma \equiv \sigma_{11}$. We get

$$E_1^{(n)} \frac{\partial^2 u^{(n)}}{\partial x^2} + E_2^{(n)} \frac{\partial u^{(n)}}{\partial x} \frac{\partial^2 u^{(n)}}{\partial x^2} = \rho^{(n)} \frac{\partial^2 u^{(n)}}{\partial t^2}. \tag{4.9}$$

Here $E_1^{(n)}$ stands for linear modulus of elasticty, whereas $E_2^{(n)}$ is nonlinear modulus. In equation (4.8) for $E_2^{(n)}$, the first term $3\left(\lambda^{(n)} + 2\mu^{(n)}\right)$ describes the geometric nonlinearity, whereas the second term $2\left(A^{(n)} + 3B^{(n)} + C^{(n)}\right)$ exhibits input of the physical nonlinearity. In majority of the physically nonlinear materials, the Landau coefficients are negative and they exceed the Lamé coefficients (Table 4.1) with regard to their absolute values. It means that the coefficient $E_2^{(n)}$ is negative, which corresponds to the case of soft nonlinearity. If a material is physically linear ($A^{(n)} = B^{(n)} = C^{(n)} = 0$), whereas nonlinearity is originated from geometric factors, then the coefficient $E_2^{(n)}$ becomes positive and this corresponds to the case of hard nonlinearity.

On the boundary of the components $\partial\Omega$, we employ the conditions of the ideal contact

$$u^{(1)} = u^{(2)}, \quad \sigma^{(1)} = \sigma^{(2)} \quad \text{at} \quad \partial\Omega, \tag{4.10}$$

where $\sigma^{(n)} = E_1^{(n)} \frac{\partial u^{(n)}}{\partial x} + \frac{E_2^{(n)}}{2} \left(\frac{\partial u^{(n)}}{\partial x}\right)^2$.

Observe that according to the degree of accuracy of the truncated series (4.4) in the equation (4.9) the terms of order $O\left[\left(\frac{\partial u^{(n)}}{\partial x}\right)^2 \frac{\partial^2 u^{(n)}}{\partial x^2}\right]$ are neglected.

4.3 MACROSCOPIC WAVE EQUATION

Employment of the asymptotic method of homogenization allows for transition from the boundary value problems for nonhomogeneous media (4.9), (4.10), to the wave equation for homogeneous media.

Let us introduce the non-dimensional variables $\bar{u} = u/U$, $\bar{x} = x/L$, where U stands for the displacements amplitude, and L is the wave length.

PDEs (4.9), (4.10) are recast to the following form

$$E_1^{(n)} \frac{\partial^2 \bar{u}^{(n)}}{\partial \bar{x}^2} + \delta E_2^{(n)} \frac{\partial \bar{u}^{(n)}}{\partial \bar{x}} \frac{\partial^2 \bar{u}^{(n)}}{\partial \bar{x}^2} = \rho^{(n)} L^2 \frac{\partial^2 \bar{u}^{(n)}}{\partial t^2}, \tag{4.11}$$

$$\bar{u}^{(1)} = \bar{u}^{(2)},$$

$$E_1^{(1)} \frac{\partial \bar{u}^{(1)}}{\partial \bar{x}} + \delta \frac{E_2^{(1)}}{2} \left(\frac{\partial \bar{u}^{(1)}}{\partial \bar{x}} \right)^2 = E_1^{(2)} \frac{\partial \bar{u}^{(2)}}{\partial \bar{x}} + \delta \frac{E_2^{(2)}}{2} \left(\frac{\partial \bar{u}^{(n)}}{\partial \bar{x}} \right)^2 \quad \text{at} \quad \partial\Omega, \tag{4.12}$$

where $\delta = U/L$ presents a small parameter.

We introduce two spatial scales associated with the size l of the periodicity cell (microlevel) and with the wave length L (macrolevel). Assuming $l < L$, we introduce the small parameter $\eta = l/L$ characterizing the inhomogenity of the composite materials.

We employ the "slow" $\bar{x} = \bar{x}$ and "fast" $y = \eta^{-1}\bar{x}$ variables. In the co-ordinate y, the cell size is equal to 1. The derivatives are computed in the following way

$$\frac{\partial}{\partial \bar{x}} = \frac{\partial}{\partial \bar{x}} + \eta^{-1} \frac{\partial}{\partial y}, \qquad \frac{\partial^2}{\partial \bar{x}^2} = \frac{\partial^2}{\partial \bar{x}^2} + 2\eta^{-1} \frac{\partial^2}{\partial \bar{x} \partial y} + \eta^{-2} \frac{\partial^2}{\partial y^2}. \tag{4.13}$$

A solution to the boundary value problems (4.11), (4.12) is searched in the following form

$$\bar{u}^{(n)} = u_0(\bar{x}) + \eta u_1^{(n)}(\bar{x}, y) + \eta^2 u_2^{(n)}(\bar{x}, y) + \ldots, \tag{4.14}$$

where first term u_0 presents the averaged part of the solution; it slowly changes on the macrolevel and does not depend on the fast co-ordinates. The following terms $u_i^{(n)}, i = 1, 2, 3, ...,$ introduce corrections of the order η^i and describe the local oscillations of the solution in microlevel.

Splitting of relations (4.11), (4.12) with respect to the parameter η yields the following set of recurrent sequence of the nonlinear equations

$$E_1^{(n)} \left(\frac{\partial^2 u_{i-2}^{(n)}}{\partial \bar{x}^2} + \frac{\partial^2 u_{i-1}^{(n)}}{\partial \bar{x} \partial y} + \frac{\partial^2 u_i^{(n)}}{\partial y^2} \right) + \delta E_2^{(n)} \sum_{j=0}^{i-1} \left[\left(\frac{\partial u_j^{(n)}}{\partial \bar{x}} + \frac{\partial u_{j+1}^{(n)}}{\partial y} \right) \times \right.$$
$$\left. \left(\frac{\partial^2 u_{i-j-2}^{(n)}}{\partial \bar{x}^2} + 2 \frac{\partial^2 u_{i-j-1}^{(n)}}{\partial \bar{x} \partial y} + \frac{\partial^2 u_{i-j}^{(n)}}{\partial y^2} \right) \right] = \rho^{(n)} L^2 \frac{\partial^2 u_{i-2}^{(n)}}{\partial t^2} \tag{4.15}$$

and the boundary conditions

$$u_i^{(1)} = u_i^{(2)} \quad \text{at} \quad \partial\Omega, \tag{4.16}$$

$$
\begin{aligned}
&E_1^{(1)}\left(\frac{\partial u_{i-1}^{(1)}}{\partial \bar{x}}+\frac{\partial u_i^{(1)}}{\partial y}\right)+\\
&\delta\frac{E_2^{(1)}}{2}\sum_{j=0}^{i-1}\left[\left(\frac{\partial u_j^{(1)}}{\partial \bar{x}}+\frac{\partial u_{j+1}^{(1)}}{\partial y}\right)\left(\frac{\partial u_{i-j-1}^{(1)}}{\partial \bar{x}}+\frac{\partial u_{i-j}^{(1)}}{\partial y}\right)\right]=\\
&E_1^{(2)}\left(\frac{\partial \bar{u}_{i-1}^{(2)}}{\partial \bar{x}}+\frac{\partial \bar{u}_i^{(2)}}{\partial y}\right)+\\
&\delta\frac{E_2^{(2)}}{2}\sum_{j=0}^{i-1}\left[\left(\frac{\partial u_j^{(2)}}{\partial \bar{x}}+\frac{\partial u_{j+1}^{(2)}}{\partial y}\right)\left(\frac{\partial u_{i-j-1}^{(2)}}{\partial \bar{x}}+\frac{\partial u_{i-j}^{(2)}}{\partial y}\right)\right] \quad \text{at} \quad \partial\Omega,
\end{aligned}
\tag{4.17}
$$

where $i = 1, 2, 3, \ldots$, $u_{-1}^{(n)} = 0$.

Owing to periodicity of the microstructure of the composite material, the functions $u_i^{(n)}$ also satisfy the periodicity conditions:

$$u_i^{(n)}(\bar{x}, y) = u_i^{(n)}(\bar{x}, y \pm 1), \tag{4.18}$$

and hence the boundary value problem (4.15)–(4.17) can be considered in the interval of one separated cell (Fig. 4.2).

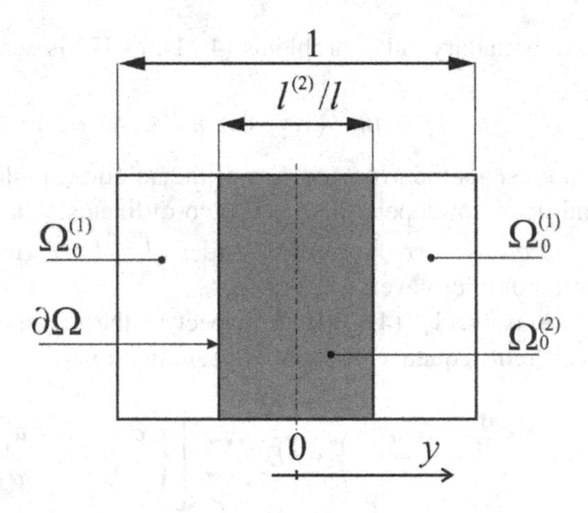

Figure 4.2 Cell of periodicity of the layered composite material.

Relation (4.18), without loss of generality, can be substituted by the conditions of zero average value with regard to the period, i.e.

$$\int\limits_{\Omega_0^{(1)}} u_i^{(1)} dy + \int\limits_{\Omega_0^{(2)}} u_i^{(2)} dy = 0. \tag{4.19}$$

Algorithm of the used homogenization method is as follows. Solving sequently the boundary value problems defined on the cell (4.15)–(4.17), (4.19) for $i = 1, 2, \ldots, k$, we find $u_1^{(n)}$, $u_2^{(n)}$, ..., $u_k^{(n)}$.

Then we proceed with the averaging operator $\int_{-1/2}^{1/2} (\cdot)\, dy$ on the $(k+1)$ equation of (4.15). The term $u_{k+1}^{(n)}$ is cancelled due the boundary condition (4.17) and relations (4.18). In result, one gets the macroscopic wave equation of order η^{k-1}.

In this section we carry out computations for $k = 3$. Owing to the given accuracy of the model (4.4), the terms $u_2^{(n)}$ and $u_3^{(n)}$ can be taken into account in the linear approximation. The solution $u_1^{(n)}$ is found based on the following series with regard to the small parameter δ:

$$u_1^{(n)} = u_{10}^{(n)} + \delta u_{11}^{(n)} + O\left(\delta^2\right), \tag{4.20}$$

where

$$u_{10}^{(2)} = A_{10} y \frac{\partial u_0}{\partial \bar{x}}, \qquad u_{10}^{(1)} = (B_{10} y + C_{10}) \frac{\partial u_0}{\partial \bar{x}},$$

$$u_{11}^{(2)} = A_{11} y \left(\frac{\partial u_0}{\partial \bar{x}}\right)^2, \qquad u_{11}^{(1)} = (B_{11} y + C_{11}) \left(\frac{\partial u_0}{\partial \bar{x}}\right)^2,$$

$$A_{10} = -\frac{c^{(1)}\left(E_1^{(2)} - E_1^{(1)}\right)}{c^{(2)} E_1^{(1)} + c^{(1)} E_1^{(2)}}, \qquad B_{10} = -A_{10} \frac{c^{(2)}}{c^{(1)}},$$

$$C_{10} = -\frac{B_{10}}{2} \quad \text{for} \quad y > 0, \qquad C_{10} = \frac{B_{10}}{2} \quad \text{for} \quad y < 0,$$

$$A_{10} = -\frac{c^{(1)}\left[\left(E_1^{(1)}\right)^2 E_2^{(2)} - \left(E_1^{(2)}\right)^2 E_2^{(1)}\right]}{2\left(c^{(2)} E_1^{(1)} + c^{(1)} E_1^{(2)}\right)^3}, \qquad B_{11} = -A_{11} \frac{c^{(2)}}{c^{(1)}},$$

$$C_{11} = -\frac{B_{11}}{2} \quad \text{for} \quad y > 0, \qquad C_{11} = \frac{B_{11}}{2} \quad \text{for} \quad y < 0,$$

and $c^{(n)} = l^{(n)}/l$ stand for the volume fractions of the components, $c^{(1)} + c^{(2)} = 1$.

The macroscopic nonlinear wave equation of the order η^2 takes the following form

$$E_1 \frac{\partial^2 u_0}{\partial \bar{x}^2} + \delta E_2 \frac{\partial u_0}{\partial \bar{x}} \frac{\partial^2 u_0}{\partial \bar{x}^2} + \eta^2 E_3 \frac{\partial^4 u_0}{\partial \bar{x}^4} \left[1 + O(\delta)\right] + O\left(\eta^4\right) = \rho L^2 \frac{\partial^2 u_0}{\partial t^2}. \tag{4.21}$$

Carrying out the transition into dimensional variables $u = \bar{u}U$, $x = \bar{x}L$, we recast equation (4.21) to the following form

$$E_1 \frac{\partial^2 u}{\partial x^2} + E_2 \frac{\partial u}{\partial x} \frac{\partial^2 u}{\partial x^2} + \eta^2 L^2 E_3 \frac{\partial^4 u}{\partial x^4} = \rho \frac{\partial^2 u}{\partial t^2}, \tag{4.22}$$

where: ρ is the averaged density, $\rho = c^{(1)}\rho^{(1)} + c^{(2)}\rho^{(2)}$; E_1, E_2, E_3 are the effective linear stiffness coefficients.

The following analytical formulas for effective elastic coefficients are obtained

$$E_1 = \frac{E_1^{(1)} E_1^{(2)}}{c^{(1)} E_1^{(2)} + c^{(2)} E_1^{(1)}}, \tag{4.23}$$

$$E_2 = \frac{c^{(1)} E_2^{(1)} \left(E_1^{(2)}\right)^3 + c^{(2)} E_2^{(2)} \left(E_1^{(1)}\right)^3}{\left(c^{(1)} E_1^{(2)} + c^{(2)} E_1^{(1)}\right)^3}, \tag{4.24}$$

$$E_3 = E_1 \frac{\left(c^{(1)} c^{(2)}\right)^2 v_0^4}{12\left(v^{(1)} v^{(2)}\right)^2} \left(\frac{z^{(1)}}{z^{(2)}} - \frac{z^{(2)}}{z^{(1)}}\right)^2, \tag{4.25}$$

where: $z^{(n)}$ is the acoustic impedance, $z^{(n)} = \sqrt{E_1^{(n)} \rho^{(n)}}$; $v^{(n)}$ is the velocity of propagation of the wave through component (n) in the linear case, $v^{(n)} = \sqrt{E_1^{(n)} / \rho^{(n)}}$; v_0 is the effective phase velocity in the linear homogeneous medium (for $E_2 \to 0$, $l/L \to 0$), $v_0 = \sqrt{E_1/\rho}$.

Formulas (4.23)–(4.25) present the dependence of the effective characteristics of the composite material on properties and volume fraction of the component. It should be mentioned that the analytical results for E_1, E_3 have been known earlier, whereas the relations (4.24) for E_2 has been obtained for the first time.

The first term in the left hand side of the wave equation (4.22) corresponds to the model of linear homogeneous medium, whereas the second term accounts of the nonlinear effect. For majority of the physically nonlinear materials $E_2^{(n)} < 0$, and therefore the effective coefficient E_2 will be also negative, $E_2 < 0$ (soft nonlinearity). In the case of the physically linear medium, when the nonlinearity is caused only by geometric reasons, we get $E_2^{(n)} > 0$ and $E_2 > 0$ (hard nonlinearity).

The third term in equation (4.22) describes the dispersion effect caused by energy dissipation on the medium nonhomogenuities. The coefficient E_3 is equal to zero and dispersion vanishes: (i) in the case of homogeneous material ($c^{(1)} = 1$, $c^{(2)} = 0$ or $c^{(1)} = 0$, $c^{(2)} = 1$) and (ii) in the case of equality of the acoustic impedances of the components ($z^{(1)} = z^{(2)}$), when local reflections of the signal on the separation boundary $\partial\Omega$ are absent (it follows from formulas (4.25), $E_3 \geq 0$). Therefore, occurence of microstructure implies negative (normal) dispersion which yields decrease of the phase velocity under increase of the frequency. The given conclusion holds only in frame of the long-wave approximation governed by equation (4.22). For the high frequencies, when the wave length is comparable with the size of the periodicity cell, the nonhomogeneous materials may exhibit either normal or anomalous dispersion.

It should be mentioned that the homogenized model (4.22) is obtained for the case of spatially infinite medium. If one considers structure of the finite size, then solutions depend on the boundary conditions defined on the macrolevel. The mentioned dependence will introduce occurrence of the additional correcting terms in formulas (4.23)–(4.25). Therefore, the formal transition of the results obtained for the infinite medium for the case of finite size may introduce errors in the boundary conditions.

The macroscopic wave equation (4.22) has the asymptotic origin and is useful when the wave length L is larger then the characteristic size l of the microstructure such that $\eta = l/L < 1$. In order to estimate the area of applicability of this model we compare the dispersive dependence which is obtained from equation (4.22) with the exact solution found for the linear case with the help of the Floquet-Bloch theory.

We consider the linear wave ($E_2 = 0$):

$$u = U\exp(ikx)\exp(i\omega t), \tag{4.26}$$

where k is the wave number, $k = 2\pi/L$; ω is the frequency.

We substitute (4.26) into equation (4.22), and obtain the asymptotic formula for the dispersive dependence

$$\omega^2 = \omega_0^2\left[1 - 4\pi^2\frac{E_3}{E_1}\eta^2 + O\left(\eta^4\right)\right], \tag{4.27}$$

where: $\omega_0 = v_0 k$; $\eta = l/L = kl/(2\pi)$.

The formula (4.27) describes the acoustic branch of the dispersive curve. Keeping the larges number of terms we may increase the numerical accuracy of the solution. However, in principle it is impossible to model the successive branches in the upper spectrum part. With decreasing the wave length (increasing the frequency) in the composite material the pass and stop bands appear [98]. If a frequency goes into the stop band, there appears standing wave, the group velocity of which is equal to zero.

Following theory of Floquet-Bloch [98], the exact dependence between ω and k can be found by solving the transcendent equation of the following form

$$\cos(kl) = \cos\left(k^{(1)}l^{(1)}\right)\cos\left(k^{(2)}l^{(2)}\right) - \frac{1}{2}\left(\frac{z^{(1)}}{z^{(2)}} - \frac{z^{(2)}}{z^{(1)}}\right)\sin\left(k^{(1)}l^{(1)}\right)\sin\left(k^{(2)}l^{(2)}\right), \tag{4.28}$$

where $k^{(n)} = \omega/v^{(n)}$.

Table 4.2
Properties of the components of the composite materials.

Material	λ [GPa]	μ [GPa]	ρ [kg/m^3]	v [m/s]	$z \cdot 10^{-6}$
Steel Helca 37	111	82.1	7800	5940	46.3
Aluminium D16T	57	27.6	2700	6450	17.4
Carbonplastic	5.18	3.45	1600	2750	4.40

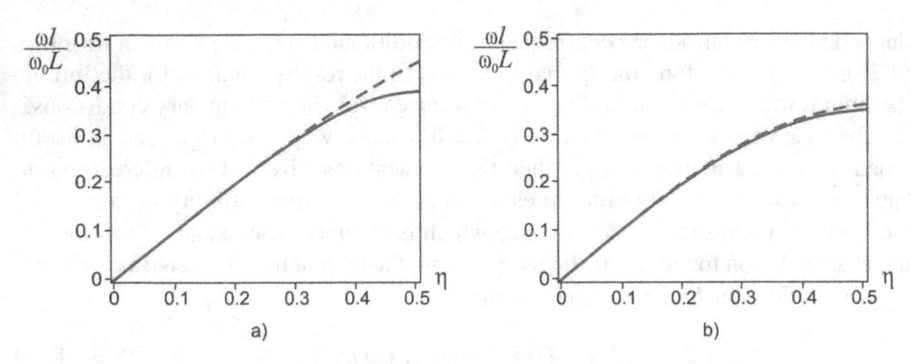

Figure 4.3 Acoustic branches of the dispersive curves in linear case: (a) steel-aluminium; (b) steel-carboplastic [reprinted with permission from Elsevier].

Fig. 4.3 presents the acoustic branches of the dispersive curves localized in the intervals of the first pass band $0 < \eta < 1/2$. The threshold value $\eta = 1/2$ $(L = 2l)$ corresponds to the first stop band. The dashed curves correspond to the asymptotic formula (4.27), whereas solid curves represent the exact solution (4.28). The computations have been carried out for the cases of low-contrast "steel-aluminium" and high-contrast "steel-carboplastic" composite materials. Properties of the components are shown in Table 4.2, based on references [286, 370]; volume fractions of components are the same, $c^{(1)} = c^{(2)} = 0.5$.

The obtained results show that the convergence of the asymptotic solution (4.27) depend on the material properties. In the considered examples the homogenized wave equation (4.22) guarantees the efficient accuracy for $\eta = l/L < 0.4$.

4.4 ANALYTICAL SOLUTION FOR STATIONARY WAVES

We consider the stationary wave propagating with a constant velocity without changing of its form. In the latter case the solution satisfies the following conditions

$$u(x,t) = u(\xi), \tag{4.29}$$

where: ξ is the running wave variable, $\xi = x - vt$, v is the phase velocity.

We introduce the following non-dimensional deformation of the wave profile:

$$f = \frac{du}{d\xi}. \tag{4.30}$$

Substituting relations (4.29), (4.30) into equation (4.22) one gets

$$\left(1 - \frac{v^2}{v_0^2}\right)\frac{df}{d\xi} + \frac{E_2}{E_1} f \frac{df}{d\xi} + \frac{E_3}{E_1} l^2 \frac{d^3 f}{d\xi^3} = 0. \tag{4.31}$$

Integration with respect to the variable ξ yields the following equations of the harmonic oscillator with squared nonlinearity

$$\frac{d^2 f}{d\xi^2} + af + bf^2 + c = 0, \tag{4.32}$$

where: $a = E_1 \left(1 - v^2/v_0^2\right) / \left(E_3 l^2\right)$, $b = E_2 / \left(2E_3 l^2\right)$, c is the constant of integration.

A few researchers did not take into account the constant c while investigating the propagation of nonlinear waves in solid media [158]. The obtained results were erroneous from the physical standpoint. It is clear that the displacements u cannot increase unboundedly. It means that the average deformation with regard to the wave period should be equal zero:

$$\frac{1}{L}\int_0^L f(\xi)\,d\xi = 0. \tag{4.33}$$

Condition (4.33) allows to find constant c. In order to find the exact analytical solution, we multiply equation (4.32) by $df/d\xi$, and after integration one gets

$$\frac{1}{2}\left(\frac{df}{d\xi}\right)^2 + \frac{a}{2}f^2 + \frac{b}{3}f^3 + f = W_0. \tag{4.34}$$

Relation (4.34) can be interpreted as the energy conservation principle of the anharmonic oscillator. The integration constant W_0 presents the initial total energy of the system, $W_0 > 0$; $W_k = (1/2)(df/d\xi)^2$ is the kinetic energy; $W_p = (a/2)f^2 + (b/3)f^3 + f$ is the potential energy; $W_0 = W_k + W_p$.

Equation (4.34) can be recast to the following form:

$$\sqrt{2}d\xi = \frac{df}{\sqrt{W_0 - cf - \frac{a}{2}f^2 - \frac{b}{3}f^3}}. \tag{4.35}$$

A solution exists if the kinetic energy satisfies the following inequality

$$W_k = W_0 - W_p = W_0 - cf - \frac{a}{2}f^2 - \frac{b}{3}f^3 > 0. \tag{4.36}$$

The cubic term $W_0 - cf - (a/2)f^2 - (b/3)f^3$, depending on the values of the coefficients, is responsible either for one or three real roots. Since the case of one real root corresponds to the unbounded solution it is not further considered, the solution is bounded and periodic if the function $W_k = W_0 - cf - (a/2)f^2 - (b/3)f^3$ takes values on the interval between its real roots.

Let us present the polynom $W_0 - cf - (a/2)f^2 - (b/3)f^3$ through its roots f_1, f_2, f_3:

$$W_0 - cf - \frac{a}{2}f^2 - \frac{b}{3}f^3 = -\frac{b}{3}(f - f_1)(f - f_2)(f - f_3), \tag{4.37}$$

where

$$f_1 + f_2 + f_3 = -\frac{3a}{2b}, \qquad f_1f_2 + f_2f_3 + f_1f_3 = \frac{3c}{b}, \qquad f_1f_2f_3 = \frac{3W_0}{b}. \tag{4.38}$$

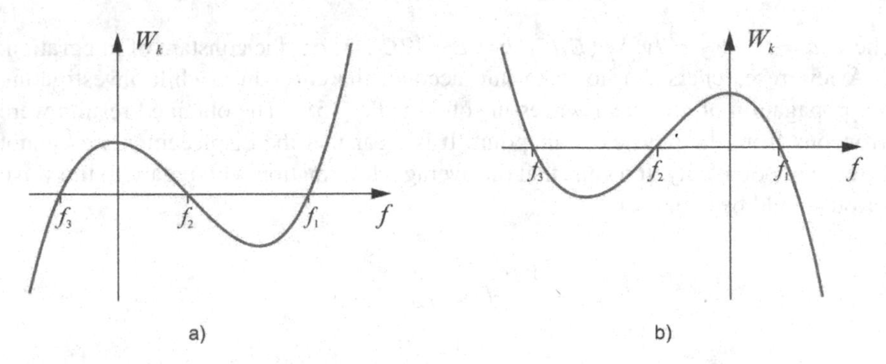

Figure 4.4 Kinetic energy of the oscillator with square nonlinearity: (a) $b < 0$, (b) $b > 0$ [reprinted with permission from Elsevier].

Equation (4.35) is recast to the following form

$$\sqrt{-\frac{2}{3}b d\xi} = \frac{df}{\sqrt{(f-f_1)(f-f_2)(f-f_3)}}. \tag{4.39}$$

Assume the following order of the roots: $f_1 \geq f_2 \geq f_3$. Form of the graph of the kinetic energy function depends on the soft ($E_2 < 0$, $b < 0$) or hard ($E_2 > 0$, $b > 0$) nonlinearity of the composite material (Fig. 4.4). In what follows we briefly consider both cases.

Soft nonlinearity. If $b < 0$, then the periodic solution exists for $f_2 \geq f \geq f_3$. Let us carry out the following change of the variables

$$g^2 = \frac{f-f_3}{f_2-f_3}, \qquad 0 \leq g^2 \leq 1, \tag{4.40}$$

$$s^2 = \frac{f_2-f_3}{f_1-f_3}, \qquad 0 \leq s^2 \leq 1. \tag{4.41}$$

Equation (4.39) takes the following form:

$$\sqrt{-\frac{b}{6}(f_1-f_3)d\xi} = \frac{dg}{\sqrt{(1-g^2)(1-s^2g^2)}}. \tag{4.42}$$

Integration of equation (4.42) yields

$$\sqrt{-\frac{b}{6}(f_1-f_3)}\xi = \int_0^g \frac{dg}{\sqrt{(1-g^2)(1-s^2g^2)}}. \tag{4.43}$$

We carry out inversion of the elliptic integral occurring on the right hand side of equation (4.43) [2]:

$$f = f_3 + (f_2-f_3)\,\mathrm{sn}^2\left[\sqrt{-\frac{b}{6}(f_1-f_3)}\xi, s\right]. \tag{4.44}$$

Let us introduce the following notations: $F = f_2 - f_3$, $\kappa = 2\sqrt{-b(f_1 - f_3)/6} = \sqrt{-2bF/(3s^2)}$, where F is the amplitude, $F > 0$, κ is the constant of propagation (analog of frequency).

Solution (4.44) has the period $L = 4K(s)/\kappa$. Substituting (4.44) into condition (4.33) yields

$$f_3 = -F\frac{1 - E(s)/K(s)}{s^2}, \tag{4.45}$$

where $K(s)$ and $E(s)$ stand for complete elliptic integrals of the first and second kind, respectively [2].

Finally, we have

$$f_1 = \frac{F}{s^2} + f_3 = F\frac{E(s)}{s^2K(s)}, \qquad f_2 = F + f_3 = F\left[1 - \frac{1 - E(s)/K(s)}{s^2}\right], \tag{4.46}$$

and solution (4.44) is recast to the following form

$$f = -F\frac{1 - E(s)/K(s)}{s^2} + F\,\mathrm{sn}^2\left(\frac{\kappa}{2}\xi, s\right). \tag{4.47}$$

Modulus s of the elliptic function stands for solution to the following transcendental equation

$$s^2K(s)^2 = -\frac{FE_2}{48\eta^2E_3}. \tag{4.48}$$

The propagation constant κ is coupled with wave length L in the following way

$$\kappa = \frac{4K(s)}{L}. \tag{4.49}$$

Velocity of the wave propagation is found from equation (4.38), and it reads

$$\frac{v^2}{v_0^2} = 1 - 16\left[3E(s)/K(s) - 2 + s^2\right]K(s)^2\frac{E_3}{E_1}\eta^2. \tag{4.50}$$

Hard nonlinearity. If $b > 0$, then the periodic solution exists for $f_1 \geq f \geq f_2$. Instead of relations (4.40), (4.41), we introduce the following dependencies

$$\begin{aligned} g^2 &= \frac{f_1 - f}{f_1 - f_2}, \qquad 0 \leq g^2 \leq 1, \\ s^2 &= \frac{f_1 - f_2}{f_1 - f_3}, \qquad 0 \leq s^2 \leq 1, \end{aligned} \tag{4.51}$$

and we introduce the notation

$$F = f_1 - f_2 > 0, \qquad \kappa = 2\sqrt{b(f_1 - f_3)/6} = \sqrt{2bF/(3s^2)}. \tag{4.52}$$

Carrying out the transformations in a way analogous to the previous, we get

$$f = F\frac{1-E(s)/K(s)}{s^2} - F\,\mathrm{sn}^2\left(\frac{\kappa}{2}\xi, s\right), \tag{4.53}$$

where modulus s is defined by the following transcendental equation

$$s^2 K(s)^2 = \frac{FE_2}{48\eta^2 E_3}. \tag{4.54}$$

The relations (4.49), (4.50) are the same. Observe that the solutions (4.47), (4.48) in the case of soft nonlinearity ($E_2 < 0$) and the solution (4.53), (4.54) in the case of hard nonlinearity ($E_2 > 0$) get over to each other while changing the sign of coefficient E_2.

4.5 ANALYSIS OF SOLUTION AND NUMERICAL RESULTS

The value of parameter s estimates intensity of the nonlinear effects as well as allows to define how strong the nonlinear regime of the wave propagation differs from the linear one. Fig. 4.5 presents the form of nonlinear waves of deformation for different values of s. The computations have been carried out for the composite material with soft nonlinearity based on equation (4.47).

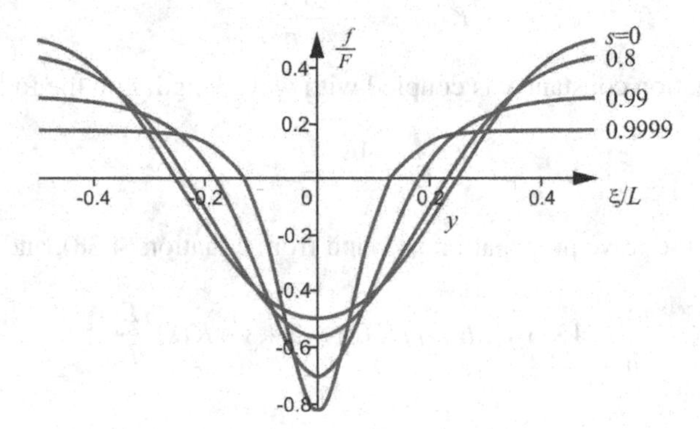

Figure 4.5 Forms of periodic nonlinear waves of deformation for $E_2 < 0$ [reprinted with permission from Elsevier].

In the case of $s = 0$ the following limiting transitions take place: $E(0) = \pi/2$, $K(0) = \pi/2$, $E(0)/K(0) \sim 1 - s^2/2$, $\mathrm{sn}(z, 0) = \sin(z)$. The found solution describes the following harmonic wave

$$f = -\frac{F}{2}\cos\left(\frac{2\pi}{L}\xi\right) \quad \text{for} \quad E_2 < 0, \qquad f = \frac{F}{2}\cos\left(\frac{2\pi}{L}\xi\right) \quad \text{for} \quad E_2 > 0,$$

$$\frac{v^2}{v_0^2} = 1 - 4\pi^2\eta^2\frac{E_3}{E_1}. \tag{4.55}$$

In the case of $s = 1$, the periodic nonlinear wave is transformed into the localized wave of the bell-form (soliton) - see Fig. 4.6. Having in mind that: $E(1) = 1$, $\lim_{s\to 1} K(s) = \infty$, $\operatorname{sn}(z,1)^2 = 1 - \operatorname{sech}(z)^2$, one gets

$$
\begin{aligned}
&f = -F\operatorname{sech}^2\left(\frac{\xi}{\Delta}\right), \quad \frac{\Delta^2}{l^2} = -\frac{12E_3}{FE_2} \quad \text{for} \quad E_2 < 0,\\
&f = F\operatorname{sech}^2\left(\frac{\xi}{\Delta}\right), \quad \frac{\Delta^2}{l^2} = \frac{12E_3}{FE_2} \quad \text{for} \quad E_2 > 0,\\
&\frac{v^2}{v_0^2} = 1 + 4\frac{E_3 l^2}{E_1 \Delta^2},
\end{aligned}
\tag{4.56}
$$

where the parameter Δ stands for the soliton width.

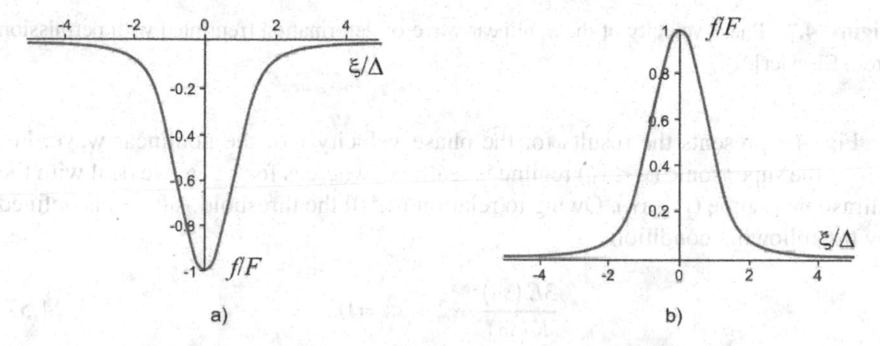

Figure 4.6 Localized nonlinear waves of deformations (solitons): (a) soft nonlinearity ($E_2 < 0$); (b) hard nonlinearity ($E_2 > 0$) [reprinted with permission from Elsevier].

Analysis of solution (4.56) allows to achieve important conclusions regarding properties of the localized nonlinear waves propagation in composite materials. Since $F > 0$, then in the case of soft nonlinearity ($E_2 < 0$) we have $f < 0$. It means that in such material the localized waves of compression propagate. On contrary, in the case of hard nonlinearity ($E_2 > 0$) we have $f > 0$, and consequently only localized waves of extension may exist.

Increase of the soliton amplitude F yields decrease of its width Δ and increase of its velocity v. Therefore, the "high" solitons have the small width and propagate faster than "low" solitons.

The velocity of solitons is higher than the velocity of waves propagating in linear homogeneous medium: $v > v_0$. It is the ultrasonic regime. The analogous effect is observed while studying nonlinear waves of deformation in homogeneous solid [158, 398].

Further numerical examples is devoted to the case of the steel-aluminium composite. Properties of the components are the same: $c^{(1)} = c^{(2)} = 0.5$. Owing to the obtained solution (4.23)–(4.25), we compute the effective elastic coefficients: $E_1 = 160$ GPa, $E_2 = -2385$ GPa (soft nonlinearity) for $E_3 = 2.73$ GPa.

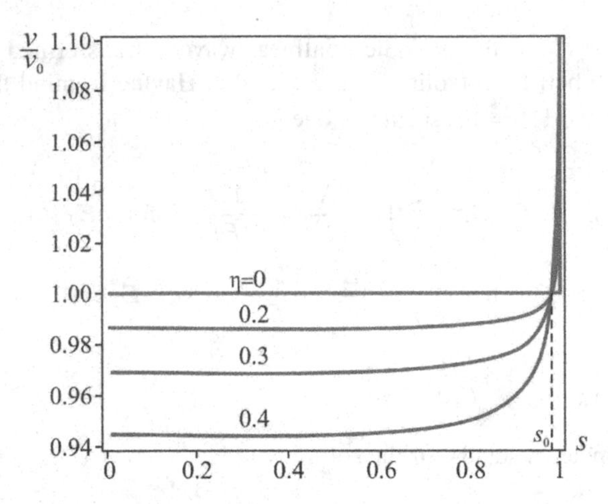

Figure 4.7 Phase velocity of the nonlinear wave of deformation [reprinted with permission from Elsevier].

Fig. 4.7 presents the results for the phase velocity v of the nonlinear wave. For $s < s_0$ the super sonic ($v < v_0$) regime is realized, whereas for $s > s_0$ we deal with the ultrasonic regime ($v > v_0$). Owing to relation (4.50) the threshold value s_0 is defined by the following condition

$$\frac{3E(s_0)}{K(s_0)} - 2 + s_0^2 = 0. \tag{4.57}$$

Solving numerically equation (4.57) yields $s_0 = 0.9803\ldots$. It should be noticed that the given value is defined only by the character of the obtained analytical solutions and does not depend on the properties of the composite material.

The obtained numerical results (Figs. 4.5, 4.7) yield conclusion that influence of nonlinearity on the form and velocity of the wave become important for $s > 0.6\ldots0.8$. In the case of smaller values of s, the wave form almost does not differ from harmonic one, whereas the regime of propagation is very close to the linear regime.

We consider how the value of parameter s depends on the wave amplitude F and on the ratio of the wave length L and the size of the material microstructure l. Note that the parameter $\eta = l/L$ characterizes intensity of the effect of dispersion. The higher value η, the bigger is influence of the microstructure on the dissipation of the wave energy. The parametric dependencies of the modulus s on F and η are shown in Fig. 4.8 (they have been found by solving numerically equation (4.48)).

The obtained data illustrate how phenomena of nonlinearity and dispersion compensate each other. Increase of the amplitude F(with constant η) implies increase of s which yields increase of intensity of the nonlinear effects. In contrary, decrease of the wave length and increase of η (with constant amplitude F) yields decrease of the parameter s; the solution becomes more close to its linear counterpart.

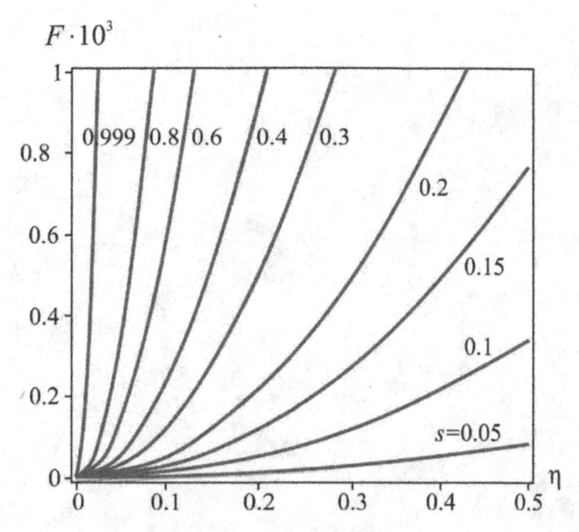

Figure 4.8 Modulus s characterizing intensity of the nonlinear effects [reprinted with permission from Elsevier].

Analysis of the results presented in Fig. 4.8 allows to estimate areas of applibility of various approximate theories, which are employed for modeling of elastic waves propagating in solids. For the majority of materials used for mechanical constructions, the area of elastic deformations is bounded by the value of $F \leq 10^{-3}$. Nonlinear effects are exhibited for $s > 0.6$ (in the latter case $\eta < 0.13$). Consequently, modeling of nonlinear waves can be carried out in the scope of the length wave linear approximation (for instance by employing of the higher order continuous or homogenize models).

Occurrence of dispersion plays a key role when the wave length is close to the size of the internal structure of material, i.e. for $\eta > 0.3$ (see Fig. 4.3). Then we get $s < 0.28$, which means that the nonlinear effects can be neglected. Therefore, the problem of propagation of short and strongly dispersive waves can be analyzed in the scope of linear theory. The solution can be obtained with a help of the Floquet-Bloch method [98].

The so far given statements are not valid if the model allows for large values of elastic deformations ($F > 10^{-2} \ldots 10^{-1}$). In the latter case, the nonlinear and dispersive effects may appear simulteneously. They are exhibited by the rubber-type materials and elastomers, molecular and atomic chains, nanotubes, etc.

5 Antiplane Shear Waves in Fiber Composites with Structural Nonlinearity

The antiplane shear waves propagating in a fiber composite with an account of structural nonlinearity are studied. Boundary value problem for imperfect bonding conditions are given and discussed. Then, the macroscopic wave equation is derived and analytical solution for stationary waves is presented and analyzed in the case of soft and hard nonlinearities.

5.1 BOUNDARY VALUE PROBLEM FOR IMPERFECT BONDING CONDITIONS

We consider the fiber composite consisting of the matrix $\Omega^{(1)}$ and the square lattice of cylindrical inclusions $\Omega^{(2)}$ (Fig. 5.1). We assume that physical and geometric nonlinearity can be neglected, and the non-linear properties of the model are originated from imperfect contact conditions on the boundary of the components. The latter effect stands for the structural nonlinearity which is associated with nonhomogeneity of the internal structure of the composite material.

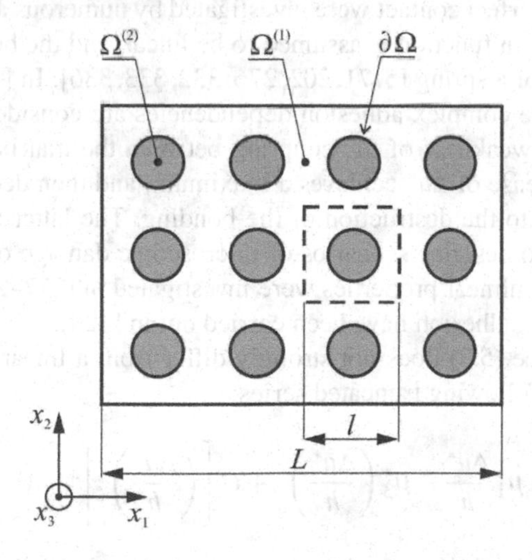

Figure 5.1 Fiber composite material.

We investigate the propagation of the elastic waves in plane perpendicular to the fibers axis. In this case the components of the stress-strain state depend only on two spatial co-ordinates x_1, x_2, and the system of equations of motion is split into two uncoupled problems governing propagation of plane waves and antiplane shear waves. Our further considerations are limited to consider only antiplane transverse waves. The governing wave equation has the following form:

$$\mu^{(n)}\left(\frac{\partial^2 u^{(n)}}{\partial x_1^2}+\frac{\partial^2 u^{(n)}}{\partial x_2^2}\right)=\rho^{(n)}\frac{\partial^2 u^{(n)}}{\partial t^2}, \tag{5.1}$$

where: $\mu^{(n)}$ is the Young's modulus; $u^{(n)}$ is the displacement in direction x_3, $n=1,2$.

We consider the model of imperfect contact between fibers and the matrix. We suppose the following bonding conditions concerning stresses

$$\sigma^* = \sigma^{(1)} = \sigma^{(2)} \quad \text{at} \quad \partial\Omega, \tag{5.2}$$

where: $\sigma^{(n)} = \mu^{(n)}(\partial u^{(n)}/\partial n)$, $\partial/\partial n$ – derivation along normal n to the boundary $\partial\Omega$.

The violation of the mechanical coupling between the fibers and the matrix supplies occurrence of discontinuities in the displacement fields. One of the most popular approach aimed on modeling of the imperfect contact consists of employment of the dependence between the displacements jump $\Delta u^* = u^{(1)} - u^{(2)}$ and the stresses σ^* on the separation boundary:

$$\Delta u^* = f(\sigma^*) \quad \text{at} \quad \partial\Omega, \tag{5.3}$$

where $f(\sigma^*)$ stands for the function of adhesion.

Problems of imperfect contact were investigated by numerous authors. In the simple cases the adhesion function is assumed to be linear, and the boundary of separation plays the role of a spring [5, 71, 202, 276, 332, 333, 336]. In [159, 160, 436, 437] a few kinds of more complex adhesion dependencies are considered. Owing to the employed models, weakness of the coupling between the matrix and inclusion σ^* increases with increase of Δu^* achieves a maximum, and then decreases up to zero, which corresponds to the destruction of the bonding. The latter approach has been successively used to describe scenarios of microscopic damage of composite materials. Bonds with nonlinear properties were investigated in [277–279]. Experimental investigations of the adhesion have been carried out in [428].

If the dependence (5.3) does not strongly differ from a linear one, it can be approximated by the following truncated series:

$$\sigma^* = \mu_1^* \frac{\Delta u^*}{h} + \mu_3^* \left(\frac{\Delta u^*}{h}\right)^3 + O\left[\left(\frac{\Delta u^*}{h}\right)^5\right] \quad \text{at} \quad \partial\Omega, \tag{5.4}$$

where h stands for the thickness of the boundary of separation between the components.

Deformation exhibits symmetry and hence the adhesion function is odd, i.e. we have $f(\sigma^*) = -f(-\sigma^*)$, and only odd powers of the quantity $\Delta u^*/h$ appear in the expression (5.4). The coefficients μ_1^*, μ_3^* represent the linear and nonlinear stiffness in the bonding of components, respectively. The series (5.4) yields

$$\frac{\Delta u^*}{h} = \frac{\sigma^*}{\mu_1^*} - \frac{\mu_3^*}{\mu_1^*}\left(\frac{\sigma^*}{\mu_1^*}\right)^3 + O\left[\left(\frac{\sigma^*}{\mu_1^*}\right)^5\right] \quad \text{at} \quad \partial\Omega. \tag{5.5}$$

In order to model effects of the imperfect contact, the following nondimensional adhesion parameters are introduced

$$\alpha = \frac{h u^{(1)}}{l u_1^*}, \qquad \beta = \frac{\mu_3^*}{\mu_1^*}\left(\frac{\mu^{(1)}}{\mu_1^*}\right)^2, \tag{5.6}$$

where the thickness of the boundary h is rescaled with regard to the size of the cell of periodicity l, whereas the stiffnesses μ_1^*, μ_3^* are rescaled with respect to the modulus $\mu^{(1)}$.

Let the thickness and stiffness of the bonding tends to zero: $h \to 0$, $\mu_1^* \to 0$, $\mu_3^* \to 0$. Then in the asymptotic limit, depending on the values of parameters α and β, one may takes into account various kinds of adhesion between the components. Formula (5.5) is recast to the following form

$$u^{(1)} - u^{(2)} = \alpha l \frac{\sigma^*}{\mu^{(1)}} - \alpha\beta l\left(\frac{\sigma^*}{\mu^{(1)}}\right)^3 \quad \text{at} \quad \partial\Omega. \tag{5.7}$$

The value $\alpha = 0$ corresponds to the ideal contact, whereas the transition $\alpha \to \infty$ stands for lack of the contact between fibers and the matrix. In the case $\beta = 0$ the elastic properties of the bonding are linear while increase of $|\beta|$ implies increase of the nonlinearity. In the case $\beta < 0$ ($\beta > 0$) the nonlinearity is soft (hard).

Relations (5.1), (5.2) and (5.6) present the input boundary value problem. The proposed model of the imperfect bonding is correct in the case of soft nonlinearity. Equation (5.6) has an asymptotic character and can be used if in its right hand side the ratio of the second to first term does not exceed 10^{-1}. The ratio of stress σ^* to the matrix stiffness $\mu^{(1)}$ is approximately equal to the elastic deformation ε. Therefore, one gets the following estimation: $|\beta|\varepsilon^2 \leq 10^{-1}$. The allowed values of β are presented in Table 5.1.

Table 5.1
Maximum allowed values of the adhesion parameter β vs. the values of elastic deformations ε.

ε	10^{-6}	10^{-5}	10^{-4}	10^{-3}
$\|\beta\|$	10^{11}	10^{9}	10^{7}	10^{5}

In the engineering practice the effect of the imperfect bonding can be caused by various reasons either through: defects and imperfection while fabricating the material, or via separation of a boundary in the exploitation process. Besides, there is a possibility when the adhesion between inclusions and the matrix is decreased, which implies decrease of the local stress concentrations. The proposed model of imperfect bonding does not depend on the physical circumstances causing damage of the coupling between components, and hence they can be used for description of a wide class of problems. There is only one requirement to be satisfied: weak nonlinear behavior of adhesion $f(\sigma^*)$ allows for employment of the truncated series (5.4). The parameters α and β have the phenomenological character. It is assumed that for real materials their values can be found experimentally [428].

The boundary value problem (5.1), (5.2), (5.7) stands for various physical interpretations. Besides of the considered example of the elastic wave, it can also describe processes of propagation of electromagnetic waves in the composites with the dielectric inclusions.

5.2 MACROSCOPIC WAVE EQUATION

Let us introduce the following nondimensional variables:

$$\bar{u} = \frac{u}{U}, \quad \bar{n} = \frac{n}{L}, \quad \bar{x}_k = \frac{x_k}{L}, \quad k = 1,2, \quad U = \max u. \tag{5.8}$$

Taking into account that

$$\sigma^* = \mu^{(1)}(\partial u^{(1)}/\partial \mathbf{n}) \quad \text{at} \quad \partial\Omega, \tag{5.9}$$

the boundary value problem (5.1), (5.2), (5.7) can be written in the following way

$$\mu^{(n)}\left(\frac{\partial^2 \bar{u}^{(n)}}{\partial \bar{x}_1^2} + \frac{\partial^2 \bar{u}^{(n)}}{\partial \bar{x}_2^2}\right) = \rho^{(n)} L^2 \frac{\partial^2 \bar{u}^{(n)}}{\partial t^2}, \tag{5.10}$$

$$\mu^{(1)} \frac{\partial \bar{u}^{(1)}}{\partial \bar{n}} = \mu^{(2)} \frac{\partial \bar{u}^{(2)}}{\partial \bar{n}} \quad \text{at} \quad \partial\Omega, \tag{5.11}$$

$$\bar{u}^{(1)} - \bar{u}^{(2)} = \alpha\eta \frac{\partial \bar{u}^{(1)}}{\partial \bar{n}} - \alpha\eta\delta \left(\frac{\partial \bar{u}^{(1)}}{\partial \bar{n}}\right)^3 \quad \text{at} \quad \partial\Omega, \tag{5.12}$$

where $\eta = l/L$, $\delta = \beta(U/L)^2$ stands for the small parameter.

In the order to obtain the macroscopic wave equation, we employ the asymptotic method of homogenization. We introduce "slow" $\bar{x}_k$ and "fast" $y_k = \eta^{-1}\bar{x}_k$ coordinates. The solution has the following asymptotic form

$$\bar{u}^{(n)} = u_0(\bar{x}_k) + \eta u_1^{(n)}(\bar{x}_k, y_k) + \eta^2 u_2^{(n)}(\bar{x}_k, y_k) + \ldots, \tag{5.13}$$

where

$$u_i^{(n)}(\bar{x}_k, y_k) = u_i^{(n)}(\bar{x}_k, y_k \pm 1), \quad i = 1,2,3,\ldots \tag{5.14}$$

The recurrent system of boundary value problems on a cell takes the form

$$\mu^{(n)}\left(\frac{\partial^2 u_{i-2}^{(n)}}{\partial \bar{x}_1^2}+\frac{\partial^2 u_{i-2}^{(n)}}{\partial \bar{x}_2^2}+2\frac{\partial^2 u_{i-1}^{(n)}}{\partial \bar{x}_1 \partial y_1}+2\frac{\partial^2 u_{i-1}^{(n)}}{\partial \bar{x}_2 \partial y_2}+\frac{\partial^2 u_{i}^{(n)}}{\partial y_1^2}+\frac{\partial^2 u_{i}^{(n)}}{\partial y_2^2}\right)= \rho^{(n)}L^2\frac{\partial^2 u_{i-2}^{(n)}}{\partial t^2}, \tag{5.15}$$

$$\mu^{(1)}\left(\frac{\partial u_{i-1}^{(1)}}{\partial \bar{n}}+\frac{\partial u_{i}^{(1)}}{\partial \mathrm{m}}\right)=\mu^{(2)}\left(\frac{\partial u_{i-1}^{(2)}}{\partial \bar{n}}+\frac{\partial u_{i}^{(2)}}{\partial \mathrm{m}}\right) \quad \text{at} \quad \partial\Omega, \tag{5.16}$$

$$u_i^{(1)}-u_i^{(2)}=\alpha\left(\frac{\partial u_{i-1}^{(1)}}{\partial \bar{n}}+\frac{\partial u_{i}^{(1)}}{\partial \mathrm{m}}\right)-\alpha\delta\left(\frac{\partial u_{i-1}^{(1)}}{\partial \bar{n}}+\frac{\partial u_{i}^{(1)}}{\partial \mathrm{m}}\right)^3 \quad \text{at} \quad \partial\Omega, \tag{5.17}$$

where: $i = 1, 2, 3\ldots$, $\bar{u}_{-1}^{(n)} = 0$, and $\partial/\partial m$ is the derivative along the normal to $\partial\Omega$ expressed in "fast" co-ordinates.

The periodicity conditions (5.14) can be substituted by the condition of zero average value along the period

$$\iint_{\Omega_0^{(1)}} u_i^{(1)} dy_1 dy_2 + \iint_{\Omega_0^{(2)}} u_i^{(2)} dy_1 dy_2 = 0. \tag{5.18}$$

The boundary value problems (5.15)–(5.18) can be considered in the interval of the separated cell of periodicity of the composite material (Fig. 5.2). In "fast" coordinates the cell size is equal to unit, and the fiber radius is $\sqrt{c^{(2)}/\pi}$, where $c^{(2)}$ stands for the volume fraction of the fibers, and $0 \leq c^{(2)} \leq \pi/4$.

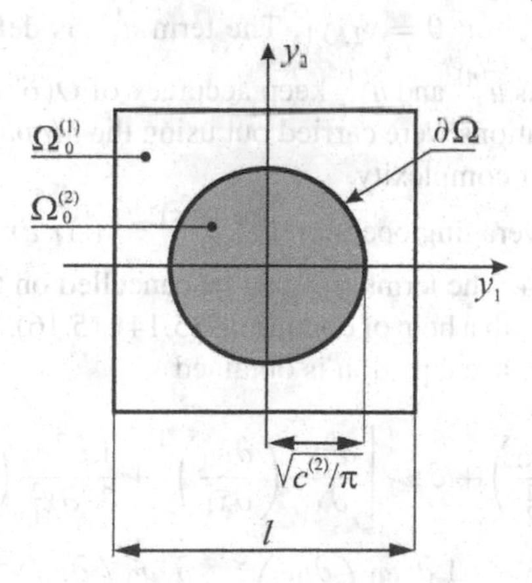

Figure 5.2 Cell of periodicity of the fiber composite material.

In order to find the approximate analytical solution we reduce the problem on cell into its axially symmetric form. For this purpose we replace the square cell by

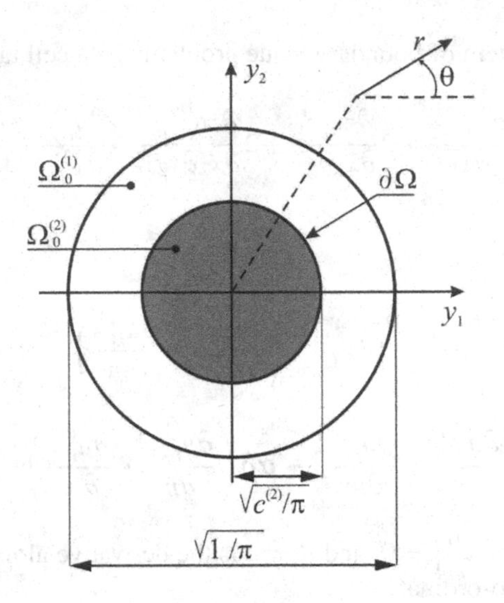

Figure 5.3 Simplification of the geometric form of the cell.

a circular one with the same area (Fig. 5.3). The accuracy of this approach increase with decreasing the volume fraction of fibers.

The solution to the nonlinear problems on a cell can be approximated by the following expansion:

$$u_i^{(n)} = u_{i0}^{(n)} + \delta u_{i1}^{(n)} + O\left(\delta^2\right). \tag{5.19}$$

The coefficients of the series (5.19) are found using the polar system of coordinates $r = \sqrt{y_1^2 + y_2^2}$, $tan\ \theta = y_2/y_1$. The term $u_1^{(n)}$ is defined with accuracy of $O(\delta)$, whereas the terms $u_2^{(n)}$ and $u_3^{(n)}$ keep accuracy of $O(\delta^0)$.

Symbolic transformations were carried out using the *Maple* package, but they are omitted here due to their complexity.

We employing the averaging operator $\int_0^{2\pi} \int_0^{(4\pi)^{-1/2}} (\cdot)\, r\, dr\, d\theta$ with respect to the equation (5.15) for $i = 4$. The terms $\bar{u}_4^{(n)}$ can be cancelled on the basis of the Gauss-Ostrogradsky theorem with a help of conditions (5.14), (5.16). In result, the following macroscopic nonlinear wave equation is obtained

$$\begin{aligned}
&\mu_1\left(\frac{\partial^2 u_0}{\partial \bar{x}_1^2} + \frac{\partial^2 u_0}{\partial \bar{x}_2^2}\right) + \delta\mu_2\left[\frac{\partial^2 u_0}{\partial \bar{x}_1^2}\left(\frac{\partial u_0}{\partial \bar{x}_1}\right)^2 + \frac{1}{3}\frac{\partial^2 u_0}{\partial \bar{x}_1^2}\left(\frac{\partial u_0}{\partial \bar{x}_2}\right)^2 + \right.\\
&\left.\frac{4}{3}\frac{\partial u_0}{\partial \bar{x}_1}\frac{\partial u_0}{\partial \bar{x}_2}\frac{\partial^2 u_0}{\partial \bar{x}_1 \partial \bar{x}_2} + \frac{1}{3}\frac{\partial^2 u_0}{\partial \bar{x}_2^2}\left(\frac{\partial u_0}{\partial \bar{x}_1}\right)^2 + \frac{\partial^2 u_0}{\partial \bar{x}_2^2}\left(\frac{\partial u_0}{\partial \bar{x}_2}\right)^2\right] + \\
&\eta^2\mu_3\left(\frac{\partial^4 u_0}{\partial \bar{x}_1^4} + 2\frac{\partial^4 u_0}{\partial \bar{x}_1^2 \partial \bar{x}_2^2} + \frac{\partial^4 u_0}{\partial \bar{x}_2^4}\right) + O\left(\delta^2 + \eta^4\right) = \rho L^2 \frac{\partial^2 u_0}{\partial t^2}.
\end{aligned} \tag{5.20}$$

Transition to dimensional variables $u = u_0 U$, $x_k = \bar{x}_k L$ allows for recasting equations (5.20) to the following form

$$\begin{gathered}
\mu_1\left(\frac{\partial^2 u}{\partial x_1^2}+\frac{\partial^2 u}{\partial x_2^2}\right)+\beta\mu_2\left[\frac{\partial^2 u}{\partial x_1^2}\left(\frac{\partial u}{\partial x_1}\right)^2+\frac{1}{3}\frac{\partial^2 u}{\partial x_1^2}\left(\frac{\partial u}{\partial x_2}\right)^2+\right.\\
\left.\frac{4}{3}\frac{\partial u}{\partial x_1}\frac{\partial u}{\partial x_2}\frac{\partial^2 u}{\partial x_1\partial x_2}+\frac{1}{3}\frac{\partial^2 u}{\partial x_2^2}\left(\frac{\partial u}{\partial x_1}\right)^2+\frac{\partial^2 u}{\partial x_2^2}\left(\frac{\partial u}{\partial x_2}\right)^2\right]+\\
\eta^2L^2\mu_3\left(\frac{\partial^4 u}{\partial x_1^4}+2\frac{\partial^4 u}{\partial x_1^2\partial x_2^2}+\frac{\partial^4 u}{\partial x_2^4}\right)+O\left(\delta^2+\eta^4\right)=\rho\frac{\partial^2 u}{\partial t^2},
\end{gathered} \tag{5.21}$$

where μ_1, μ_2, μ_3 are effective moduli of stiffness, obtained analytically:

$$\begin{aligned}
\mu_1 = &\left[\mu^{(1)}\left(c^{(2)}\right)^2\pi^{3/2}\left(\mu^{(2)}\right)^2+2\left(c^{(2)}\right)^2\pi^{3/2}\mu^{(2)}\left(\mu^{(1)}\right)^2-\right.\\
&\mu^{(1)}\left(c^{(2)}\right)^4\pi^{3/2}\left(\mu^{(2)}\right)^2+2\left(c^{(2)}\right)^4\pi^{(3/2)}\mu^{(2)}\left(\mu^{(1)}\right)^2-\\
&\left(c^{(2)}\right)^4\pi^{3/2}\left(\mu^{(1)}\right)^3+\left(c^{(2)}\right)^2\pi^{3/2}\left(\mu^{(1)}\right)^3+\\
&2\mu^{(1)}\alpha\left(\mu^{(2)}\right)^2\pi^2\left(c^{(2)}\right)^{3/2}+2\alpha\mu^{(2)}\pi^2\left(c^{(2)}\right)^{3/2}\left(\mu^{(1)}\right)^2-\\
&\mu^{(1)}\alpha^2\left(\mu^{(2)}\right)^2\pi^{5/2}\left(c^{(2)}\right)^3+\mu^{(1)}\alpha^2\left(\mu^{(2)}\right)^2c^{(2)}\pi^{5/2}+\\
&\left.2\mu^{(1)}\alpha\left(\mu^{(2)}\right)^2\pi^2\left(c^{(2)}\right)^{7/2}-2\alpha\mu^{(2)}\left(c^{(2)}\right)^{7/2}\pi^2\left(\mu^{(1)}\right)^2\right]/\\
&\left[2\alpha\mu^{(2)}\pi^2\left(c^{(2)}\right)^{3/2}\mu^{(1)}+2\alpha\mu^{(2)}\left(c^{(2)}\right)^{7/2}\pi^2\mu^{(1)}-\right.\\
&2\left(\mu^{(2)}\right)^2\left(c^{(2)}\right)^3\pi^{3/2}+2\left(c^{(2)}\right)^2\pi^{3/2}\mu^{(2)}\mu^{(1)}-\\
&2\left(c^{(2)}\right)^4\pi^{3/2}\mu^{(2)}\mu^{(1)}+2\alpha\left(\mu^{(2)}\right)^2\pi^2\left(c^{(2)}\right)^{3/2}+\\
&\alpha^2\left(\mu^{(2)}\right)^2\pi^{5/2}\left(c^{(2)}\right)^3+\alpha^2\left(\mu^{(2)}\right)^2c^{(2)}\pi^{5/2}-2\alpha\left(\mu^{(2)}\right)^2\pi^2\left(c^{(2)}\right)^{7/2}+\\
&2\left(\mu^{(1)}\right)^2\left(c^{(2)}\right)^3\pi^{3/2}+\left(c^{(2)}\right)^2\pi^{3/2}\left(\mu^{(2)}\right)^2+\\
&\left(c^{(2)}\right)^4\pi^{3/2}\left(\mu^{(2)}\right)^2+\left(c^{(2)}\right)^4\pi^{3/2}\left(\mu^{(1)}\right)^2+\left(c^{(2)}\right)^2\pi^{3/2}\left(\mu^{(1)}\right)^2+\\
&\left.4\alpha\mu^{(2)}\pi^2\left(c^{(2)}\right)^{5/2}\mu^{(1)}+2\alpha^2\left(\mu^{(2)}\right)^2\pi^{5/2}\left(c^{(2)}\right)^2\right],\\
\mu_2 = &\,36\alpha\mu^{(1)}\pi\left(c^{(2)}\right)^{5/2}\left(\mu^{(2)}\right)^4/\\
&\left[-4\left(c^{(2)}\right)^5\pi^{1/2}\left(\mu^{(2)}\right)^4+\left(c^{(2)}\right)^6\pi^{1/2}\left(\mu^{(1)}\right)^4+\right.
\end{aligned} \tag{5.22}$$

$$
\begin{aligned}
&6\left(c^{(2)}\right)^{4}\pi^{1/2}\left(\mu^{(1)}\right)^{4}-4\left(c^{(2)}\right)^{3}\pi^{1/2}\left(\mu^{(2)}\right)^{4}+\left(c^{(2)}\right)^{2}\pi^{1/2}\left(\mu^{(1)}\right)^{4}+\\
&\left(c^{(2)}\right)^{6}\pi^{1/2}\left(\mu^{(2)}\right)^{4}+12c^{(2)}\left(\mu^{(2)}\right)^{3}\alpha^{2}\pi^{3/2}\mu^{(1)}+6c^{(2)}\left(\mu^{(2)}\right)^{2}\alpha^{2}\pi^{3/2}\left(\mu^{(1)}\right)^{2}+\\
&24\left(c^{(2)}\right)^{2}\left(\mu^{(2)}\right)^{2}\alpha^{2}\pi^{3/2}\left(\mu^{(1)}\right)^{2}+36\left(c^{(2)}\right)^{3}\left(\mu^{(2)}\right)^{2}\alpha^{2}\pi^{3/2}\left(\mu^{(1)}\right)^{2}+\\
&24\left(c^{(2)}\right)^{2}\left(\mu^{(2)}\right)^{3}\alpha^{2}\pi^{3/2}\mu^{(1)}-24\left(c^{(2)}\right)^{4}\left(\mu^{(2)}\right)^{3}\alpha^{2}\pi^{3/2}\mu^{(1)}+\\
&4\left(c^{(2)}\right)^{5}\pi^{1/2}\left(\mu^{(1)}\right)^{4}+6\left(c^{(2)}\right)^{4}\pi^{1/2}\left(\mu^{(2)}\right)^{4}-\\
&12\left(c^{(2)}\right)^{4}\pi^{1/2}\left(\mu^{(1)}\mu^{(2)}\right)^{2}-8\left(c^{(2)}\right)^{5}\pi^{1/2}\left(\mu^{(1)}\right)^{3}\mu^{(2)}+8\left(c^{(2)}\right)^{3}\pi^{1/2}\mu^{(2)}\left(\mu^{(1)}\right)^{3}-\\
&4\left(c^{(2)}\right)^{6}\pi^{1/2}\left(\mu^{(2)}\right)^{3}\mu^{(1)}-4\left(c^{(2)}\right)^{6}\pi^{1/2}\mu^{(2)}\left(\mu^{(1)}\right)^{3}+\left(c^{(2)}\right)^{2}\pi^{1/2}\left(\mu^{(2)}\right)^{4}+\\
&\alpha^{4}\left(\mu^{(2)}\right)^{4}\pi^{5/2}+6\left(c^{(2)}\right)^{2}\pi^{1/2}\left(\mu^{(1)}\mu^{(2)}\right)^{2}+24\left(c^{(2)}\right)^{4}\left(\mu^{(2)}\right)^{2}\alpha^{2}\pi^{3/2}\left(\mu^{(1)}\right)^{2}-\\
&12\left(c^{(2)}\right)^{5}\left(\mu^{(2)}\right)^{3}\alpha^{2}\pi^{3/2}\mu^{(1)}+6\left(c^{(2)}\right)^{5}\left(\mu^{(2)}\right)^{2}\alpha^{2}\pi^{3/2}\left(\mu^{(1)}\right)^{2}+\\
&12\left(\mu^{(2)}\right)^{3}\pi\alpha\left(c^{(2)}\right)^{3/2}\mu^{(1)}+16\left(\mu^{(2)}\right)^{3}\pi^{2}\alpha^{3}\left(c^{(2)}\right)^{3/2}\mu^{(1)}+\\
&12\left(\mu^{(2)}\right)^{2}\pi\alpha\left(c^{(2)}\right)^{3/2}\left(\mu^{(1)}\right)^{2}+4\mu^{(2)}\pi\alpha\left(c^{(2)}\right)^{3/2}\left(\mu^{(1)}\right)^{3}+\\
&4\left(\mu^{(2)}\right)^{3}\pi^{2}\alpha^{3}\left(c^{(2)}\right)^{9/2}\mu^{(1)}-24\left(\mu^{(2)}\right)^{2}\pi\alpha\left(c^{(2)}\right)^{9/2}\left(\mu^{(1)}\right)^{2}+\\
&16\mu^{(2)}\pi\alpha\left(c^{(2)}\right)^{9/2}\left(\mu^{(1)}\right)^{3}+24\left(\mu^{(2)}\right)^{3}\pi^{2}\alpha^{3}\left(c^{(2)}\right)^{5/2}\mu^{(1)}+\\
&16\mu^{(2)}\pi\alpha\left(c^{(2)}\right)^{5/2}\left(\mu^{(1)}\right)^{3}+24\left(\mu^{(2)}\right)^{2}\pi\alpha\left(c^{(2)}\right)^{5/2}\left(\mu^{(1)}\right)^{2}+\\
&4\pi\mu^{(2)}\alpha\left(c^{(2)}\right)^{11/2}\left(\mu^{(1)}\right)^{3}-12\pi\left(\mu^{(2)}\right)^{2}\alpha\left(c^{(2)}\right)^{11/2}\left(\mu^{(1)}\right)^{2}+\\
&12\pi\left(\mu^{(2)}\right)^{3}\alpha\left(c^{(2)}\right)^{11/2}\mu^{(1)}+16\left(\mu^{(2)}\right)^{3}\pi^{2}\alpha^{3}\left(c^{(2)}\right)^{7/2}\mu^{(1)}-\\
&24\left(\mu^{(2)}\right)^{3}\pi\alpha\left(c^{(2)}\right)^{7/2}\mu^{(1)}+24\mu^{(2)}\pi\alpha\left(c^{(2)}\right)^{7/2}\left(\mu^{(1)}\right)^{3}+\\
&8\left(c^{(2)}\right)^{5}\pi^{1/2}\mu^{(1)}\left(\mu^{(2)}\right)^{3}+6\left(c^{(2)}\right)^{6}\pi^{1/2}\left(\mu^{(1)}\mu^{(2)}\right)^{2}+4\left(c^{(2)}\right)^{3}\pi^{1/2}\left(\mu^{(1)}\right)^{4}+\\
&6\left(c^{(2)}\right)\left(\mu^{(2)}\right)^{4}\alpha^{2}\pi^{3/2}-12\left(c^{(2)}\right)^{3}\left(\mu^{(2)}\right)^{4}\alpha^{2}\pi^{3/2}+6\left(c^{(2)}\right)^{5}\left(\mu^{(2)}\right)^{4}\alpha^{2}\pi^{3/2}+\\
&8\left(\mu^{(2)}\right)^{4}\pi^{2}\alpha^{3}\left(c^{(2)}\right)^{3/2}+4\left(\mu^{(2)}\right)^{4}\pi\alpha\left(c^{(2)}\right)^{3/2}-4\left(\mu^{(2)}\right)^{4}\pi^{2}\alpha^{3}\left(c^{(2)}\right)^{9/2}+\\
&8\left(\mu^{(2)}\right)^{4}\pi\alpha\left(c^{(2)}\right)^{9/2}+4\alpha^{4}\left(\mu^{(2)}\right)^{4}\pi^{5/2}c^{(2)}+6\alpha^{4}\left(\mu^{(2)}\right)^{4}\pi^{5/2}\left(c^{(2)}\right)^{2}+\\
&4\alpha^{4}\left(\mu^{(2)}\right)^{4}\pi^{5/2}\left(c^{(2)}\right)^{3}+\alpha^{4}\left(\mu^{(2)}\right)^{4}\pi^{5/2}\left(c^{(2)}\right)^{4}-8\left(\mu^{(2)}\right)^{4}\pi\alpha\left(c^{(2)}\right)^{5/2}-\\
&4\pi\left(\mu^{(2)}\right)^{4}\alpha\left(c^{(2)}\right)^{11/2}-8\left(\mu^{(2)}\right)^{4}\pi^{2}\alpha^{3}\left(c^{(2)}\right)^{7/2}+4\left(c^{(2)}\right)^{2}\pi^{1/2}\left(\mu^{(2)}\right)^{3}\mu^{(1)}+
\end{aligned}
$$

$$4\left(c^{(2)}\right)^2\pi^{1/2}\mu^{(2)}\left(\mu^{(1)}\right)^3-8\left(c^{(2)}\right)^3\pi^{1/2}\mu^{(1)}\left(\mu^{(2)}\right)^3+$$
$$4\pi^2\left(\mu^{(2)}\right)^4\alpha^3\left(c^{(2)}\right)^{1/2}+4\pi^2\left(\mu^{(2)}\right)^3\alpha^3\left(c^{(2)}\right)^{1/2}\mu^{(1)}\Big].$$

The values of the effective modulus μ_3 were defined through numerical integration along the area of the cell of periodicity using the standard procedure of the *Maple* package.

The obtained results are based on the simplification of the geometric form of the cell. In order to estimate accuracy of the proposed approximation, we consider the effective linear modulus μ_1 in the case of the ideal contact between components ($\alpha = 0$). The solution takes the following form:

$$\mu_1 = \mu^{(1)}\frac{\left(1-c^{(2)}\right)\mu^{(1)}+\left(1+c^{(2)}\right)\mu^{(2)}}{\left(1+c^{(2)}\right)\mu^{(1)}+\left(1-c^{(2)}\right)\mu^{(2)}}. \tag{5.23}$$

Observe that formula (5.23) can be obtained with a help of the self-consistent approach and coincides with the lower Hashin-Strickman bound [123]. A comparison of the analytical solution (5.23) with the theoretical results achieved in [358] with a help of the Rayleigh method for the composite with infinitely stiff fibers ($\mu^{(2)}/\mu^{(1)} \to \infty$) is shown in Table 5.2. The given example is characterized by large concentrations of the local stresses, and hence the error of approximate analytical solution achieves a maximum. Decreasing of the fibers stiffness decreases the error.

Table 5.2
Effective linear shear modulus $\mu_1/\mu^{(1)}$ of the composite material with infinitely stiff fibers

$c^{(2)}$	Formula (5.23)	Data [358]	Error %
0.1	1.222	1.222	0.0
0.2	1.50	1.500	0.0
0.3	1.857	1.860	0.2
0.4	2.333	2.351	0.8
0.5	3.000	3.080	2.6
0.6	4.000	4.342	7.9
0.7	5.667	7.433	24.8
0.74	6.692	11.01	39.2
0.76	7.333	15.44	52.5
0.77	-	20.43	-
0.78	-	35.93	-

Formula (5.23) guarantees high accuracy for $c^{(2)} < 0.5$. Error increases with increase of the volume fraction of fibers ($c^{(2)} > 0.6$) when the distance between neighboring fibers is small. It is associated with occurrence of essential concentration of local stresses. In the latter case we need to take into account the space packing of fibers. It should be mentioned that for majority of real composite materials the volume fraction of the fibers does not achieve the value 0.4...0.5. It means that the obtained approximate solutions can be employed in real world situation.

As a case study, we consider the composite material composed of the aluminum matrix ($\mu^{(1)} = 27.9$ GPa, $\rho^{(1)} = 2700$ kg/m^3) and nickel fibers ($\mu^{(2)} = 75.4$ GPa, $\rho^{(2)} = 8940$ kg/m^3). Figures 5.4–5.6 report the results obtained for the effective elastic moduli μ_1, μ_2, μ_3 for different values of volume fractions of the fibers $c^{(2)}$ and the adhesion parameter α. Figure 5.4 presents a comparison of the obtained solution with the results obtained in [358] in the case of ideal contact ($\alpha = 0$, circles) and in the case of material with voids ($\mu^{(2)} = 0$, dashed curve).

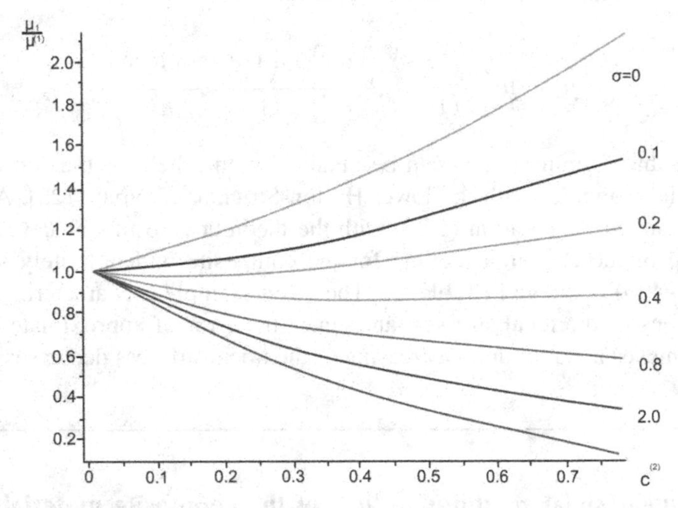

Figure 5.4 Effective linear shear modulus of the composite [reprinted with permission from Elsevier].

The effective moduli μ_1, μ_3 take maximum values in the case of ideal contact of components ($\alpha = 0$). The weakness of the coupling between fibers and the matrix (increase of the parameter α) yields decrease of μ_1 and μ_3.

The effective coefficient μ is equal zero for $\alpha = 0$. It means that in the case of ideal contacting material the studied behavior is linear. In the interval $\alpha = 0.05...0.1$ a rapid increase of μ_2 is observed, and influence of nonlinearity achieves maximum. Further, decrease of the coupling between the components implies decrease of the coefficient μ_2 and decrease of intensity of the nonlinear effects.

On the other hand, on increase of volume fraction of inclusions $c^{(2)}$ (with fixed α) yields increase of the effective modulus μ_2. It can be explained in the following way: in the considered example nonlinear behavior of material is mainly governed by the properties on the separation boundary $\partial\Omega$. Therefore, increase of the area $\partial\Omega$ implies increase of the nonlinearity.

The effective modulus μ_3 is characterized by intensity of the dispersion effect. For $c^{(2)} = 0$ the material is homogeneous and there is no dispersion ($\mu_3 = 0$). In the case

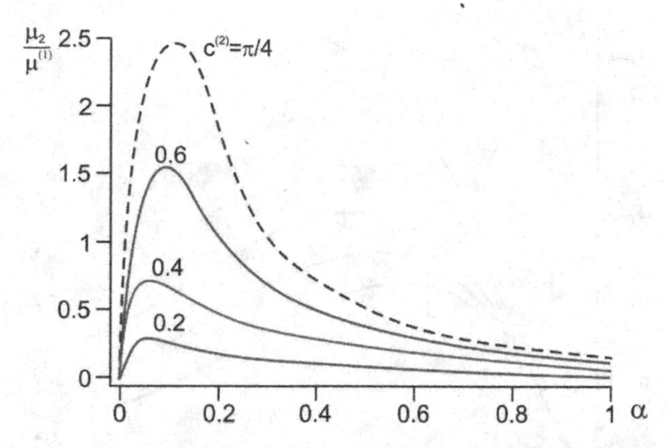

Figure 5.5 Effective nonlinear shear modulus of the composite material aluminum-nickel [reprinted with permission from Elsevier].

of the average values of the volume fraction of fibers we have $c^{(2)} = 0.25...0.45$. It means that nonhomogeneity of the internal structure of composite is expressed strongly, and the coefficient μ_3 achieves its maximum.

Observe that the coefficients μ_1, μ_2 and μ_3 are always considered as positive.

The macroscopic wave equation (5.21) holds for the case of long waves, when $\eta = l/L < 1$. In order to estimate the area of applicability of the homogenized model, we compare the dispersive dependences obtained based on equation (5.21) with the solution found with a help of the Floquet-Bloch approach.

We consider the linear case ($\mu_2 = 0$). The harmonic wave is governed by the following equation

$$u = U\exp(i\mathbf{k}\cdot\mathbf{x})\exp(i\omega t), \tag{5.24}$$

where: $\mathbf{k}$ is the wave vector, $\mathbf{k} = k_1\mathbf{e}_1 + k_2\mathbf{e}_2$; $\mathbf{e}_1, \mathbf{e}_2$ are the unit basis vectors in the rectangular system of co-ordinates; $\mathbf{x} = x_1\mathbf{e}_1 + x_2\mathbf{e}_2$.

Projections of the vector $\mathbf{k}$ to the axes of the co-ordinates are as follows: $k_1 = k\cos\varphi$, $k_2 = k\sin\varphi$, where $k = |\mathbf{k}| = \sqrt{(k_1^2 + k_2^2)} = 2\pi/L$, and φ stands for the angle between axes x_1 and direction of the vector $\mathbf{k}$.

Substituting of ansatz (5.24) into equation (5.21) yields the following dispersive relation

$$\omega^2 = \omega_0^2\left[1 - 4\pi^2\left(\sin^4\varphi + \cos^4\varphi\right)\frac{\mu_3}{\mu_1}\eta^2 + O(\eta^4)\right], \tag{5.25}$$

where: $\omega_0 = v_0 k$ (v_0 stands for the effective phase velocity in the linear homogeneous medium) and $v_0 = \sqrt{\mu_1/\rho}$, $\eta = \frac{l}{L} = \frac{kl}{2\pi}$.

Fig. 5.7 presents a comparison of the solution (5.25) (dashed curves) with the computational results obtained using a Floquet-Bloch method. The computations are carried out for the composite composed of aluminum matrix and nickel fibers. Physical properties of components have been given in the above, whereas the volume fraction of the fibers is $c^{(2)} = 0.4$.

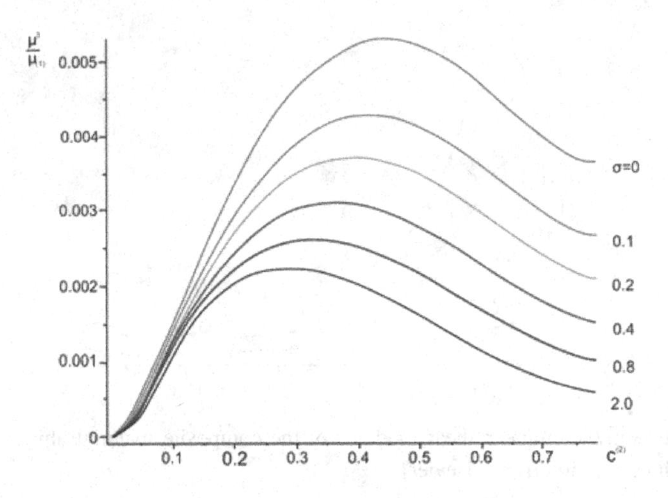

Figure 5.6 Effective shear modulus of order η^2 [reprinted with permission from Elsevier].

Dispersive diagram is composed of two parts separated by dotted vertical line. The right hand side correspond to the orthogonal ($\varphi = 0$), whereas the left hand side corresponds to diagonal ($\varphi = \pi/4$) direction of the wave propagation. In the long wave interval ($\omega \to 0$) the solution becomes isotropic and does not depend on the angle φ. However, decrease of the wave length (and increase of the frequency ω) yields exhibition of the anisotropic properties of the composite material.

Based on the obtained results one may conclude that the homogenized wave equation (5.21) preserves high accuracy for $\eta < 0.4$ (for $\varphi = 0$) and for $\eta < 0.5$ (for $\varphi = \pi/4$).

5.3 ANALYTICAL SOLUTION FOR STATIONARY WAVES

Let a stationary wave is propagated in direction defined by the wave vector $\mathbf{k}$. In this case the solution satisfies the following condition

$$u(\mathbf{x}, t) = u(\xi), \tag{5.26}$$

where: $\xi = \mathbf{e}_k \cdot \mathbf{x} - vt$, $\mathbf{e_k}$ is the unit wave vector, $\mathbf{e}_k = \mathbf{e}_1 \cos\varphi + \mathbf{e}_2 \sin\varphi$.

Let us introduce nondimensional deformation of wave profile:

$$f = \frac{du}{d\xi}. \tag{5.27}$$

Substituting relations (5.27), (5.26) into equation (5.21) and carrying out the integration with respect to ξ, the following equation of the anharmonic oscillator with a cubic nonlinearity is obtained

$$\frac{d^2 f}{d\xi^2} + af + bf^3 + = 0, \tag{5.28}$$

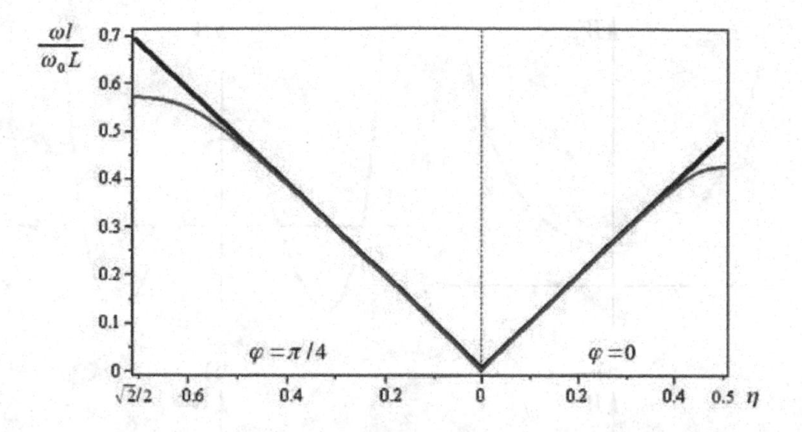

Figure 5.7 Acoustic branches of the dispersion curves of the composite material "aluminum-nickel" (linear case) [reprinted with permission from Elsevier].

where: $a = \mu_1 (1 - v^2/v_0^2) / (\mu_3 l^2)$, $b = \beta \mu_2 / (3\mu_3 l^2)$ and c is the constant of integration.

The condition of equality to zero of the averaged deformation along period of the wave in the considered case yields $c = 0$.

We multiply equation (5.28) by $df/d\xi$, and the carried out integration yields

$$\frac{1}{2}\left(\frac{df}{d\xi}\right)^2 + \frac{a}{2}f^2 + \frac{b}{4}f^4 = W_0, \tag{5.29}$$

where W_0 is the total energy of the oscillator with $W_0 > 0$; $W_k = (1/2)(d\,f/d\xi)^2$ is the kinetic energy; $W_p = (a/2)f^2 + (b/4)f^4$ is the potential energy, and $W_0 = W_k + W_p$.

Separation of variables in equation (5.29) gives:

$$\sqrt{2}d\xi = \frac{df}{\sqrt{W_0 - \frac{a}{2}f^2 - \frac{b}{4}f^4}}. \tag{5.30}$$

The solution is bounded and periodic if the kinetic energy

$$W_k = W_0 - W_p = W_0 - (a/2)f^2 - (b/4)f^4 \tag{5.31}$$

takes the positive values on interval between its two real roots. The form of the solution depends on the signs of the coefficients a and b. We consider separately the case of soft ($\beta < 0, b < 0$) and hard ($\beta > 0, b > 0$) nonlinearity.

Soft nonlinearity. If $a < 0, b < 0$, the polynomial W_k does not have real roots (Fig. 5.8a). Bounded periodic solution of equation (5.28) does not exist.

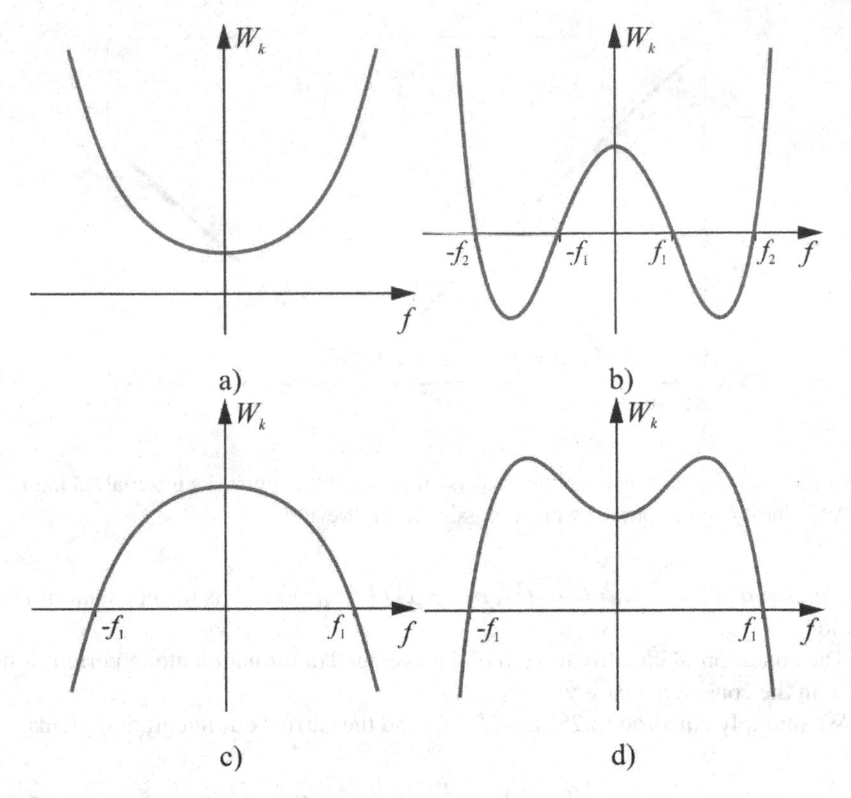

Figure 5.8 Kinetic energy of oscillator with cubic nonlinearity; a) $a < 0, b < 0$; b) $a > 0, b < 0$; c) $a > 0, b > 0$; d) $a < 0, b > 0$ [reprinted with permission from Elsevier].

If $a > 0, b < 0$, the polynomial W_k has four real roots $\pm f_1$, $\pm f_2$ (Fig. 5.8b), where

$$f_1^2 = \frac{a - \sqrt{a^2 + 4bW_0}}{-b}, \quad f_2^2 = \frac{a + \sqrt{a^2 + 4bW_0}}{-b}, \quad f_1^2 < f_2^2. \tag{5.32}$$

The periodic solutions exist for $f_1 \geq f \geq -f_1$. Let us express the polynomial W_k by its roots $W_k = -(b/4)\left(f_1^2 - f^2\right)\left(f_2^2 - f^2\right)$, and let us carry out the following change of variables

$$g^2 = f^2/f_1^2, \qquad 0 \leq g^2 \leq 1, \tag{5.33}$$

$$s^2 = f_1^2/f_2^2, \qquad 0 \leq s^2 \leq 1. \tag{5.34}$$

Then equation (5.30) takes the following form

$$\sqrt{-\frac{b}{2}f_2^2}d\xi = \frac{dg}{\sqrt{(1-g^2)(1-s^2g^2)}}. \tag{5.35}$$

Integration of equation (5.35) and inverse of the elliptic integral yields the following function

$$f = f_1 \operatorname{sn}\left(\sqrt{-\frac{b}{2} f_2^2}\xi, s\right), \tag{5.36}$$

where

$$F = 2f_1, \qquad \kappa = \sqrt{-\frac{b}{2} f_2^2}, \tag{5.37}$$

and F stands for the amplitude ($F > 0$, κ - constant).

Then, formula (5.36) takes the following form

$$f = \frac{F}{2} \operatorname{sn}(\kappa\xi, s). \tag{5.38}$$

The wave length is equal $L = 4K(s)/\kappa$.

Based on the relations (5.32), (5.34), (5.37) and after the trivial transformations, the following transcendental equation is obtained, which allows to find the module s:

$$s^2 K(s)^2 = -\frac{\beta F^2 \mu_2}{384\eta^2 \mu_3 (\sin^4\varphi + \cos^4\varphi)}, \tag{5.39}$$

The wave velocity is described by the following formula

$$\frac{v^2}{v_0^2} = 1 - 16(1+s^2)K(s)^2 \frac{\mu_3}{\mu_1}\eta^2(\sin^4\varphi + \cos^4\varphi). \tag{5.40}$$

Hard nonlinearity. Let $b > 0$. Graphs of the kinetic energies are shown in Fig. 5.8c (for $a > 0$) and in Fig. 5.8d (for $a < 0$). In both cases the polynomial W_k has two real $\pm f_1$ and two imaginary roots $\pm i f_2$ of the following forms

$$f_1^2 = \frac{-a + \sqrt{a^2 + 4bW_0}}{b}, \qquad f_2^2 = \frac{a + \sqrt{a^2 + 4bW_0}}{b}. \tag{5.41}$$

The periodic solutions exist if $f_1 \geq f \geq -f_1$.

We substitute W_k in the form $W_k = (b/4)(f_1^2 - f^2)(f_2^2 + f^2)$ and introduce the following variables: $g^2 = 1 - \frac{f^2}{f_1^2}$, $0 \leq g^2 \leq 1$; $s^2 = \left(1 + f_2^2/f_1^2\right)^{-1}$. Here $0 \leq s^2 \leq 1/2$ for $a > 0$ for $1/2 \leq s^2 \leq 1$ for $a < 0$ and $F = 2f_1 > 0$, $\kappa = \sqrt{(b/2)\left(f_1^2 + f_2^2\right)}$.

We carry out the transformations described earlier, and in result we find

$$f = \frac{F}{2} \operatorname{cn}(\kappa\xi, s), \tag{5.42}$$

$$s^2 K(s)^2 = \frac{\beta F^2 \mu_2}{384\eta^2 \mu_3 (\sin^4\varphi + \cos^4\varphi)}, \tag{5.43}$$

$$\frac{v^2}{v_0^2} = 1 - 16(1 - 2s^2)K(s)^2 \frac{\mu_3}{\mu_1}\eta^2(\sin^4\varphi + \cos^4\varphi). \tag{5.44}$$

The wave length is equal to $L = 4K(s)/\kappa$. Equations (5.43) and (5.39) turn to each other while changing the sign of the parameter β.

5.4 ANALYSIS OF SOLUTION AND NUMERICAL RESULT

In the considered case the nonlinearity type (soft for $\beta < 0$ and hard for $\beta > 0$) has a principal influence on the solution character. The form of periodic waves of deformation is shown in Fig. 5.9. The computations are carried out based on the formulas (5.38) and (5.42).

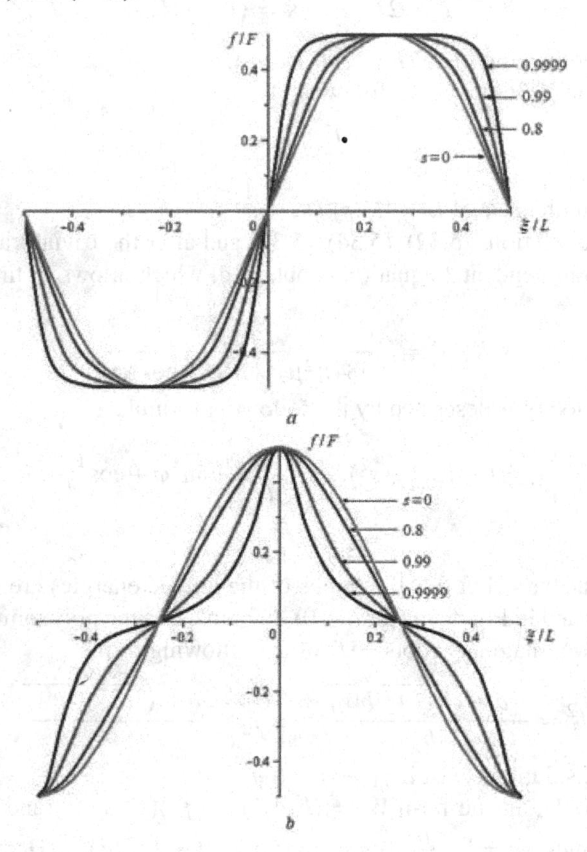

Figure 5.9 Forms of periodic nonlinear waves of deformations: (a) soft nonlinearity ($\beta < 0$) and (b) hard nonlinearity ($\beta > 0$) [reprinted with permission from Elsevier].

In the case of $s = 0$, we have two limiting transitions to the linear solution:

$$f = \frac{F}{2}\sin\left(\frac{2\pi}{L}\xi\right) \quad \text{for} \quad \beta < 0,$$

$$f = \frac{F}{2}\cos\left(\frac{2\pi}{L}\xi\right) \quad \text{for} \quad \beta > 0, \tag{5.45}$$

$$\frac{v^2}{v_0^2} = 1 - 4\pi^2\frac{\mu_3}{\mu_1}\eta^2(\sin^4\varphi + \cos^4\varphi).$$

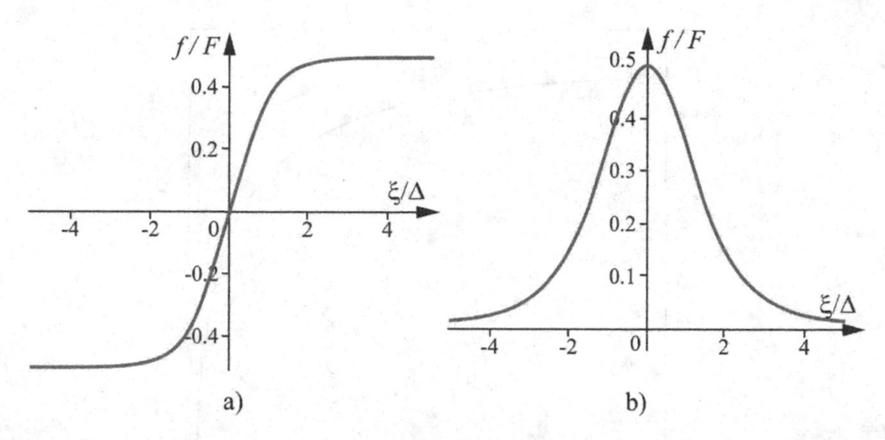

Figure 5.10 Localized nonlinear waves of deformations: (a) soft nonlinearity ($\beta < 0$) and (b) hard nonlinearity ($\beta > 0$) [reprinted with permission from Elsevier].

In the case $s = 1$ in the composite material the localized waves deformation appears. If the nonlinearity is soft, the shock wave is created (kink in Fig. 5.10a):

$$f = \frac{F}{2}\tanh\left(\frac{\xi}{\Delta}\right), \quad \frac{\Delta^2}{l^2} = -\frac{24\mu_3(\sin^4\varphi + \cos^4\varphi)}{F^2\beta\mu_2},$$
$$\frac{v^2}{v_0^2} = 1 - \frac{2\mu_3 l^2}{\mu_1\Delta^2}(\sin^4\varphi + \cos^4\varphi), \quad \beta < 0. \tag{5.46}$$

Kink is propagated in the under sonic regime ($v < v_0$). The increase of the wave amplitude F yields decrease of its width Δ and velocity v.

If the nonlinearity is hard, the localized solutions takes the form of soliton (Fig. 5.10b):

$$f = \frac{F}{2\cosh(\xi/\Delta)}, \quad \frac{\Delta^2}{l^2} = \frac{24\mu_3(\sin^4\varphi + \cos^4\varphi)}{F^2\beta\mu_2},$$
$$\frac{v^2}{v_0^2} = 1 + \frac{\mu_3 l^2}{\mu_1\Delta^2}(\sin^4\varphi + \cos^4\varphi), \quad \beta > 0. \tag{5.47}$$

In the latter case, the propagation regime is supersonic ($v > v_0$). Increase of the soliton amplitude implies decrease of its width with increase of its velocity.

Therefore, in the considered problem the type of nonlinearity defines the form of the localized wave of deformation, but not the mechanical properties of material.

We analyze the numerical results for the composite composed on the aluminum matrix ($\mu^{(1)} = 27.9$ GPa, $\rho^{(1)} = 2700$ kg/m^3) and fibers made from nickel ($\mu^{(2)} = 75.4$ GPa, $\rho^{(2)} = 8940$ kg/m^3). The volume fraction of fiber is $c^{(2)} = 0.4$. The following values of the adhesion parameters are fixed: $\alpha = 0.1$, $|\beta| = 10^5$. The effective elastic coefficients are: $\mu_1 = 33.1$ GPa, $\mu_2 = 21.6$ GPa, $\mu_3 = 0.119$ GPa. The wave propagates in direction orthogonal with respect to the sizes of the periodicity cell: $\varphi = 0$.

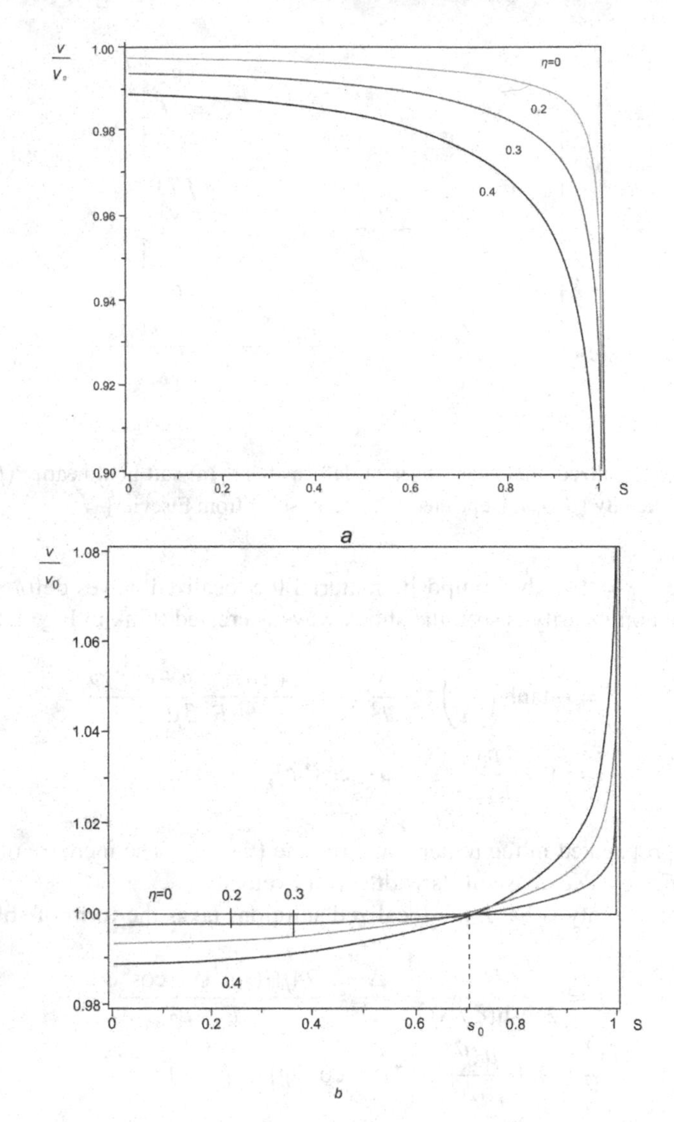

Figure 5.11 Phase velocity of nonlinear wave of deformations: (a) soft nonlinearity ($\beta < 0$) and (b) hard nonlinearity ($\beta > 0$) [reprinted with permission from Elsevier].

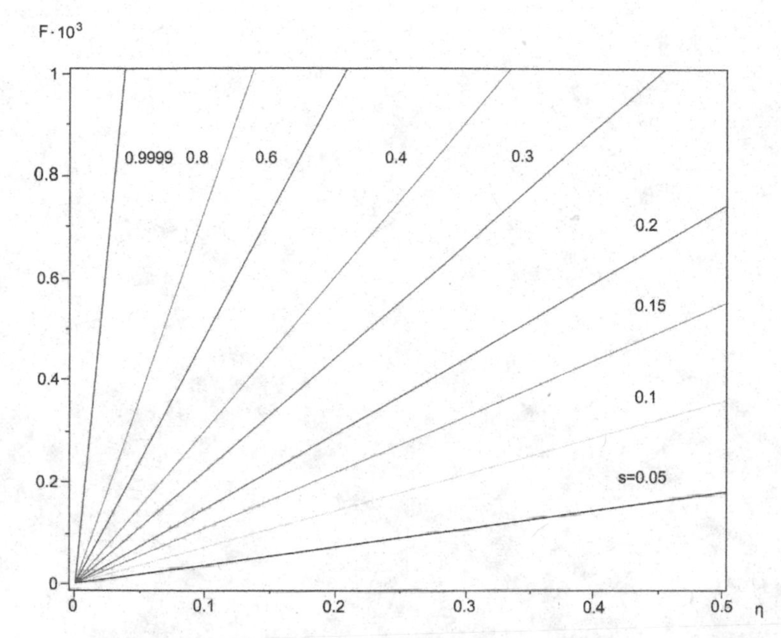

Figure 5.12 Modulus s characterizing intensity of nonlinear effects [reprinted with permission from Elsevier].

The dependence of the phase velocity v versus modulus s is reported in Fig. 5.11. In the case of soft nonlinearity one deals with the subsonic regime for which $v < v_0$. In the case of the hard nonlinearity one has supersonic regime and $v < v_0$ if $s^2 < 1/2$ and $v > v_0$ if $s^2 > 1/2$. The threshold value $s^2 = 1/2$, for which the change of the velocity regime takes place, is defined by the qualitative character of the analytical solution (5.44) and does not depend neither on the properties of the composite material nor on direction of the wave propagation.

Fig. 5.12 presents the parametric dependence of the modulus s versus F obtained based on the numerically found solution to equation (5.39) (or (5.43)). The area of elastic deformation is bounded by the value $F \leq 10^{-3}$ being characteristic for the majority of the materials used in real constructions.

As it follows from Fig. 5.9, 5.11, influence of the nonlinearity on the form and velocity of the wave becomes important for $s > 0.6$. For $\eta < 0.2$, the solution can be obtained with a help of the employed method of homogenization. Accuracy of the homogenized model (5.21) is decreased for $\eta > 0.4$ (see Fig. 5.7) (in this case $s < 0.34$), and the nonlinear effects can be neglected. Therefore propagation of nonlinear waves can be investigated in scope of the developed long-wave models, whereas the short waves with strong dispersion effects can be approximated by linear models.

6 Formation of Localized Nonlinear Waves in Layered Composites

This chapter deals with nonstationary wave propagation and formation of localized nonlinear waves in layered composite materials. First, the mathematical model is given and the pseudo-spectral method is described. Then, the Fourier-Padé approximation is used to decrease the Gibbs-Wilbraham effects. The process of formation of the localized nonlinear waves originated from the initial impulses of deformation is modeled numerically. The following special cases are considered: (i) generation of a single localized wave; (ii) generation of a train of the localized waves; (iii) scattering of the initial tension impulse.

6.1 INITIAL MODEL AND PSEUDO-SPECTRAL METHOD

Solutions obtained in the previous describe stationary waves which propagate without change of their forms and velocities. However, important role plays analysis of non-stationary dynamic processes. In this section we study evolution of the initial impulses of deformations and formation of the localized nonlinear waves. Here we employ numerical methods.

Consider longitudinal waves propagating in a layered composite material in a direction perpendicular to the layers arrangement. Let us define deformation of the wave profile as $f = du/dx$ and introduce the non-dimensional variable $\bar{f} = f/F$, $\bar{x} = x/l, \bar{t} = t/T$, where f is the deformation amplitude, T is the time needed by wave propagating in linear homogenous medium to move along the size of periodicity cell l, and $T = l/v_0$, $v_0 = \sqrt{E/\rho}$. Then wave equation (4.22) can be presented in the following form

$$\frac{\partial^2 \bar{f}}{\partial \bar{t}^2} = P\left(\bar{f}\right), \qquad P\left(\bar{f}\right) = \frac{\partial^2 \bar{f}}{\partial \bar{x}^2} + \frac{FE_2}{2E_1}\frac{\partial^2\left(\bar{f}^2\right)}{\partial \bar{x}^2} + \frac{E_3}{E_1}\frac{\partial^4 \bar{f}}{\partial \bar{x}^4}. \tag{6.1}$$

We take the following initial conditions

$$\bar{f} = f_0(\bar{x}), \qquad \frac{d\bar{f}}{d\bar{t}} = 0 \quad \text{at} \quad \bar{t} = 0, \tag{6.2}$$

where the function $f_0 = (\bar{x})$ defines the form of initial impulse of deformation.

In order to solve the Cauchy problem (6.1), (6.2), the following numerical procedure is employed. Integration in time is carried out by the 4th order Runge-Kutta method. We present the searched solutions $\overline{f}(\overline{x},\overline{t})$ in the form of the discrete sequence of the approximating functions

$$\bar{f}(\bar{x},\bar{t}) = f_k(\bar{x}) \quad \text{for} \quad \bar{t} = k\Delta t, \quad k = 0,1,2\ ..., \tag{6.3}$$

where Δt stands for the step of integration.

The function $f_k(\bar{x})$ is estimated based on the following recurrent relations [245]

$$\begin{aligned}
&f_{k+1} = f_k + \dot{f}_k \Delta t + \frac{1}{6}(m_1 + m_2 + m_3)\Delta t, \\
&\dot{f}_0 = 0, \quad \dot{f}_{k+1} = \dot{f}_k + \frac{1}{6}(m_1 + 2m_2 + 2m_3 + m_4), \\
&m_1 = P(f_k)\Delta t, \quad m_2 = P\left(f_k + \dot{f}_k \Delta t/2\right)\Delta t, \\
&m_3 = P\left(f_k + \dot{f}_k \Delta t/2 + m_1 \Delta t/4\right)\Delta t, \\
&m_4 = P\left(f_k + \dot{f}_k \Delta t + m_2 \Delta t/2\right)\Delta t.
\end{aligned} \tag{6.4}$$

In order to approximate the searched solution in the spatial variable we use the Fourier series. The wave equation (6.1) is symmetric with regard to the co-ordinate's origin $\bar{x} = 0$. If we consider the case when the impulse $f_0(\bar{x})$ is described by an even function then the solution should be also represented by an even function, $\dot{f}_k(\bar{x}) = f_k(-\bar{x})$. Each of the functions $f_k(\bar{x})$ can be presented by the following series forms

$$f_k(\bar{x}) = \sum_{n=0}^{N} c_{n,k} \cos\left(n\frac{2\pi\bar{x}}{L_0}\right), \qquad k = 0,1,2,..., \tag{6.5}$$

where L_0 stands for the non-dimensional size of the spatial domain.

The coefficients $c_{n,0}$, $c_{n,1}$ are defined through the initial conditions (6.2), and they take the following form

$$c_{n,0} = \frac{2}{L_0}\int_{-L_0/2}^{L_0/2} f_0(\bar{x})\cos\left(n\frac{2\pi\bar{x}}{L_0}\right), \qquad c_{n,1} = c_{n,0}. \tag{6.6}$$

The remaining coefficients $c_{n,k}$, $k = 2,3,4,...$, are computed with the use of formulas (6.4). The series (6.5) takes into account the boundary conditions of the periodic extension for $\bar{x} = L_0/2$. Since we consider the waves in infinite medium, we consider the solution evolution in the interval $0 < \bar{x} < L_0/2$.

We have known that the wave equation (6.1) can be used if $\eta = l/L < 0.4$. The minimal period of harmonic in series (6.5) is equal to $L = L_0 l/N$. Then we get the enequality

$$N/L < 0.4. \tag{6.7}$$

Condition (6.7) implies the restriction on the number of the terms of the series (6.5).

Series (6.5) present a continuous approximation where the Runge-Kutta method is based on the discrete difference scheme. Combination of both approaches is named the pseudo-spectral method [175, 393]. The pseudo-spectral method has found the wide application while solving numerically nonlinear wave equations [69, 243, 395].

6.2 THE FOURIER-PADÉ APPROXIMATION

In order to improve convergence of the series (6.5) and to decrease the Gibbs-Wilbraham effects [96, 143, 220], we employ the method of Fourier-Padé approximation. It allows to decrease the number of terms in the series (6.5) and to increase the minimum harmonic period. The latter circumstance is of particular importance. It allows to extend the area of applicability of the wave equation (6.1) the rational Padé approximation of the function $g(z)$ and it can be defined in the following way. Let us present $g(z)$ in the form of the series

$$g(z) = c_0 + c_1 z + c_2 z^2 + \ldots + c_N z^N + O\left(z^{N+1}\right). \tag{6.8}$$

Then the Padé approximation takes the following form:

$$g(z) = \frac{a_0 + a_1 z + a_2 z^2 + \cdots + a_K z^K}{1 + b_1 z + b_2 z^2 + \cdots + b_M z^M} + O\left(z^{N+1}\right), \tag{6.9}$$

where $K + M = N$.

The coefficients are defined by the following rule: development of (6.9) into the Maclaurin series should coincide with the series (6.8) up to the term $O(z^N)$ inclusively.

Transformation (6.9) allows essentially increase convergence of the series (6.8). In practice mostly the diagonal Padé approximation are used, where $K = M = N/2$. In what follow we consider the diagonal approximations.

There are a few algorithms devoted to computation of the coefficients while using the Padé approximation [47]. In this section we employ the following procedure. The coefficients of that denominator b are defined through the solution of the system of linear algebraic equation of the following form:

$$\begin{pmatrix} c_1 & c_2 & c_3 & \cdots & c_{N/2} \\ c_2 & c_3 & c_4 & \cdots & c_{N/2+1} \\ c_3 & c_4 & c_5 & \cdots & c_{N/2+2} \\ \cdots & \cdots & \cdots & \cdots & \cdots \\ c_{N/2} & c_{N/2+1} & c_{N/2+2} & \cdots & c_{N-1} \end{pmatrix} \begin{pmatrix} b_{N/2} \\ b_{N/2-1} \\ b_{N/2-2} \\ \cdots \\ b_1 \end{pmatrix} = \begin{pmatrix} c_{N/2+1} \\ c_{N/2+2} \\ c_{N/2+3} \\ \cdots \\ c_N \end{pmatrix} \tag{6.10}$$

and then the numerator coefficients a_i are defined using the following recurrent formula

$$a_0 = c_0, \quad a_i = c_i + \sum_{k=1}^{i} b_k c_{i-k}, \quad i = 1,\ 2,\ 3,\ \ldots,\ N/2. \tag{6.11}$$

In order to solve system (6.10), the Gauss-Jordan method is employed (it belongs to the one of the most effective methods from the computational point of view). The given algorithm requires $O[(N/2)^3]$ arithmetic operations, where $N/2$ stands for the number of the system equations.

We use the Fourier series based on the following exponential form

$$f_k(\bar{x}) = \sum_{n=0}^{N} c_{k,n} \cos\left(n\frac{2\pi\bar{x}}{L_0}\right) = \mathrm{Re}\left(\sum_{n=0}^{N} c_{k,n} e^{i\frac{2\pi\bar{x}}{L_0}n}\right), \tag{6.12}$$

and we carry out the replacement $e^{i\frac{2\pi\bar{x}}{L_0}n} \to z$. Then, the diagonal Fourier-Padé approximation can be presented in the following form

$$f_k(\bar{x}) = \mathrm{Re}\left(\frac{a_0 + a_1 z + a_2 z^2 + \ldots + a_{N/2} z^{N/2}}{1 + b_1 z + b_2 z^2 + \ldots + b_{N/2} z^{N/2}}\right), \tag{6.13}$$

where $z = \exp(2\pi\bar{x}i/L_0)$, and the coefficients a_i, b_i are yielded by relations (6.10), (6.11).

As the case study, we consider the piece-wise continuous function

$$f(\bar{x}) = \begin{cases} 0 & \text{for} \quad -50 < \bar{x} < -10, \\ 1 & \text{for} \quad -10 < \bar{x} < 10, \\ 0 & \text{for} \quad 10 < \bar{x} < 50. \end{cases} \tag{6.14}$$

Fig. 6.1 presents a part of the Fourier series sum (solid curve) being compared with the Fourier-Padé approximation (dashed curve) for $N = 32$. The absolute error is shown in Fig. 6.2 in the logarithmic scale for $N = 32$ and $N = 64$. Employment of the Fourier-Padé approximation allows for essential improvement of the convergence of the considered series. In the points of discontinuity there are local oscillation of the Fourier series observed, yielded by the Gibbs-Wilbraham phenomena.

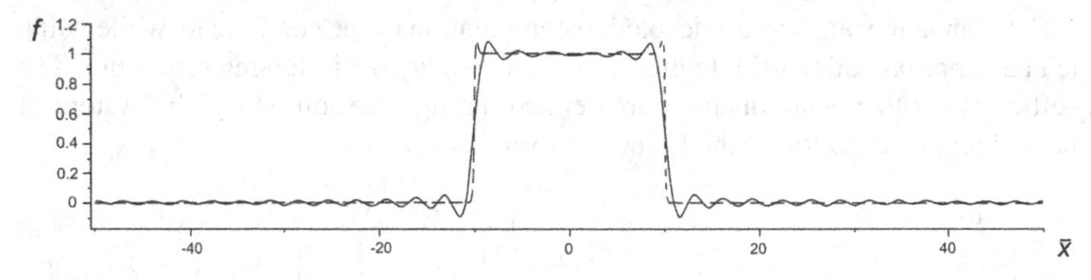

Figure 6.1 Approximation of the piece-wise discontinuous function $N = 32$ [reprinted with permission from Elsevier].

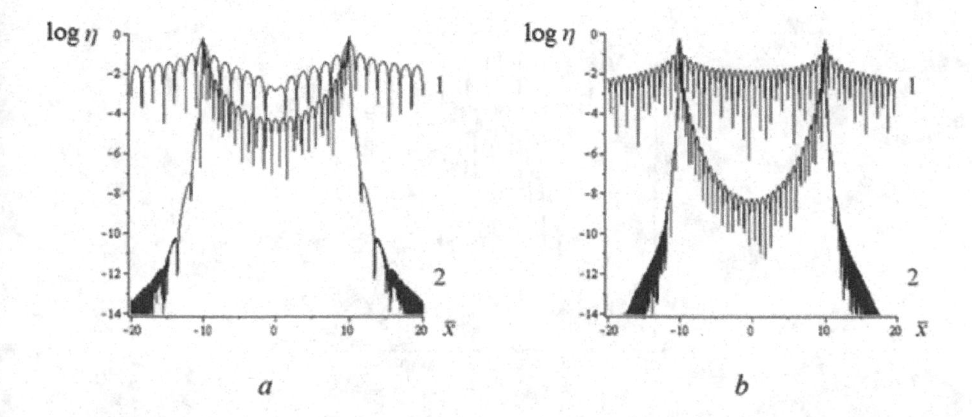

Figure 6.2 Absolute error of approximations of the piece-wise discontinuous function: (a) $N = 32$; (b) $N = 64$. 1-Fourier series; 2- Fourier-Padé approximation [reprinted with permission from Elsevier].

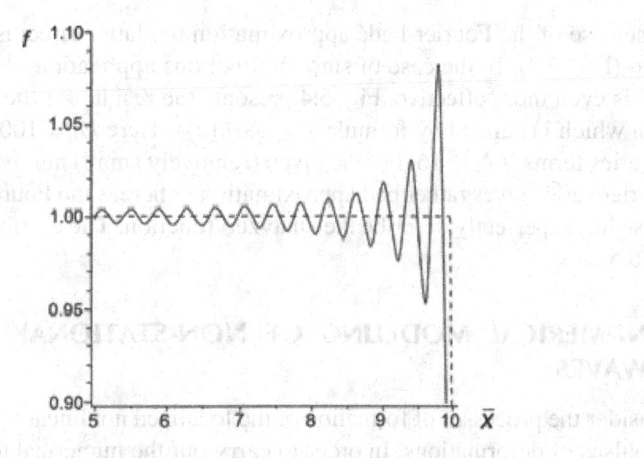

Figure 6.3 Approximations of the piece-wise discontinuous function near the point of discontimaty by Fourier series (solid curve) and Fourier-Padé (dashed curve) approximations ($N = 256$) [reprinted with permission from Elsevier].

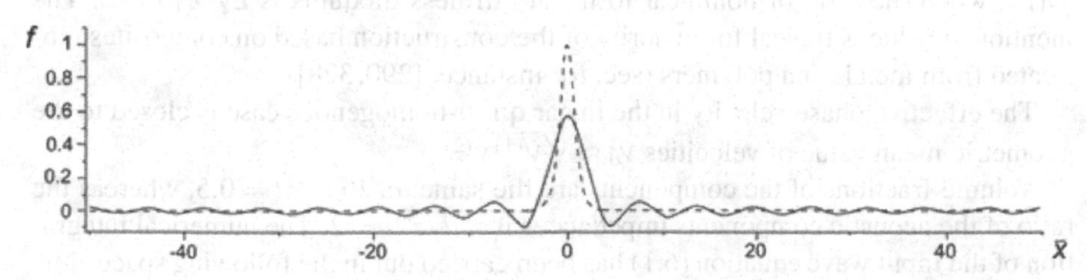

Figure 6.4 Approximation of the localized wave by Fourier series (solid curve) and Fourier-Padé (dashed curve) approximation ($N = 16$) [reprinted with permission from Elsevier].

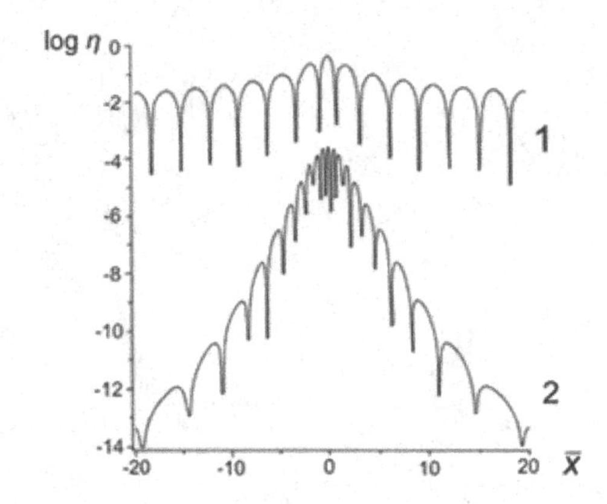

Figure 6.5 Absolute error of approximations of the localized wave by Fourier series (curve 1) and Fourier-Padé (curve 2) approximations ($N = 16$) [reprinted with permission from Elsevier].

In the case of the Fourier-Padé approximation the latter effect is more weak represented (Fig. 6.3). In the case of smooth functions application of the Fourier-Padé method is even more effective. Fig. 6.4 presents the results for the case of localized solution which is defined by formula $f = \sec h(\overline{x})^2$. Here $L_0 = 100$, and the number of the series terms is $N = 16$. For the given (relatively small) number of series terms, the Fourier series gives rather bad approximation, whereas the Fourier-Padé approximations almost perfectly describe the analyzed function. The absolute error is shown in Fig. 6.5.

6.3 NUMERICAL MODELING OF NON-STATIONARY NONLINEAR WAVES

We consider the processes of formation of the localized nonlinear waves from the initial impulses of deformations. In order to carry out the numerical tests, the program code in C language has been developed. With the use of *GNU Compiler Collection* we have employed, while carrying out the computations, the values of physical properties of the composite material. We have considered the case of soft nonlinearity, when the ratio of nonlinear to linear stiffness modulus is $E_2/E_1 \approx 10$. The mentioned value is typical for majority of the construction based on composites fabricated from metals and polymers (see, for instance, [290, 328]).

The effective phase velocity in the linear quasi-homogenous case is closed to the geometric mean value of velocities $v_0 \approx \sqrt{v^{(1)}v^{(2)}}$.

Volume fractions of the components are the same: $c^{(1)} = c^{(2)} = 0.5$, whereas the ratio of the acoustic components impedances is $z^{(2)}/z^{(1)} = 2$. The numerical integration of the input wave equation (6.1) has been carried out in the following space-time area: $-L_0/2 \leq \overline{x} \leq L_0/2$, $L_0 = 10000$, $0 \leq t \leq 5000$. The number of the included Fourier series terms (Fig. 6.5) is $N = 4000$, and the time integration step $\Delta t = 0.5$.

We consider the evolution of the initial impulse of the rectangular form

$$f_0(\bar{x}) = \begin{cases} 0 & \text{for} \quad -\frac{L_0}{2} < \bar{x} < -\frac{\Delta}{2}, \\ 1 & \text{for} \quad -\frac{\Delta}{2} \le \bar{x} \le \frac{\Delta}{2}, \\ 0 & \text{for} \quad \frac{\Delta}{2} < \bar{x} < \frac{L_0}{2}, \end{cases} \tag{6.15}$$

where Δ stands for the nondimensional impulse width.

The numerical computation is carried out for the three various scenarios: generation of the separate localized wave, generation of a sequence of the localized waves and scattering of the initial impulse.

1. Generation of a single localized wave.

Let the initial excitation forms are relatively narrow impulse of width $\Delta = 5$. The deformation amplitude is $F = 10^{-2}$. Observe that such a relatively high value of the elastic deformations is characteristic for the elastomers and rubber-type materials, as well as for polymers and organic molecular chains.

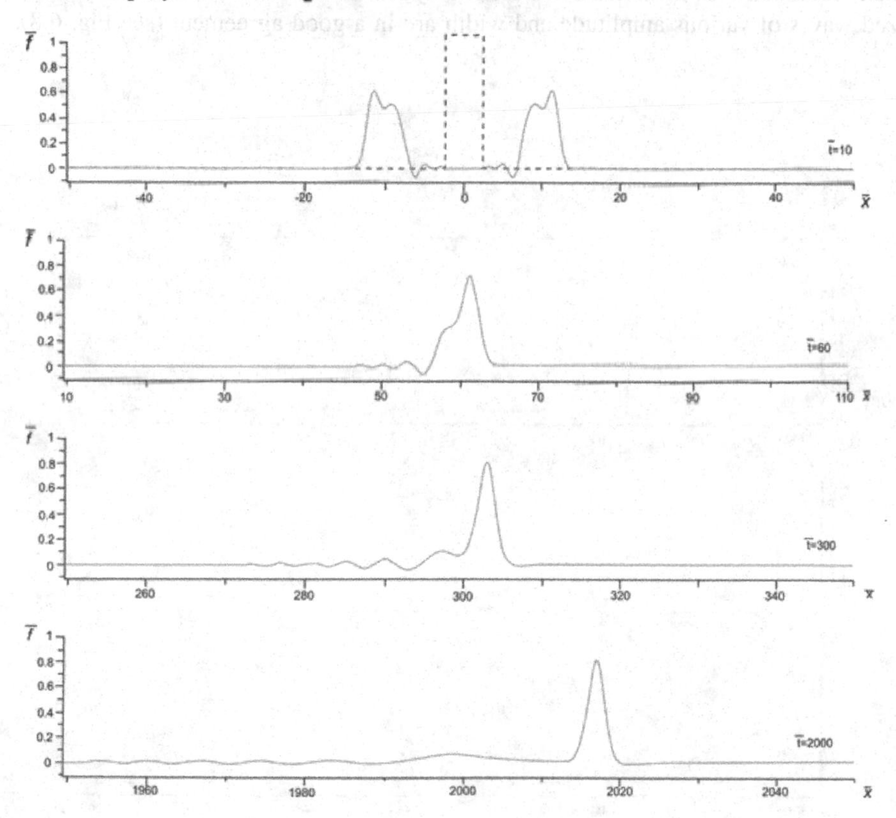

Figure 6.6 Evolution of the "narrow" impulse [reprinted with permission from Elsevier].

The evolution of the nonlinear wave is shown in Fig. 6.6. Here and further the dashed curves correspond to the solution for time instant $t = 0$. The rectangular impulse is collapsed into two symmetric parts, which move into opposite directions. We consider the interval $0 < \bar{x} < L_0/2$, since in the interval $-L_0/2 < \bar{x} < 0$ the solution is the same. The initial impulse is transformed into the bell-shape localized wave (soliton) with the fast oscillating tail. It should be emphasized that the formation of the localized wave, is carried out in the interval of order $x \sim 10^3 l$. The characteristic size of the internal structure of the composite materials is typically in intervals $l \sim 10^{-6} - 10^{-3}$ m. Consequently, the solitons can be observed in the real constructions of the size $l \sim 10$ m and more.

2. Generation of a train of the localized waves.

The initial excitation in the form of the wide impulse implies formation of the sequence of solitons. The evolution of the rectangular impulse for $\Delta = 20$, $F = -10^{-2}$ is reported in Fig. 6.7. "High" solitons have smaller width and are propagated faster than "low" solitons. Both numerical and analytical solutions of the stationary localized waves of various amplitude and width are in a good agreement (see Fig. 6.8).

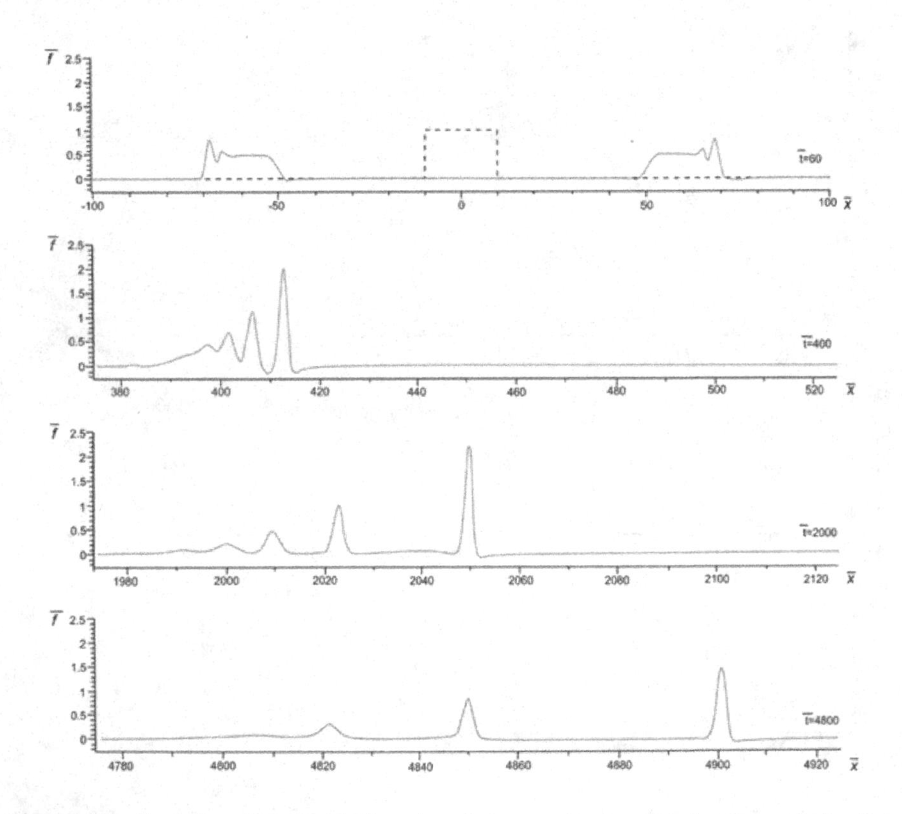

Figure 6.7 Evolution of the "wide" impulse of compression [reprinted with permission from Elsevier].

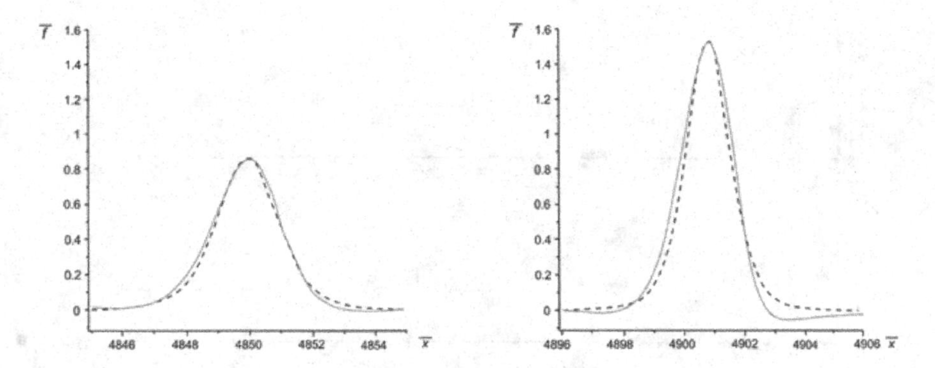

Figure 6.8 Stationary localized waves for $\bar{t} = 4800$; numerical/analytical solution are denoted by solid/dashed curve, respectively [reprinted with permission from Elsevier].

3. Scattering of an initial impulse of tension.

Impulse of tension $\Delta = 5$, $F = 10^{-2}$ does not imply formation of the localized waves. Instead, we observe the scattering of energy, which yields the delocalization of the initial excitation (Fig. 6.9). The reported results are in the agreement with conclusion carried out in section 4 while analyzing the analytical solution for the material with nonlinearity where only localized waves of compression are propagated.

The results obtained in this section are also in a qualitative agreement with the results reported by Porubov [370] and Samsonov [398] regarding properties of the localized nonlinear waves propagating in solid bodies. It should be also pointed out that we illustrated here the phenomena associated with occurrence of the structural dispersion caused by non-homogeneity of the internal microstructure of the composite materials. The obtained solutions allow to establish a dependence between parameters of the microstructure and characteristics of nonlinear waves of deformation which are propagated in materials on the macrolevel.

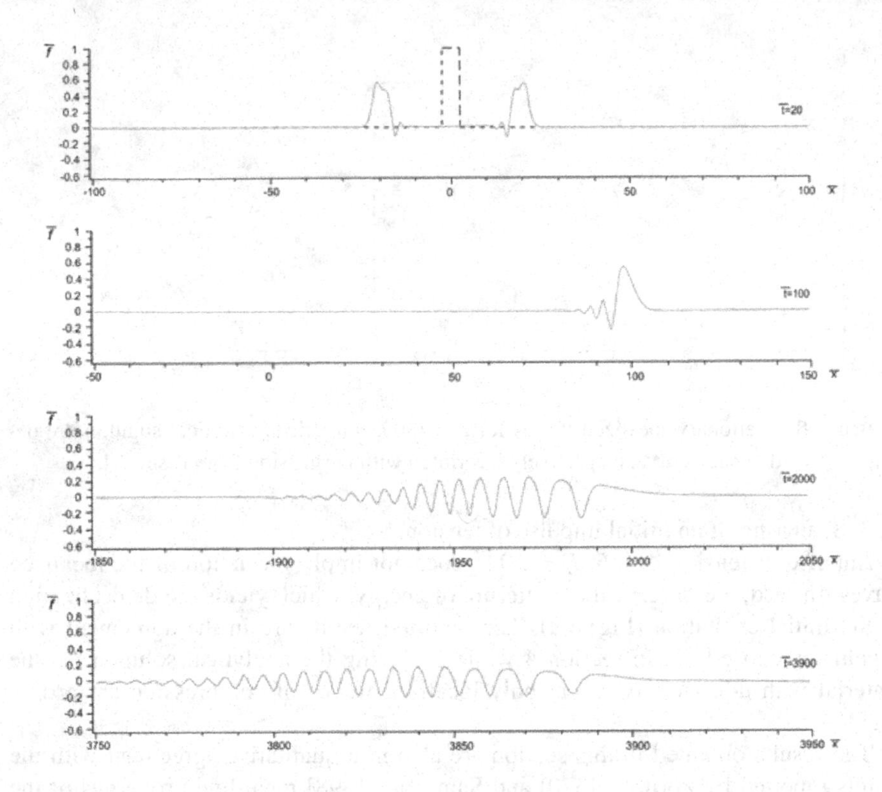

Figure 6.9 Delocalization of the initial impulse of tension [reprinted with permission from Elsevier].

7 Vibration Localization in 1D Linear and Nonlinear Lattices: Discrete and Continuous Models

This chapter deals with the localization of vibrations in one-dimensional linear and nonlinear lattices. The vibration localization can occur in the case of inhomogeneity under the following conditions: (i) the frequency spectrum of the periodic structure includes stopbands (ii) a perturbation of periodicity is present, and (iii) the eigenfrequency of the perturbed element falls into a stopband. Under these conditions, the energy can be spatially localized in the vicinity of the defect with an exponential decay in the infinity. The influence of nonlinearity can shift frequency into the stopband zone.

The localization frequencies are determined and the attenuation factors are calculated. Discrete and continuous models are developed and compared. The limits of the applicability of the continuous models are established. Analysis of the linear problem has allowed a better understanding of particular features of the nonlinear problem and has led to developing a new approach for the analysis of nonlinear lattices alternative to the method of continualization.

7.1 INTRODUCTION

The effect of spatial localization of oscillations plays an important role in dynamics of materials and structures. Local perturbation of properties of the dynamical system (presence of defects, cracks, holes, inclusions, etc.) leads to concentration of energy. Under certain conditions it may cause localized modes in form of standing waves with zero group velocity. In this case all the energy is concentrated in the vicinity of the perturbed element and decays exponentially with distance from it. A nonlinearity of the system can cause a similar effect on its behavior leading to localized vibrations [441].

Localization of vibrations can lead to local loss of stability of structures and result in their failure. At the same time, this effect allows the designer to control the distribution of dynamic loads that can be used to create the vibration dampers and to reduce the vibration amplitude in the most critical parts of the system. Localized modes play an important role in the acoustic diagnostics methods, which are used to detect various defects.

Localization effects in homogeneous and near-homogeneous structures are investigated in [218, 280]. The problems of localization of oscillations in linear and nonlinear mass lattices are reviewed in [295], where attention is paid to the construction of asymptotic solutions in the continuous approximations. Homogenization and high-frequency asymptotics are used to describe localization in one-dimensional lattices in [131, 339]. Problems of localization of oscillations in two-dimensional lattices are considered in [127].

Mode localization phenomenon in infinite periodic or bi-periodic mass-spring systems with disorders was investigated in [104, 106] using so-called U-transformation method. This method is based on the expansion normal mode.

Many authors used Green's functions to describe the effect of localization. The vibrational spectra of the defected polymers are analyzed in [254]. Localizations in the linear continuous systems with defects (beam and plate on an elastic foundation; non-uniform strings, rods and beams), as well as in discrete systems (a discrete lattice on a discrete elastic foundation; bi-periodic discrete lattice) are studied in [324].

The perturbation method was used in [363] to analyze the localization of vibrations in irregular structures.

7.2 MONATOMIC LATTICE WITH A PERTURBED MASS

The spectrum of infinite lattice in linear approximation does not contain localized modes because of symmetry restrictions. Such a solution can be valid for the lattice with broken symmetry. The detailed analysis of the influence of symmetry breaking on the vibration localization of lattice is given in [299–301].

Periodic systems are sensitive to certain types of periodically-breaking disorder, resulting in a phenomenon known as mode localization. Localization of periodic oscillations in a discrete system implies that some amplitude substantially exceed the rest. As a result the energy of the system is found to be predominantly confined on such a part [441]. The majority of existing studies of this phenomenon relate to linear systems; in particular, it was shown that in a linear regular system localization can occur when weak perturbations of the periodicity are introduced [104].

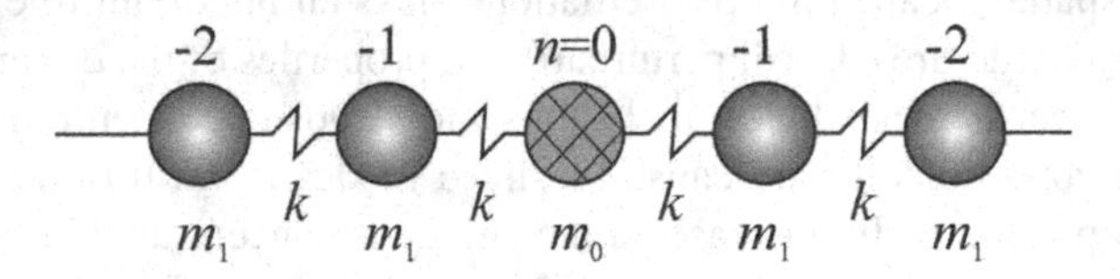

Figure 7.1 Monatomic lattice with a perturbed mass [reprinted with permission from Springer Nature].

Consider a monatomic lattice in which the mass of an element $n = 0$ differs from all the other elements and is equal to m_0, while masses of all other elements are equal to m_1, see Fig. 7.1. The equations of motion are

$$m_0 \ddot{u}_0 + k(2u_0 - u_1 - u_{-1}) = 0, \tag{7.1}$$

$$m_0 \ddot{u}_0 + k(2u_0 - u_1 - u_{-1}) = 0. \tag{7.2}$$

We seek a spatially localized solution in the following form:

$$u_0 = A_0 e^{i\omega t}, \qquad u_n = A_1 e^{-\alpha|n|} e^{i\beta n} e^{i\omega t}, \qquad n = \pm 1, \pm 2, ..., \tag{7.3}$$

where α is the coefficient of localization, $\alpha > 0$.

Conditions for existence of solution (7.3) and the numerical values of its parameters are determined from the compatibility conditions for equations of motion.

Substituting Eq. (7.3) into Eqs. (7.1) and (7.2), we obtain

$$-m_0\omega^2 A_0 + 2k[A_0 - e^{-\alpha}\cosh(i\beta)A_1] = 0, \tag{7.4}$$

$$-m_1\omega^2 A_1 + k(2A_1 - A_0 e^{\alpha} e^{-i\beta} - A_1 e^{-\alpha} e^{i\beta}) = 0, \quad n = 1, \tag{7.5}$$

$$-m_1\omega^2 A_1 + k(2A_1 - A_0 e^{\alpha} e^{i\beta} - A_1 e^{-\alpha} e^{-i\beta}) = 0, \quad n = -1, \tag{7.6}$$

$$-m_1\omega^2 + k(2 - e^{-\alpha} e^{i\beta} - e^{\alpha} e^{-i\beta}) = 0, \quad n \geq 2, \tag{7.7}$$

$$-m_1\omega^2 + k(2 - e^{-\alpha} e^{-i\beta} - e^{\alpha} e^{i\beta}) = 0, \quad n \leq -2. \tag{7.8}$$

From Eqs. (7.5), (7.7) or (7.6), (7.8) we find $A_0 = A_1$, and from Eqs. (7.7), (7.8) we obtain $\sinh(i\beta) = 0$. Hence, $\beta = 0$ or $\beta = \pi$. The first root corresponds to a trivial solution $\omega = 0$ and can be discarded. For $\beta = \pi$ we obtain a perturbed π-waveform

$$u_n = (-1)^n A_1 e^{-\alpha|n|} e^{i\omega t}. \tag{7.9}$$

Equations (7.4) and (7.7) for $\beta = \pi$ allow to determine the coefficient of localization:

$$\alpha = \ln\left(\frac{1-\delta}{1+\delta}\right), \tag{7.10}$$

where $\delta = m_0/m_1 - 1$ is a dimensionless parameter that determines the degree of disturbance.

Vibration localization is possible only if $\alpha > 0$, and therefore Eq. (7.9) yields: $-1 < \delta < 0$. Otherwords, localization is possible only for the perturbed mass which is smaller than other masses.

For small values of δ we obtain

$$\alpha \approx -2\delta. \tag{7.11}$$

From Eq. (7.7) we find the frequency at which the localization takes place

$$\frac{\omega^2}{\omega_1^2} = \frac{2}{1-\delta^2}. \tag{7.12}$$

The numerical results are shown in Figs. 7.2 and 7.3. Decreasing mass m_0 leads to a shift in the frequency of localization to the stop band zone: $\omega > \sqrt{2}\omega_1$.

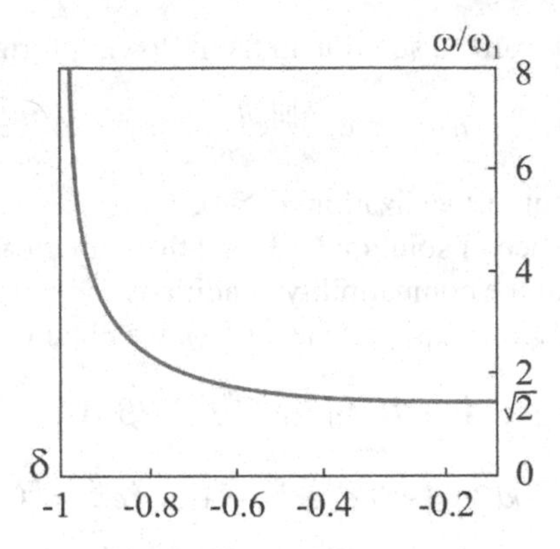

Figure 7.2 Frequency of localization in the monatomic lattice [reprinted with permission from Springer Nature].

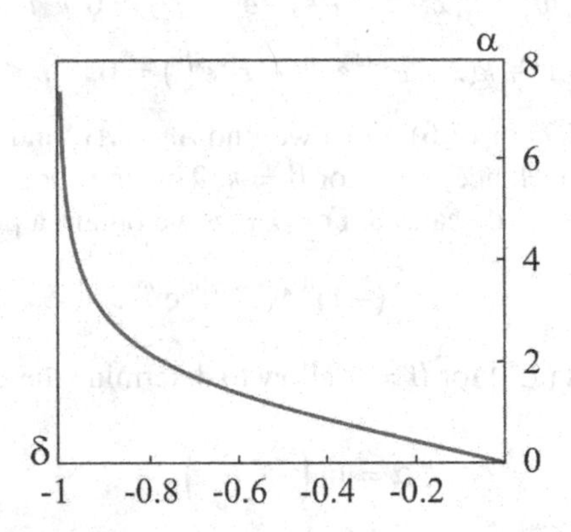

Figure 7.3 Coefficient of localization in the monatomic lattice [reprinted with permission from Springer Nature].

7.3 MONATOMIC LATTICE WITH A PERTURBED MASS – THE CONTINUOUS APPROXIMATION

The basic equations in the framework of envelope continualization approximation (semi discrete approach) [247, 250] and section 10.5 can be written as follows

$$m_1 u^{\pm}_{tt} + 4ku^{\pm} + kh^2 u^{\pm}_{xx} = 0, \tag{7.13}$$

$$\text{at} \quad x = 0, \quad u^+ = u^- = u, \tag{7.14}$$

$$kh(u^+_x - u^-_x) = -\delta m_1 u_{tt}. \tag{7.15}$$

Here, functions $u^+(x,t)$, $u^-(x,t)$ describe the envelope of the saw-tooth vibrations to the right and from the left of zero, respectively, and h is the distance between the masses.

The solution is sought in the form

$$u^{\pm} = C^{\pm} e^{-\alpha|x|} e^{\omega t}, \tag{7.16}$$

where

$$\alpha = \sqrt{\frac{m_1\omega^2 - 4k}{kh^2}}. \tag{7.17}$$

Equation (7.14) yields $C^+ = C^-$. Assume that

$$m_1\omega^2 = 4k + \omega_1^2, \qquad \omega_1^2 << 4k. \tag{7.18}$$

Then Eq. (7.15) gives

$$\alpha h \approx -2\delta, \tag{7.19}$$

which coincides with the first term of expansion of the solution for the discrete system in the series of a small parameter δ (7.11).

The expression for the frequency is given by

$$\omega^2 = \frac{4k}{m_1}(1+\delta^2). \tag{7.20}$$

It is interesting to note that the rearrangement of expression (7.20) in the Padé approximant (see, e.g., [47]) leads to the exact expression for the frequency of the discrete model (7.12).

7.4 DIATOMIC LATTICE

Consider now an infinite lattice consisting of the alternating masses m_1 and m_2, see Fig. 7.4.

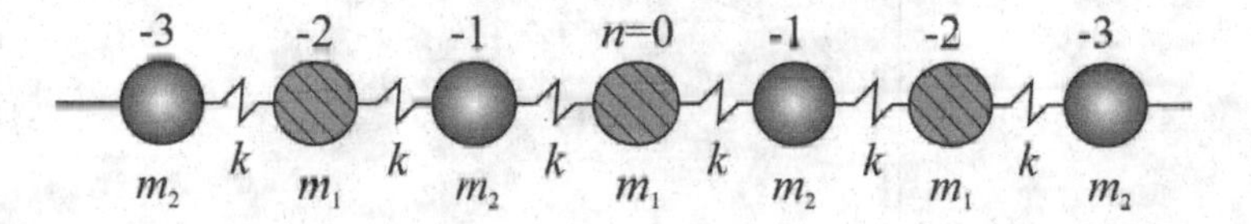

Figure 7.4 Diatomic lattice [reprinted with permission from Springer Nature].

The equations of motion are

$$m_1\ddot{u}_n + k(2u_n - u_{n+1} - u_{n-1}) = 0 \quad \text{for} \quad n = \pm1, \pm3..., \tag{7.21}$$

$$m_2\ddot{u}_n + k(2u_n - u_{n+1} - u_{n-1}) = 0 \quad \text{for} \quad n = 0, \pm2, \pm4, ... \tag{7.22}$$

Substituting expressions

$$u_n = A_1 e^{i\beta|n|} e^{i\omega t} \quad \text{for} \quad n = \pm1, \pm3, \pm5, ... \tag{7.23}$$

$$u_n = A_2 e^{i\beta|n|} e^{i\omega t} \quad \text{for} \quad n = 0, \pm2, \pm4, ... \tag{7.24}$$

into the Eqs. (7.21), (7.22) and eliminating the amplitudes A_1, A_2, we find the dispersion relation

$$\left(1 - \frac{\omega^2}{\omega_1^2}\right)\left(1 - \frac{\omega^2}{\omega_2^2}\right) = \cos^2\beta, \tag{7.25}$$

where $\omega_i = \sqrt{2k/m_i}$, $i = 1, 2$.

Dispersion curves are shown in Fig. 7.5. Here and below, it is assumed $m_2/m_1 = \omega_1^2/\omega_2^2 = 2$.

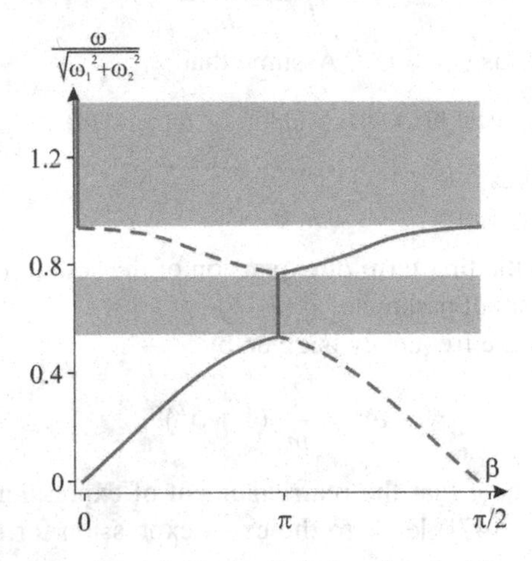

Figure 7.5 Dispersion curve for a diatomic chain [reprinted with permission from Springer Nature].

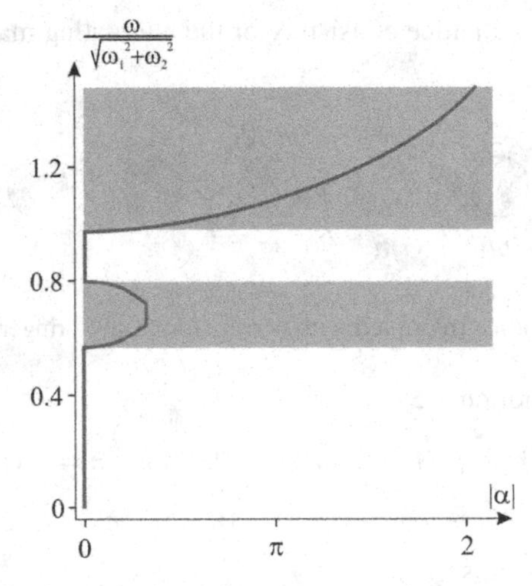

Figure 7.6 The attenuation coefficient of a diatomic chain [reprinted with permission from Springer Nature].

Note that the Eq. (7.25) has multiple roots. The dispersion dependences of normal waves in periodic structures are multivalued in nature [83]. The choice of a unique solution is one of the fundamental questions in the theory of wave propagation in periodic structures. There are different opinions in the literature on this point (see, for instance, [83, 86, 98, 232, 272, 408]). We agree with the point of view expressed in [83, 408] that a physically justified choice of a single dispersion branch for each wave requires additional information. Reducing the uncertainty in choosing dispersion curve must base on a limiting transition from an analyzed periodic structure to another one, for which dispersion dependences are known with more certainty. In the considering case, the dispersion curve is chosen from the condition of existence of the limiting transition to the monatomic chain.

In the single-mode there is a unique relationship between ω and β for a given solution (see [408]). The relevant branches of the spectrum are shown in Fig. 7.5 by solid lines, while all other branches are shown by dashed lines.

Diatomic lattice has two stop bands realized for $\beta = \pi/2$ (even and odd mass pairs are moving in opposite phases), and for $\beta = \pi$ (the neighboring masses are moving in opposite phases).

Note that the lower and upper limits of the first stop band coincide with the natural frequencies ω_2 and ω_1 of vibration of the disconnected masses, and the lower boundary of second stop band is $\sqrt{\omega_1^2 + \omega_2^2}$.

By setting the wave number equal to $\beta + i\alpha$, we find the attenuation coefficient:

$$\alpha = \text{arcsin}h\sqrt{-\left(1 - \frac{\omega^2}{\omega_1^2}\right)\left(1 - \frac{\omega^2}{\omega_2^2}\right)}$$
$$\text{at} \quad \beta = \pi/2, \quad \omega_2 < \omega < \omega_1, \tag{7.26}$$

$$\alpha = \frac{1}{2}\ln\left[1 + 2D \pm 2\sqrt{D\left(1 - \frac{\omega^2}{\omega_1^2}\right)\left(1 - \frac{\omega^2}{\omega_2^2}\right)}\right]$$
$$\text{at} \quad \beta = \pi, \quad \omega > \sqrt{\omega_1^2 + \omega_2^2}, \tag{7.27}$$

where $D = (\omega^2/\omega_1^2)(\omega^2/\omega_2^2) - \omega^2/\omega_1^2 - \omega^2/\omega_2^2$.

Numerical results for the attenuation coefficient α are shown in Fig. 7.6.

7.5 DIATOMIC LATTICE WITH A PERTURBED MASS

Consider now a diatomic lattice in which the mass of an element $n = 0$ is equal to m_0, and it is different from the masses of the rest even elements, see Fig. 7.7.

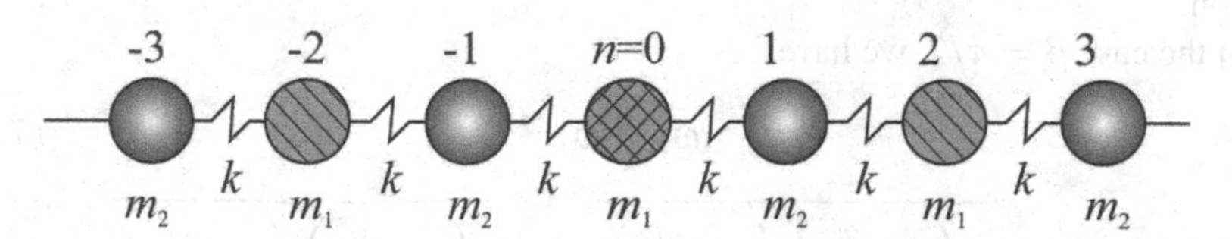

Figure 7.7 Diatomic lattice with a perturbed mass [reprinted with permission from Springer Nature].

The equations of motion are

$$m_0\ddot{u}_0 + k(2u_0 - u_1 - u_{-1}) = 0, \tag{7.28}$$

$$m_1 \ddot{u}_n + k(2u_n - u_{n+1} - u_{n-1}) = 0 \quad \text{for} \quad n = \pm 1, \pm 3..., \tag{7.29}$$

$$m_2 \ddot{u}_n + k(2u_n - u_{n+1} - u_{n-1}) = 0 \quad \text{for} \quad n = \pm 2, \pm 4, ... \tag{7.30}$$

Localized solution can be written as follows:

$$u_0 = A_0 e^{i\omega t} \quad \text{for} \quad n = 0, \tag{7.31}$$

$$u_n = A_1 e^{-\alpha|n|} e^{i\beta n} e^{i\omega t} \quad \text{for} \quad n = \pm 1, \pm 3..., \tag{7.32}$$

$$u_n = A_2 e^{-\alpha|n|} e^{i\beta n} e^{i\omega t} \quad \text{for} \quad n = \pm 2, \pm 4, ... \tag{7.33}$$

To find the values of the wave number for which there is a localization effect, consider two systems of Eqs. (7.28)–(7.30) for $n = -3, -2$ and for $n = 2, 3$. In order to eliminate the amplitude, equating to zero the determinant of matrices composed from the coefficients of A_1 and A_2

$$4\left(1 - \frac{\omega^2}{\omega_1^2}\right)\left(1 - \frac{\omega^2}{\omega_2^2}\right) = \left(e^{\alpha} e^{i\beta} + e^{-\alpha} e^{-i\beta}\right)^2, \tag{7.34}$$

$$4\left(1 - \frac{\omega^2}{\omega_1^2}\right)\left(1 - \frac{\omega^2}{\omega_2^2}\right) = \left(e^{-\alpha} e^{i\beta} + e^{\alpha} e^{-i\beta}\right)^2. \tag{7.35}$$

It follows from Eqs. (7.34) and (7.35) that either $\beta = 0$, or $\beta = \pi/2$, or $\beta = \pi$. The first root corresponds to the trivial solution $\omega = 0$. The compatibility conditions of Eqs. (7.29) and (7.30) for $n = 1, 2, 3$ yield $A_0 = A_2$.

Consider two systems of Eqs. (7.28)–(7.30): when $n = 0, 1$ and when $n = 1, 2$. Equating to zero the determinant of the corresponding matrices, we obtain

$$4\left(1 - \frac{\omega^2}{\omega_0^2}\right)\left(1 - \frac{\omega^2}{\omega_1^2}\right) = \left(e^{-2\alpha} e^{i\beta} + e^{-i\beta}\right)\left(e^{i\beta} + e^{-i\beta}\right), \tag{7.36}$$

$$4\left(1 - \frac{\omega^2}{\omega_1^2}\right)\left(1 - \frac{\omega^2}{\omega_2^2}\right) = \left(e^{-\alpha} e^{i\beta} + e^{\alpha} e^{-i\beta}\right)^2, \tag{7.37}$$

where $\omega_0 = \sqrt{2k/m_0}$, $\omega_1^2 < \omega_0^2 < \omega_2^2$.

Solving Eqs. (7.36), (7.37), we define the frequency and the coefficient of localization.

In the case $\beta = \pi/2$ we have

$$\omega^2 = \omega_0^2, \tag{7.38}$$

$$\alpha = \operatorname{arcsinh} \sqrt{-\left(1 - \frac{m_1}{m_0}\right)\left(1 - \frac{m_2}{m_0}\right)} \approx \sqrt{-\left(1 - \frac{m_1}{m_0}\right)\left(1 - \frac{m_2}{m_0}\right)}. \tag{7.39}$$

In the case $\beta = \pi$:

$$\frac{\omega^2}{\omega_1^2 + \omega_2^2} = \frac{\lambda(1 - \delta^2) + 2 + \sqrt{\lambda^2 (1 - \delta^2)^2 + 4\delta^2}}{2(1 + \lambda)(1 - \delta^2)}, \tag{7.40}$$

$$\alpha = \frac{1}{2}\ln\left[\frac{\lambda(1-\delta)^2}{2\delta + \sqrt{\lambda^2(1-\delta^2)^2 + 4\delta^2}}\right], \tag{7.41}$$

where $\lambda = m_2/m_1 = \omega_1^2/\omega_2^2$, $\delta = m_0/m_2 - 1$.

Vibration localization is possible only for $\alpha < 0$, which implies that $-1 < \delta < 0$.

For small values of δ we obtain:

in the case $\beta = \pi/2$:

$$\alpha \approx \sqrt{-\left(1 - \frac{m_1}{m_0}\right)\left(1 - \frac{m_2}{m_0}\right)}, \tag{7.42}$$

in the case $\beta = \pi$:

$$\alpha \approx -\delta\frac{m_1 + m_2}{m_2}, \tag{7.43}$$

$$\frac{\omega^2}{\omega_1^2 + \omega_2^2} \approx 1 + \delta^2\frac{m_1}{m_2}. \tag{7.44}$$

For $\lambda = 1$, the Eq. (7.40) is reduced to Eq. (7.12) and the Eq. (7.41) is reduced to Eq. (7.9).

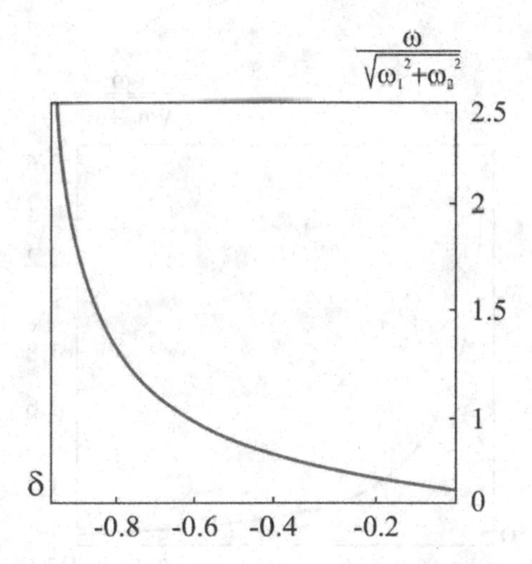

Figure 7.8 Frequency of localization in the diatomic lattice for $\beta = \pi/2$ [reprinted with permission from Springer Nature].

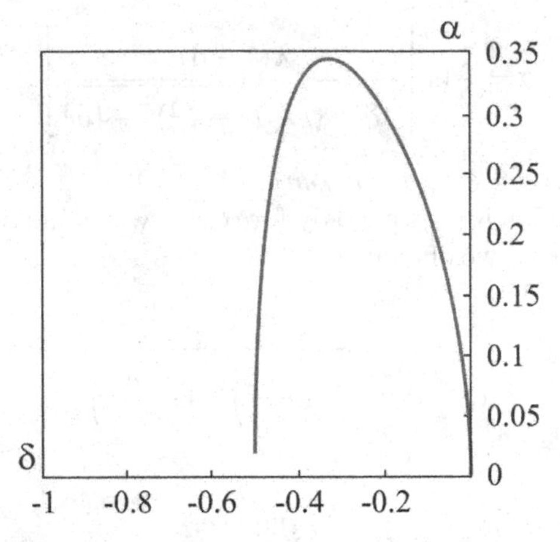

Figure 7.9 Coefficient of localization in the diatomic lattice for $\beta = \pi/2$ [reprinted with permission from Springer Nature].

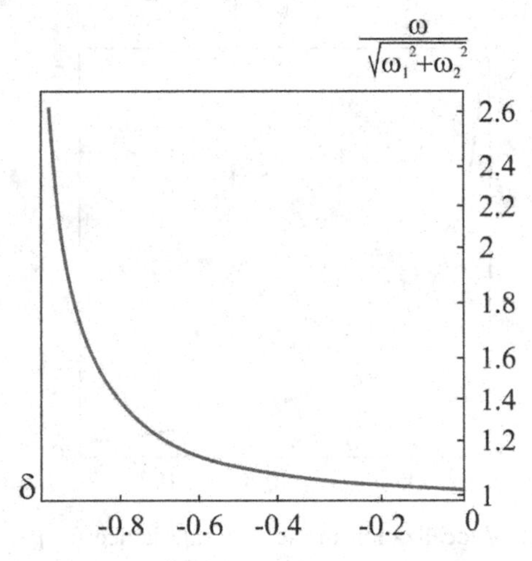

Figure 7.10 Frequency of localization in the diatomic lattice for $\beta = \pi$ [reprinted with permission from Springer Nature].

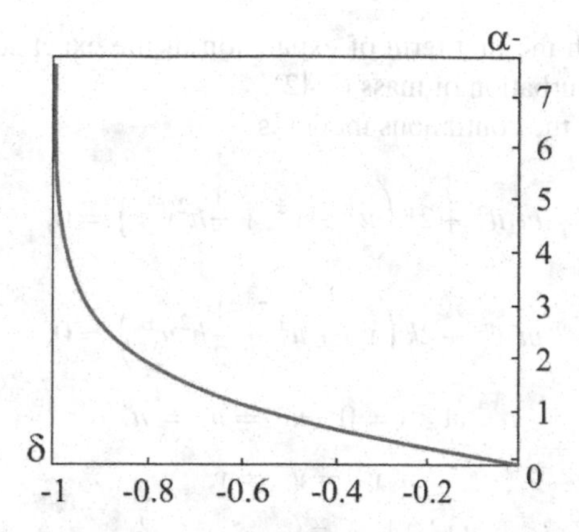

Figure 7.11 Coefficient of localization in the diatomic chain for $\beta = \pi$ [reprinted with permission from Springer Nature].

The numerical results are shown in Figs. 7.8–7.11. It was assumed in the calculations that $\lambda = 2$. In the case $\beta = \pi/2$ the localization ($\alpha > 0$) takes place in the range of values $-(\lambda-1)/\lambda < \delta < 0$. Frequency of localization coincides with the natural frequency ω_0 of the untied vibrations of the perturbed mass and is within the first stop band, $\omega_2 < \omega < \omega_1$.

In the case $\beta = \pi$ localization occurs when $\delta < 0$, which provides a shift of frequency to the second stop band: $\omega > \sqrt{\omega_1^2 + \omega_2^2}$. In both cases, the coefficient of localization α is equal in absolute value to the attenuation coefficient of the unperturbed part of chain. Substituting Eqs. (7.38) and (7.40) into Eqs. (7.26) and (7.27) we obtain the expression coinciding with the formulae (7.39) and (7.41), respectively.

7.6 DIATOMIC LATTICE WITH A PERTURBED MASS – THE CONTINUOUS APPROXIMATION

We now proceed to the continuous approximations. In the case $\beta = \pi/2$ continuous model has the following form

$$m_1\ddot{u} + 2k(u - hv_x) = 0, \tag{7.45}$$

$$m_2\ddot{v} + 2k(v - hu_x) = 0. \tag{7.46}$$

Here, $u(x,t)$ and $v(x,t)$ describe the envelope of the saw-tooth vibrations. Looking for the solution of Eqs. (7.45) and (7.46) in the form

$$u = A_1 e^{i\omega_0 t} e^{-\alpha x}, \qquad v = A_2 e^{i\omega_0 t} e^{-\alpha x}, \tag{7.47}$$

we obtain

$$\alpha h = \sqrt{-\left(1 - \frac{m_1}{m_0}\right)\left(1 - \frac{m_2}{m_0}\right)}, \tag{7.48}$$

which coincides with the first term of expansion of the exact solution (7.39) in the series of a small perturbation of mass (7.42).

In the case $\beta = \pi$ the continuous model is

$$m_1 \ddot{u}^{\pm} + 2k\left(u^{\pm} + v^{\pm} + \frac{1}{2}h^2 v^{\pm}_{xx}\right) = 0, \tag{7.49}$$

$$m_2 \ddot{v}^{\pm} + 2k\left(v^{\pm} + u^{\pm} + \frac{1}{2}h^2 u^{\pm}_{xx}\right) = 0, \tag{7.50}$$

$$\text{at} \quad x = 0 \quad u^{+} = u^{-} = u, \tag{7.51}$$

$$v^{+} = v^{-} = v, \tag{7.52}$$

$$kh^2(u^{+}_x - u^{-}_x) = \delta m_2 h \omega^2 v. \tag{7.53}$$

Here functions $u^{\pm}(x,t)$ and $v^{\pm}(x,t)$ describe the envelope of the saw-tooth vibrations to the right and left of zero respectively.

The solution of Eqs. (7.49) and (7.50) is sought in the form

$$u^{\pm} = C^{\pm}\mathrm{e}^{-\alpha|x|}, \qquad v^{\pm} = D^{\pm}\mathrm{e}^{-\alpha|x|}. \tag{7.54}$$

From the Eqs. (7.51) and (7.52) we find

$$C^{+} = C^{-} = C, \qquad D^{+} = D^{-} = D. \tag{7.55}$$

We set

$$\omega^2 = \frac{2k(m_1 + m_2)}{m_1 m_2} + \omega_1^2, \qquad \omega_1^2 << \frac{2k(m_1 + m_2)}{m_1 m_2}. \tag{7.56}$$

From Eq. (7.49) we have

$$D \approx \frac{m_1}{m_2} C. \tag{7.57}$$

From Eq. (7.53) one obtains

$$\alpha h = -\delta \frac{m_1 + m_2}{m_2}, \tag{7.58}$$

which coincides with the first term of expansion of the exact discrete solution (7.41) in the series of a small perturbation of mass (7.43).

From the Eqs. (7.49), (7.50), (7.56) and (7.58) we have

$$\omega_1^2 = \frac{2k\delta^2(m_1 + m_2)}{m_2^2}. \tag{7.59}$$

Hence

$$\omega^2 = \frac{2k(m_1 + m_2)}{m_1 m_2}\left(1 + \frac{\delta^2 m_1}{m_2}\right), \tag{7.60}$$

which coincides with the expression (7.44).

7.7 VIBRATIONS OF A LATTICE ON THE SUPPORT WITH A DEFECT

Thin-layer coats are often used as protecting or strengthening elements in the modern structures. Deformations of such structures may cause significant stresses on the interface between the base and the coat because of difference in their physical and mechanical properties. High stresses may lead to detachment of the coat and failure of the structure. Strength analysis of these structures under dynamic loading or impact is particularly important because of the possibility of localized vibrations in the neighborhood of inhomogeneities, such as inclusions, defects, etc. In [3] on the example of the detachment of a string from an elastic support, the possibility of localized vibrations is demonstrated and the effect of this localization on the growth of the detachment zone is analyzed. Below we consider vibrations of the lattice on a support with a defect.

The basic equations are

$$m_1 \ddot{u}_n + k(2u_n - u_{n+1} - u_{n-1}) + (\gamma + \Delta\gamma\delta_0^n)u_n = 0, \quad n = 0, \pm 1, \pm 2, \ldots \tag{7.61}$$

Here δ_0^n is the Kronecker delta, γ is stiffness of the support, $\Delta\gamma$ is a defect of rigidity of the support.

The solution of Eq. (7.61) is sought in the form

$$u_n = Ae^{-\alpha|n|}e^{i\beta n}e^{i\omega t}. \tag{7.62}$$

Substituting Eq. (7.62) into Eq. (7.61), we obtain

$$n = 0: -m_1\omega^2 + k(2 - e^{-\alpha}e^{i\beta} - e^{\alpha}e^{-i\beta}) + \gamma + \Delta\gamma = 0, \tag{7.63}$$

$$n > 0: -m_1\omega^2 + k[2 - e^{-\alpha}(e^{i\beta} - e^{-i\beta})] + \gamma = 0. \tag{7.64}$$

It is easy to see that the coefficient β can take values 0 or π. For $\beta = 0$ in the unperturbed state all the masses oscillate similarly, and the attenuation factor is equal to

$$\alpha = \operatorname{arcsin}h\frac{\Delta\gamma}{2k} \approx \frac{\Delta\gamma}{2k}, \quad \Delta\gamma > 0. \tag{7.65}$$

The application of the continuous approximation yields $\alpha = \frac{\Delta\gamma}{2k}$.

For $\beta = \pi$, we have a saw-tooth waveform and the attenuation factor is equal to

$$\alpha = -\operatorname{arcsin}h\frac{\Delta\gamma}{2k} \approx -\frac{\Delta\gamma}{2k}, \Delta\gamma < 0. \tag{7.66}$$

The use of envelope continualization yields $\alpha = \frac{\Delta\gamma}{2k}$.

7.8 NONLINEAR VIBRATIONS OF A LATTICE

From the latest developments in this area, we note that the use of localization in a nonlinear chain allows creating acoustic diode [282]. Also, a few of substantial surveys are published in this area [295, 410, 441].

Qualitative analysis of periodic vibrations of nonlinear n-DoF systems is carried out in [474] for both autonomous conservative and non-conservative as well as for non-autonomous systems. Sufficient conditions for strong localization are obtained. Another set of conditions provide for the existence of an one-parameter family of periodic solutions such that some amplitudes tend to infinity while the rest tend to zero as the parameter increases (asymptotic localization).

The nonlinear localized modes in a perfectly cyclic periodic system was studied in [440–442] using the multiple-scales approach.

Linear chains with nonlinear disorders were analyzed in [102, 103], where the Lindstedt-Poincaré technique and U-transformation were used for analysis of linear systems.

Many authors have used continuous approximations for analyzing the discrete systems. The pioneering work [248] should be noted here in the first place. The localized vibrations of a string with concentrated masses on the nonlinear elastic supports were studied in the similar manner in [448].

In the present section we consider the vibrations of a lattice on a nonlinear support. Basic equations can be written as follows [248, 448]:

$$m_1\ddot{u}_n + k(2u_n - u_{n+1} - u_{n-1}) + \gamma u_n - \gamma_{10}u_n^3 = 0; \;\; n = 0, \pm 1, \pm 2, ... \tag{7.67}$$

Introducing the dimensionless variable $\tau = \sqrt{\frac{k}{m_1}}t$, the Eq. (7.67) can be transferred to

$$\frac{d^2u_n}{d\tau^2} + 2u_n - u_{n+1} - u_{n-1} + \gamma_0 u_n - \gamma_1 u_n^3 = 0; \;\; n = 0, \pm 1, \pm 2, ..., \tag{7.68}$$

where $\gamma_0 = \frac{\gamma}{k}, \quad \gamma_1 = \frac{\gamma_{10}}{k}$.

Localization is possible under the shift of frequency below the minimum frequency of linear vibrations $\omega = \gamma_0$. We apply the harmonic balance method, representing the solution of Eq. (7.68) in the form

$$u_n(\tau) \approx U_n \cos(\omega\tau). \tag{7.69}$$

Substituting Eq. (7.69) into Eq. (7.68), we obtain in the first approximation of the harmonic balance method

$$\varepsilon^2 U_n + 2U_n - U_{n-1} - U_{n+1} - \frac{3}{4}\gamma_1 U_n^3 = 0, \qquad n = 0, \pm 1, \pm 2, ..., \tag{7.70}$$

where $\varepsilon^2 = \gamma_0 - \omega^2 << 1$.

The solution of Eq. (7.67) is sought in the form of perturbation series:

$$U_n = U_n^{(0)} + \varepsilon U_n^{(1)} + ... \tag{7.71}$$

In the zero-approximation we obtain

$$U_n^{(0)} = Aa^n, \tag{7.72}$$

where $a \approx 1 \pm \varepsilon$.

We consider $n \geq 0$, in this case $a \approx 1 - \varepsilon$. Using the standard procedure of the perturbation method [330], we find

$$U_n^{(1)} = \frac{3\gamma_1}{32\varepsilon^2} A^3 a^{3n}. \tag{7.73}$$

Transforming the truncated perturbation series (7.71) with the first two terms, defined by Eqs. (7.72) and (7.73), into the Padé approximants [47], we obtain

$$Aa^n + \frac{3\gamma_1}{32\varepsilon^2} A^3 a^{3n} \approx Aa^n \frac{1}{1 + \frac{3\gamma_1}{32\varepsilon^2} A^2 a^{2n}}. \tag{7.74}$$

Using the symmetry condition

$$U_1 - U_0 = 0, \tag{7.75}$$

we find

$$A = 4\varepsilon \sqrt{\frac{2}{3\gamma_1}}. \tag{7.76}$$

As a result, we obtain

$$U_n = 2\varepsilon \sqrt{\frac{2}{3\gamma_1}} \frac{2}{(1-\varepsilon)^n + (1-\varepsilon)^{-n}}. \tag{7.77}$$

Continuous approximation, described by the equation

$$\varepsilon^2 U(x) - \frac{d^2 U(x)}{dx^2} - \frac{3}{4}\gamma_1 U^3(x) = 0, \tag{7.78}$$

yields in this case [410]:

$$U(x) = 2\varepsilon \sqrt{\frac{2}{3\gamma_n}} \frac{2}{e^{\varepsilon x} + e^{-\varepsilon x}}. \tag{7.79}$$

It is of interest to compare our solution for the discrete system given by Eq. (7.77) with the solution given by Eq. (7.79) obtained in [248] in the continuous approximation of this system. It is easy to see that in the solution (7.77) the term $e^{\varepsilon x}$ is replaced by the first terms of the expansion of this exponent in the Maclaurin series. Let us analyze whether it is possible to achieve better coincidence of the solutions for the discrete and continuous approximations. Note that the expression (7.79) is the exact solution of Eq. (7.78), while the expression (7.77) is only an approximate solution of Eq. (7.70). The asymptotic method for the analysis of discrete nonlinear lattices, based on the transition in the first approximation to the integrable cases, is proposed in [124, 242]. In other words, the original systems are complemented by some terms so that the new systems have the symmetry sufficient to obtain the exact solutions. The similar idea is used in [258].

In accordance with this idea, instead of the transition to the continuous approximation, we can approximately replace the basic equations for the nonlinear nonintegrable lattice (7.67) in the following way

$$m_1 \ddot{u}_n + k(2u_n - u_{n+1} - u_{n-1}) + \gamma u_n - \frac{1}{2}\gamma_{10}(u_{n-1} + u_{n+1})\left|u_n^2\right| = 0,$$
$$n = 0, \pm 1, \pm 2, ... \tag{7.80}$$

Equation (7.80) describes a known Ablowitz-Ladik lattice (see [401], Section (5.2.1)), which has an exact solution

$$U_n = 2\varepsilon\sqrt{\frac{2}{3\gamma_1}}\frac{2}{e^{\varepsilon n} + e^{-\varepsilon n}}. \tag{7.81}$$

Continuous approximation of Eq. (7.80) coincides with the continuous approximation of Eq. (7.67), whereas the solution (7.81) coincides with the solution (7.79). Therefore, the asymptotic process proposed in [124, 242] corresponds to the continuous approximation in the first approximation.

7.9 EFFECT OF NONLINEARITY ON PASS BANDS AND STOP BANDS

In contrast to the papers [103, 248], which have examined the linear lattices with nonlinear disorders, we investigate dynamics of the nonlinear lattice.

Consider vibrations of the nonlinear lattice

$$m_1 \ddot{u}_n + k(2u_n - u_{n+1} - u_{n-1}) + \varepsilon_1\gamma_{10}[(u_n - u_{n+1})^3 + (u_n - u_{n-1})^3] = 0,$$
$$n = 0, \pm 1, \pm 2, ..., \tag{7.82}$$

where $\varepsilon_1 << 1$.

After the change of variables

$$\tau_1 = \sqrt{\frac{2k}{m_1}}t, \quad \tilde{u} = u\sqrt{\frac{\gamma_{10}}{k}} \tag{7.83}$$

we obtain the following system of equations:

$$\frac{d^2\tilde{u}}{d\tau_1^2} + \frac{1}{2}(2\tilde{u}_n - \tilde{u}_{n+1} - \tilde{u}_{n-1}) + \frac{\varepsilon_1}{2}[(\tilde{u}_n - \tilde{u}_{n+1})^3 + (\tilde{u}_n - \tilde{u}_{n-1})^3] = 0,$$
$$n = 0, \pm 1, \pm 2, ... \tag{7.84}$$

In the case of weak nonlinearity, we develop an asymptotic solution of the Eq. (7.84) using the Lindstedt-Poincaré technique [114]. For this purpose, we introduce a new variable $\tau_2 = \omega\tau_1$, and seek the solution in the form of expansions

$$\tilde{u}_n = u_n^{(0)} + \varepsilon_1 u_n^{(1)} + \varepsilon_1^2 u_n^{(2)} + \cdots \tag{7.85}$$

$$\omega^2 = \omega_0^2 + \varepsilon_1\omega_1 + \varepsilon_1^2\omega_2 + \cdots \tag{7.86}$$

In the zero approximation we obtain

$$u_n^{(0)} = A_0 e^{i\tau_2} e^{i\beta n} + c.c., \tag{7.87}$$

$$\omega_0^2 = 1 - \cos(\beta + \alpha i). \tag{7.88}$$

Here β and α are the real and imaginary parts of the wave number.

The system of equations of the first approximation is

$$\omega_0^2 \frac{d^2 u_n^{(1)}}{d\tau_2^2} + \frac{1}{2}\left(2u_n^{(1)} - u_{n+1}^{(1)} - u_{n-1}^{(1)}\right) = \omega_1 \frac{d^2 u_n^{(0)}}{d\tau_2^2} - \frac{1}{2}[\,(u_n^{(0)} - u_{n+1}^{(0)})^3 + (u_n^{(0)} - u_{n-1}^{(0)})^3], \quad n = 0, \pm 1, \pm 2, \ldots \tag{7.89}$$

From the condition of absence of secular terms we find

$$\omega_1 = 6A_0\bar{A}_0(2\cos\beta\cosh^2\alpha - \cos\beta - \cosh\alpha)(\cos\beta - \cosh\alpha), \tag{7.90}$$

where $\bar{A}_0$ is a complex conjugate of A_0.

For the pass band ($\alpha = 0$), we obtain

$$\omega_1 = 6A_0\bar{A}_0(1 - \cos\beta)^2. \tag{7.91}$$

For the stop band zone ($\beta = \pi$), we obtain

$$\omega_1 = 6A_0\bar{A}_0(2\cosh^2\alpha + \cosh\alpha - 1)(1 + \cosh\alpha). \tag{7.92}$$

The magnitude ω_{SB} of the stop band threshold ($\alpha = 0, \beta = \pi$) is

$$\omega_{SB} = \sqrt{2 + 24\varepsilon_1 A_0 \bar{A}_0}. \tag{7.93}$$

This result agrees with that obtained in [114] in a different way.

In the framework of the method of continualization of envelope, assuming $u_n = -u_{n+1}$ and afterwards passing to the continuous variable x, the original equation can be represented as follows:

$$m_1 \ddot{u}(x,t) + 4ku(x,t) + 16\varepsilon_1 \gamma_{10} u^3(x,t) = 0. \tag{7.94}$$

Using the Lindstedt-Poincaré technique we come to the above formula (7.93) for the stop band threshold.

The rate of the nonlinear effects is determined by the magnitude of the parameter ε_1 and by the amplitude modulus $|A_0|$. Figures 7.12, 7.13, and 7.14 illustrate the influence of nonlinearity on the dispersion curves, on the attenuation coefficient α, and on the stop band threshold ω_{SB}. It should be noted that a soft nonlinearity ($\varepsilon_1 < 0$) decreases the width of the pass band, while the presence of a hard nonlinearity ($\varepsilon_1 > 0$) makes the propagation zone wider.

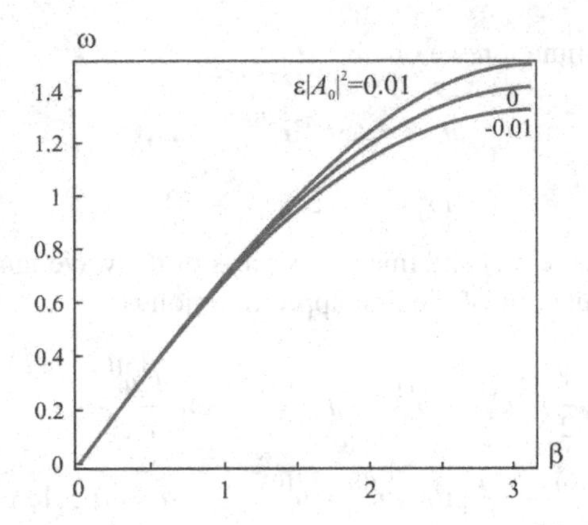

Figure 7.12 Effect of nonlinearity on the dispersion curves [reprinted with permission from Springer Nature].

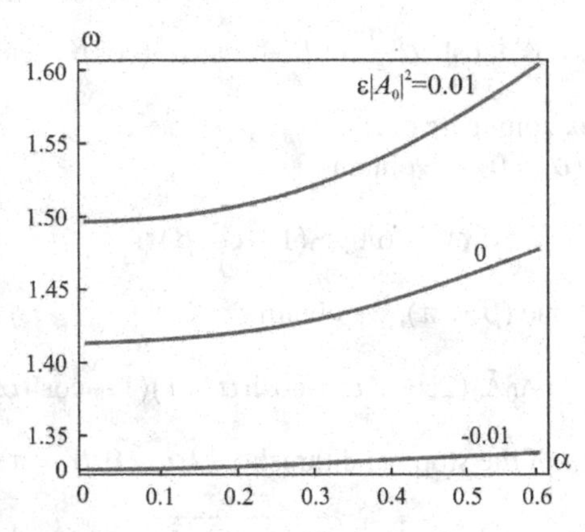

Figure 7.13 Effect of nonlinearity on the attenuation coefficient [reprinted with permission from Springer Nature].

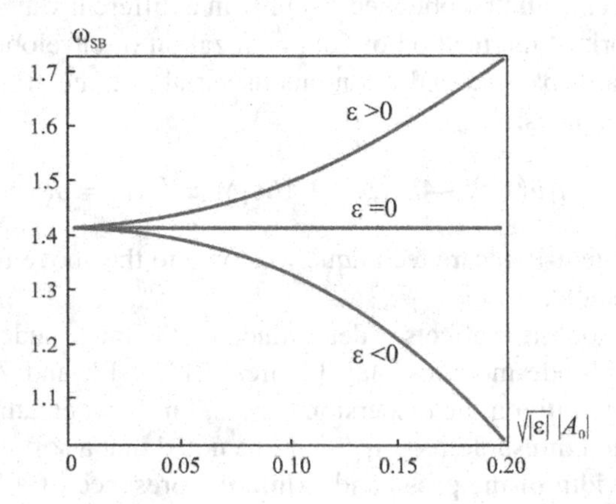

Figure 7.14 Influence of soft and hard nonlinearity [reprinted with permission from Springer Nature].

8 Spatial Localization of Linear Elastic Waves in Composite Materials with Defects

In the present chapter, the transfer-matrix approach for investigation of linear waves localization in a composite materials is used. Perturbations of the density and the volume fractions of the components are analyzed. Localization of antiplane linear shear waves in a fibrous composite is also studied. In the latter case plane-wave expansions approach gives possibility to obtain the localization frequencies and the attenuation factors. Influence of perturbation of the volume fraction of a fiber on the mentioned characteristics is investigated.

8.1 INTRODUCTION

The phenomenon of wave localization plays an important role in the dynamics of heterogeneous materials and structures. For example, linear array composite transducers being based on this effect are widely used in medical ultrasonic imaging [108]. Wave localization may be important in the study of stability of disordered trains and train-like articulated systems [394]. Wave localization in solar panels is also important from the engineering point of view [161]. Due to randomly distributed defects and manufacturing errors, composite materials designed to be periodic can never be exactly periodic in reality. The deviation from the ideal periodicity is known as disorder, it can lead to the effect of wave localization.

This effect plays an important role in dynamics of materials and structures. Local perturbation of properties of the dynamical system (presence of defects, cracks, holes, inclusions, etc.) leads to the concentration of energy. Under certain conditions, it can cause localized modes in form of standing waves with zero group velocity. In this case all the energy is concentrated in the vicinity of the perturbed element and decays exponentially with distance from it.

Wave localization can lead to a local loss of stability of the structures and results in their failure. At the same time, this effect allows the designer to control the distribution of dynamic loads that can be used to create the vibration dampers and to reduce the vibration amplitude in the most critical parts of the system. Localized modes play an important role in the acoustic diagnosis methods, which are used to detect various defects.

Wave localization in lattices is analyzed by many researches. We mention the survey paper [295] and book [441], where considerable attention is paid to the construction of asymptotic solutions in the continuous limit. Homogenization and high-frequency asymptotics are used to describe the localization in one-dimensional lattices in [131, 339]. Wave localization in 2D lattices are considered in [127]. Study the wave localization in lattices allows drawing conclusions that are common to all periodic linear systems. That is, the phenomenon of wave localization can appear under the following conditions: (i) the frequency spectrum of the periodic structure includes stop bands, (ii) the violation perturbation of periodicity (a defect) is takes place introduced, and (iii) the eigenfrequency of the defective element falls into a stop band. Then the energy can be spatially localized in the vicinity of the defect with an exponential decay at infinity.

Further, we simulate the defects by local change of the material properties or volume fractions of components of a composite material.

Wave localization in composite materials, together with the wave localization in lattices, attracted the attention of researchers. In the linear case, one of the most common approaches is the method of transfer matrices [105, 107, 240, 283, 363, 364, 464, 466]. Another possible approach is to discretize the original continuous system and then use the results obtained for the lattice structures [67, 127, 280, 281, 465]. Bloch-Floquet waves propagating in doubly-periodic composite structures with high-contrast interfaces and finite size defects are studied in [325].

In [218], the basic principles of the linear waves localization in continuous media with defects and inclusions are formulated. From a mathematical point of view, the problems considered in [218] are characterized by the discrete eigenvalues of the boundary value problems for operators.

In [3], on the example of the detachment of a string from an elastic support, the possibility of localized vibrations is demonstrated, and the effect of this localization on the increasing of the delaminated zone is analyzed. Two-scale homogenization is used in [423] for investigation of propagation and localization of elastic waves in periodic elastic composites with highly contrasting phases and highly anisotropic stiffnesses for longitudinal displacements. Various models describing damaged layer in periodically layered composites is studied in [188]. The perturbation method was used in [363] to analyze the localization of vibrations in irregular structures.

8.2 WAVE LOCALIZATION IN A LAYERED COMPOSITE MATERIAL: TRANSFER-MATRIX METHOD

Firstly, we consider the simplest problem allowing elementary numerical solution. This allows to identify the important features of the behavior of such systems and to simplify further investigations.

Let us consider an composite material consisting of alternating layers of two different components $\Omega^{(1)}$ and $\Omega^{(2)}$ (Fig. 8.1).

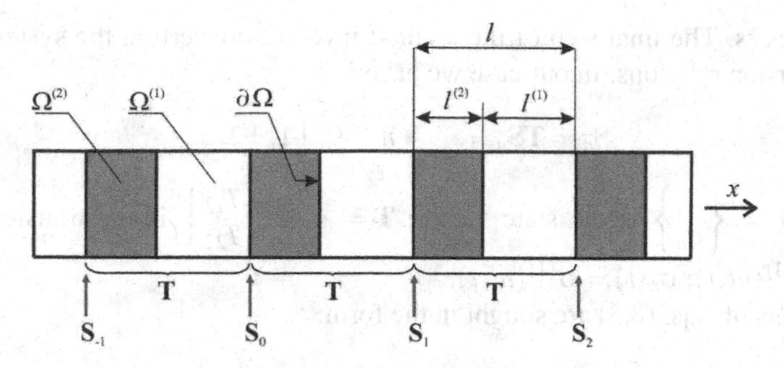

Figure 8.1 Layered composite material [reprinted with permission from John Wiley and Sons].

The original wave equation is:

$$(\lambda + 2\mu)\frac{\partial^2 u}{\partial x^2} = \rho \frac{\partial^2 u}{\partial t^2}, \tag{8.1}$$

where λ, μ are the Lamé constants, ρ is the mass density, u is the longitudinal displacement.

Due to the heterogeneity of the medium, the Lamé constants λ, μ and the mass density ρ are represented by piecewise continuous functions of co-ordinates:

$$\lambda(x) = \lambda^{(a)}(x), \quad \mu(x) = \mu^{(a)}(x), \quad \rho(x) = \rho^{(a)} \quad \text{at} \quad x \in \Omega^{(a)}. \tag{8.2}$$

Here and in the sequel the superscript (a) denotes different components of the structure, $a = 1, 2$.

Equations (8.1), (8.2) contains delta functions, so we treat a solution of this equation in a weak sense [275].

Equation (8.1) can be written in the equivalent form:

$$\left(\lambda^{(a)} + 2\mu^{(a)}\right)\frac{\partial^2 u^{(a)}}{\partial x^2} = \rho^{(a)}\frac{\partial^2 u^{(a)}}{\partial t^2}, \qquad a = 1, 2. \tag{8.3}$$

At the interface $\partial\Omega$ we assume the perfect bonding conditions that correspond to the continuity of displacements $u^{(a)}$ and stresses $\sigma^{(a)}$:

$$\left\{u^{(1)} = u^{(2)}\right\}\Big|_{\partial\Omega}, \qquad \left\{\sigma^{(1)} = \sigma^{(2)}\right\}\Big|_{\partial\Omega}, \tag{8.4}$$

where $\sigma^{(a)} = \left(\lambda^{(a)} + 2\mu^{(a)}\right)\frac{\partial u^{(a)}}{\partial x}$.

We use the transfer-matrix approach [87]. This approach is very efficient when the total system can be subdivided into a sequence of subsystems, each of them interacting with two neighbors. It is based on the following simple idea. If the displacements and stresses are known at the beginning of a layer, the displacements and stresses at the end of the layer can be derived from a simple matrix operation. A system of layers can be represented as a system matrix, which is the product of the individual layer matrices. The final step of the method involves converting the system matrix into dispersion relations. In our case we have:

$$\mathbf{S}_n = \mathbf{T}\mathbf{S}_{n-1}, \qquad n = 0, \pm 1, \pm 2, ... \tag{8.5}$$

Here $\mathbf{S}_n = \begin{Bmatrix} u_n \\ \sigma_n \end{Bmatrix}$ is the state vector; $\mathbf{T} = \begin{Bmatrix} T_{11} & T_{12} \\ T_{21} & T_{22} \end{Bmatrix}$ is the transfer matrix; $u_n(t) = u^{(1)}(nl,t)$; $\sigma_n(t) = \sigma^{(1)}(nl,t)$.

Solutions of Eqs. (8.3) are sought in the form:

$$u^{(a)} = \left[A^{(a)} \exp(-ik^{(a)}x) + B^{(a)} \exp(ik^{(a)}x)\right] \exp(i\omega t), \tag{8.6}$$

where $k^{(a)} = \omega\sqrt{\rho^{(a)}/(\lambda^{(a)} + 2\mu^{(a)})}$, $k^{(a)}$ are the wave numbers, ω is the frequency.

Using ansatz (8.6) the transfer matrix can be reduced to the following form:

$$\mathbf{T} = \mathbf{T}^{(1)}\mathbf{T}^{(2)}, \tag{8.7}$$

where

$$T_{11}^{(a)} = T_{22}^{(a)} = \cos(k^{(a)}l^{(a)}), \qquad T_{12}^{(a)} = \frac{\sin(k^{(a)}l^{(a)})}{\left(\lambda^{(a)} + 2\mu^{(a)}\right)k^{(a)}},$$
$$T_{21}^{(a)} = -\left(\lambda^{(a)} + 2\mu^{(a)}\right)k^{(a)} \sin(k^{(a)}l^{(a)}). \tag{8.8}$$

For obtaining the dispersion equation, let us consider a propagation of an effective harmonic wave. The longitudinal displacements can be represented as follows [98]:

$$u = U_0 \exp(ikx) \exp(i\omega t), \tag{8.9}$$

where k is the effective wave number.

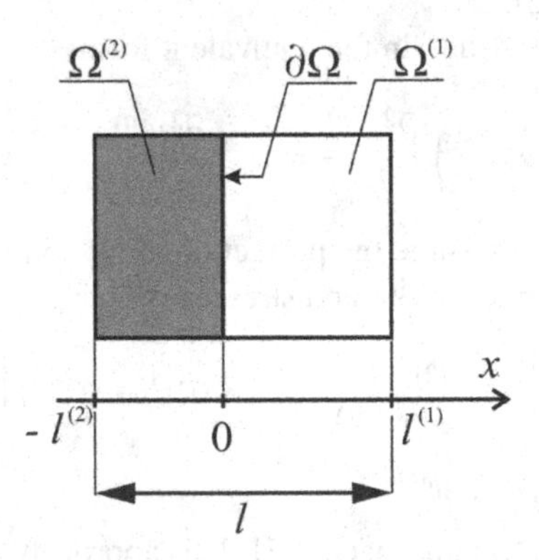

Figure 8.2 Unit cell of the layered composite material [reprinted with permission from John Wiley and Sons].

Now we determine the effective wave number k by equating expressions (8.9) and (8.6). Ratio of the displacements at the opposite sides of the periodically repeated unit cell (Fig. 8.2) can be calculated as follows:

$$u_n/u_{n-1} = \Lambda_{1,2} = \frac{1}{2}\mathrm{tr}(\mathbf{T}) \pm \sqrt{\left[\frac{1}{2}\mathrm{tr}(\mathbf{T})\right]^2 - 1}, \tag{8.10}$$

where tr(·) denotes the matrix trace, $\mathrm{tr}(\mathbf{T}) = T_{11}^{(1)}T_{11}^{(2)} + T_{12}^{(1)}T_{21}^{(2)} + T_{21}^{(1)}T_{12}^{(2)} + T_{22}^{(1)}T_{22}^{(2)}$ or

$$\mathrm{tr}(\mathbf{T}) = 2\cos(k^{(1)}l^{(1)})\cos(k^{(2)}l^{(2)}) - \sin(k^{(1)}l^{(1)})\sin(k^{(2)}l^{(2)})\left[\frac{(\lambda^{(2)}+2\mu^{(2)})k^{(2)}}{(\lambda^{(1)}+2\mu^{(1)})k^{(1)}} + \frac{(\lambda^{(1)}+2\mu^{(1)})k^{(1)}}{(\lambda^{(2)}+2\mu^{(2)})k^{(2)}}\right]. \tag{8.11}$$

The dispersion relation reads:

$$\exp(ikl) = \Lambda_{1,2}. \tag{8.12}$$

For the identification of the band gaps, one can use the following conditions:
- pass bands:

$$k \in \mathrm{Re}, \qquad |\mathrm{tr}(\mathbf{T})| < 2, \tag{8.13}$$

- stop bands:

$$k \in \mathrm{Im}, \qquad |\mathrm{tr}(\mathbf{T})| > 2. \tag{8.14}$$

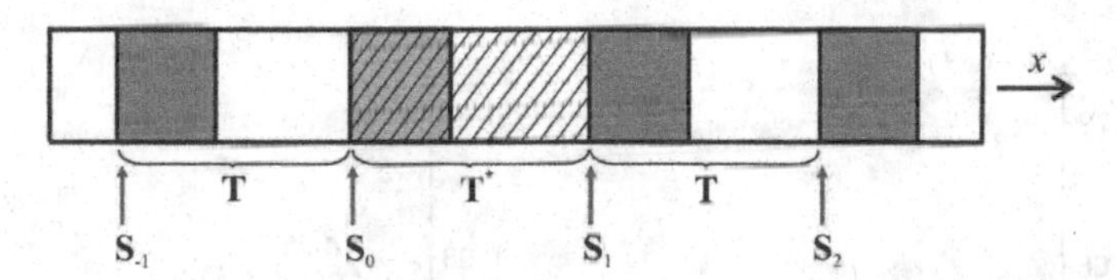

Figure 8.3 Layered composite with a defect unit cell (shaded) [reprinted with permission from John Wiley and Sons].

Now let us suppose that the layered composite has a defect unit cell (Fig. 8.3). For the determination of the localization wave number k one has the following condition:

$$2\cos(k^{(1)}l^{(1)})\cos(k^{(2)}l^{(2)}) - \sin(k^{(1)}l^{(1)})\sin(k^{(2)}l^{(2)})\left[\frac{(\lambda^{(2)}+2\mu^{(2)})k^{(2)}}{(\lambda^{(1)}+2\mu^{(1)})k^{(1)}} + \frac{(\lambda^{(1)}+2\mu^{(1)})k^{(1)}}{(\lambda^{(2)}+2\mu^{(2)})k^{(2)}}\right] = \pm 2. \tag{8.15}$$

From Eq. (8.15) one obtains: $k = 0,\ \pi/l$.
Dispersion equation for the defect unit cell is used for the determination of the localization frequency ω^*:

$$\mathrm{tr}(\mathbf{T}^*) = \pm 2 \quad \text{at} \quad k = 0, \quad \pi/l, \tag{8.16}$$

where $\mathbf{T}^*$ is the transfer matrix for the defect cell.

In the case of the wave localization, the solution for the whole structure can be written as follows:

$$u = U_0 \exp\left[i\left(k + i\alpha\right)x\right]\exp(i\omega^* t), \tag{8.17}$$

where α is the attenuation factor.

Dispersion equation for the regular unit cell is used for the determination of α:

$$\exp\left[i\left(k + i\alpha\right)l\right] = \Lambda_{1,2} \quad \text{at} \quad \omega = \omega^*. \tag{8.18}$$

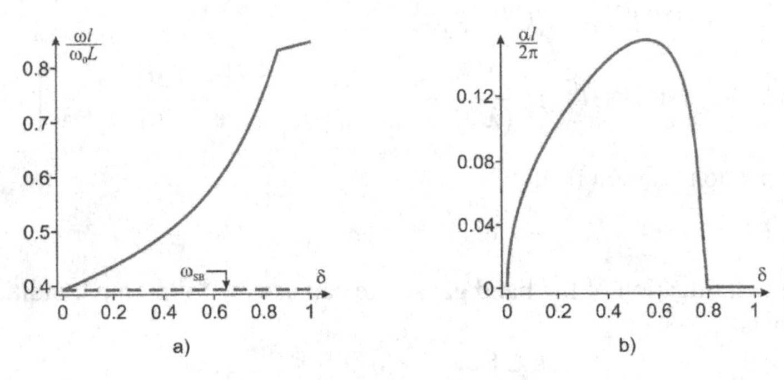

Figure 8.4 Wave localization at the beginning of the 1st stop band induced by the decrease of density of the defect cell: a) localization frequency, b) attenuation factor [reprinted with permission from John Wiley and Sons].

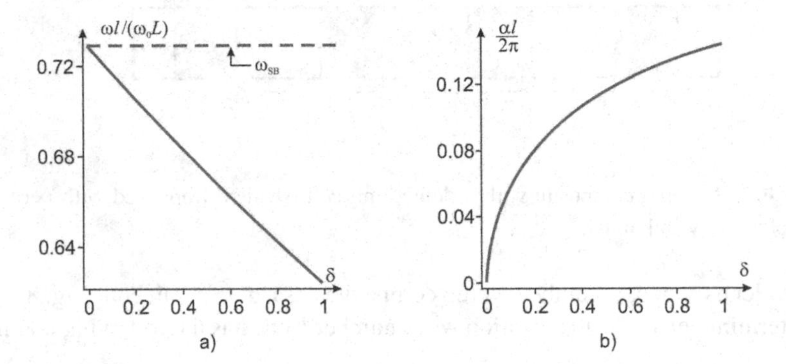

Figure 8.5 Wave localization at the end of the 1st stop band induced by the increase of density of the defect cell: a) localization frequency, b) attenuation factor [reprinted with permission from John Wiley and Sons].

Let us consider the results of calculation of Eqs. (8.16), (8.18) for the steel-aluminium composite characterized by the following values of the parameters: the volume fractions $c^{(a)} = l^{(a)}/l$ of the components are the same $c^{(1)} = c^{(2)} = 0.5$, the elastic coefficients $\lambda^{(1)} + 2\mu^{(1)} = 112$ GPa and $\lambda^{(2)} + 2\mu^{(2)} = 275$ GPa, the mass densities $\rho^{(1)} = 2700$ kg/m^3 and $\rho^{(2)} = 7800$ kg/m^3.

We will study an influence of different parameters of the defect cell on the wave localization. Firstly, we consider perturbation of the density of the defect cell:

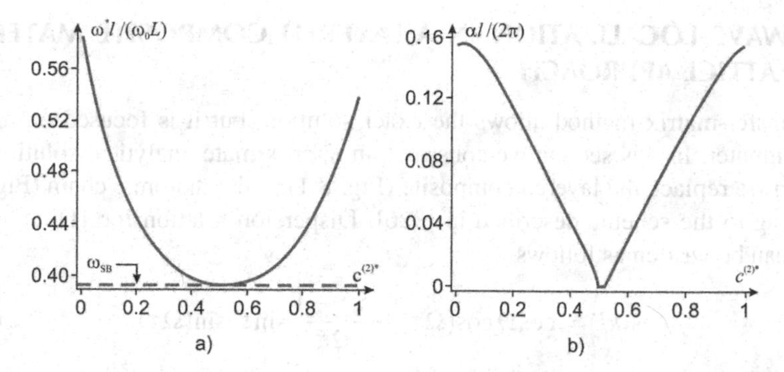

Figure 8.6 Wave localization at the beginning of the 1st stop band induced by the perturbation of the volume fraction of an inclusion: a) localization frequency, b) attenuation factor [reprinted with permission from John Wiley and Sons].

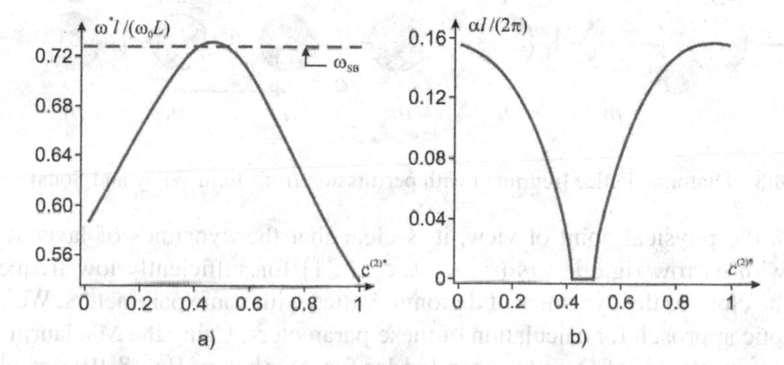

Figure 8.7 Wave localization at the end of the 1st stop band induced by the perturbation of the volume fraction of an inclusion: a) localization frequency, b) attenuation factor [reprinted with permission from John Wiley and Sons].

$\rho^{(2)*} = \rho^{(2)}(1-\delta)$, $\rho^{(2)*} < \rho^{(2)}$. The parameter $\delta \leq 1$ characterizes the measure of perturbation. Numerical results depicted in Fig. 8.4 show that decrease in mass density $\rho^{(2)}$ induces localization at the beginning of the 1st stop band with $\omega > \omega_{SB}$. Here and in the sequel ω_0 denotes the frequency of the original perfectly periodic structure; ω_{SB} denotes the frequency of the first stop band. Increase in mass density $\rho^{(2)}$ ($\rho^{(2)*} = \rho^{(2)}(1+\delta)$, $\rho^{(2)*} > \rho^{(2)}$) induces localization at the end of the 1st stop band (Fig. 8.5) with $\omega < \omega_{SB}$.

Perturbation of the volume fraction of inclusions leads to the localization at the beginning of the 1st stop band with $\omega > \omega_{SB}$ (Fig. 8.6) or at the end of the 1st stop band with $\omega < \omega_{SB}$ (Fig. 8.7). The volume fractions of the components of defect cell are defined as $c^{(1)*}, c^{(2)*}$ ($c^{(1)*} + c^{(2)*} = 1$).

8.3 WAVE LOCALIZATION IN A LAYERED COMPOSITE MATERIAL: LATTICE APPROACH

The transfer-matrix method allows the exact solution, but it is focused on the use of a computer. In this section we construct an approximate analytical solution. For this aim we replace the layered composite (Fig. 8.1) to the diatomic chain (Fig. 8.8) according to the scheme described in [326]. Dispersion relation for 1D composite (8.12) can be written as follows

$$\cos(kl) = \cos\Omega\cos(\Omega\tau) - \frac{\xi^2+1}{2\xi}\sin\Omega\sin(\Omega\tau), \tag{8.19}$$

where $\Omega = \frac{\omega l^{(1)}}{C_1}$, $\tau = \frac{l^{(2)}C_1}{l^{(1)}C_2}$, $\xi = \frac{\sqrt{(\lambda^{(1)}+2\mu^{(1)})\rho^{(1)}}}{\sqrt{(\lambda^{(2)}+2\mu^{(2)})\rho^{(2)}}}$, $C_i = \sqrt{\frac{\lambda^{(i)}+2\mu^{(i)}}{\rho^{(i)}}}$.

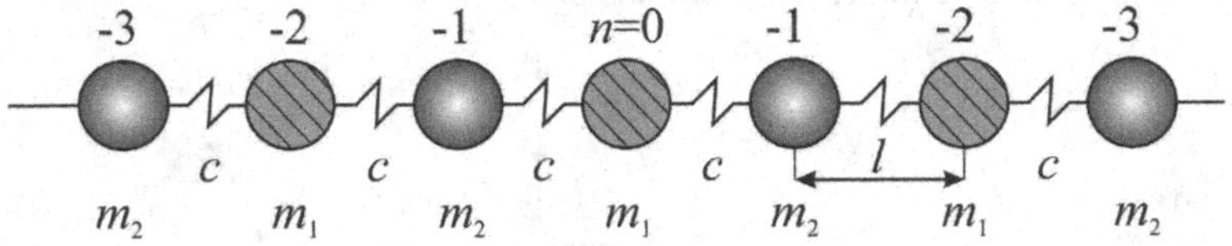

Figure 8.8 Diatomic lattice [reprinted with permission from John Wiley and Sons].

From the physical point of view, it is clear that the dynamics of layered composite with narrow rigid layers ($\xi >> 1, \tau << 1$) for sufficiently low frequencies should be close to the dynamics of diatomic lattice with some parameters. We use an asymptotic approach for calculation of these parameters. Using the Maclaurin series expansion in powers of Ω up to second order for the r.h.s. of Eq. (8.19) one obtains (ξ and τ have order 1)

$$A_1\omega^4 - A_2\omega^2 + 2[1-\cos(kl)] = 0, \tag{8.20}$$

where

$$\begin{aligned} A_1 &= \frac{1}{12}\left(\frac{l^{(1)}}{C_1}\right)^4\left(1+2\xi\tau+6\tau^2+2\xi\tau^3+2\frac{\tau}{\xi}\right), \\ A_2 &= \left(\frac{l^{(1)}}{C_1}\right)^2\left(1+\xi\tau+\tau^2+\frac{\tau}{\xi}\right). \end{aligned} \tag{8.21}$$

In expressions (8.21) we take into account the terms up to the second order concerning τ and $1/\xi$.

Now let us consider an infinite lattice consisting of the alternating masses m_1 and m_2 (Fig. 8.8). The dispersion relation for the diatomic lattice can be written as follows [326]

$$\frac{4}{\omega_1^2\omega_2^2}\omega^4 - 4\left(\frac{1}{\omega_1^2}+\frac{1}{\omega_2^2}\right)\omega^2 + 2[1-\cos(kl)] = 0, \tag{8.22}$$

where $\omega_i = \sqrt{2c/m_i}$, $i = 1,2$, c is the spring rigidity, k is the wave number.

The parameters of the lattice model are determined from the following condition: the equation (8.20) that describes the dynamic behavior of the composite material must fit the dispersion relation of the diatomic lattice.

Comparing Eqs. (8.20) and (8.22), one obtains

$$\omega_1^2\omega_2^2 = \frac{4}{A_1}, \qquad \omega_1^2 + \omega_2^2 = \frac{A_2}{A_1}, \tag{8.23}$$

where $\omega_{1(2)}^2 = \left(A_2 - \sqrt{A_2^2 - 16A_1}\right)/(2A_1)$ and $\omega_{2(1)}^2 = \left(A_2 + \sqrt{A_2^2 - 16A_1}\right)/(2A_1)$.

The lattice approach, described by Eqs. (8.22), (8.23), can be used for estimating the parameters of the wave localization in 1D composites, see Sections 7.4–7.6. For instance, we consider perturbation of the density of a defect cell: $\rho^{(2)*} = \rho^{(2)}(1-\delta)$. In this case $kl = 2\pi$, using Sections 7.4–7.6, we define the localization frequency ω^* and the attenuation factor α:

$$\frac{\omega^{*2}}{\omega_1^2 + \omega_2^2} = \frac{\lambda(1-\delta^2) + 2 + \sqrt{\lambda^2(1-\delta^2)^2 + 4\delta^2}}{2(1+\lambda)(1-\delta^2)},$$
$$\alpha = \frac{1}{2}\ln\left[\frac{\lambda(1-\delta)^2}{2\delta + \sqrt{\lambda^2(1-\delta^2)^2 + 4\delta^2}}\right], \tag{8.24}$$

where $\lambda = \omega_1^2/\omega_2^2$.

8.4 ANTIPLANE SHEAR WAVES IN A FIBER COMPOSITE

Now let us consider a 2D problem. Namely, we study transverse antiplane shear waves propagating in the x_1x_2 plane through a regular structure consisting of a spatially infinite elastic matrix $\Omega^{(1)}$ and elastic inclusions $\Omega^{(2)}$ (Fig. 8.9). The original 2D wave equation is (Section 3.1):

$$\nabla(\mu\nabla u) = \rho\frac{\partial^2 u}{\partial t^2}, \tag{8.25}$$

where $\nabla = \mathbf{e}_1\partial/\partial x_1 + \mathbf{e}_2\partial/\partial x_2$, u is the displacement in the x_3 direction, $\mathbf{e}_1$, $\mathbf{e}_2$ are the unit Cartesian vectors.

Due to the heterogeneity of the medium the Lamé constant μ and the mass density ρ is represented by piecewise continuous functions of co-ordinates:

$$\mu(\mathbf{x}) = \mu^{(a)}, \quad \rho(\mathbf{x}) = \rho^{(a)} \quad \text{for} \quad \mathbf{x} \in \Omega^{(a)}, \quad a = 1,2, \quad \mathbf{x} = x_1\mathbf{e}_1 + x_2\mathbf{e}_2. \tag{8.26}$$

Equation (8.25) can be written in the equivalent form:

$$\mu^{(a)}\nabla^2 u^{(a)} = \rho^{(a)}\frac{\partial^2 u^{(a)}}{\partial t^2}, \tag{8.27}$$

$$\left\{u^{(1)} = u^{(2)}\right\}\Big|_{\partial\Omega}, \quad \left\{G^{(1)}\frac{\partial u^{(1)}}{\partial \mathbf{n}} = G^{(2)}\frac{\partial u^{(2)}}{\partial \mathbf{n}}\right\}\Big|_{\partial\Omega}, \tag{8.28}$$

where $\nabla^2 = \partial^2/\partial x_1^2 + \partial/\partial x_2^2$, $\partial/\partial\mathbf{n}$ is the normal derivative to the contour $\partial\Omega$.

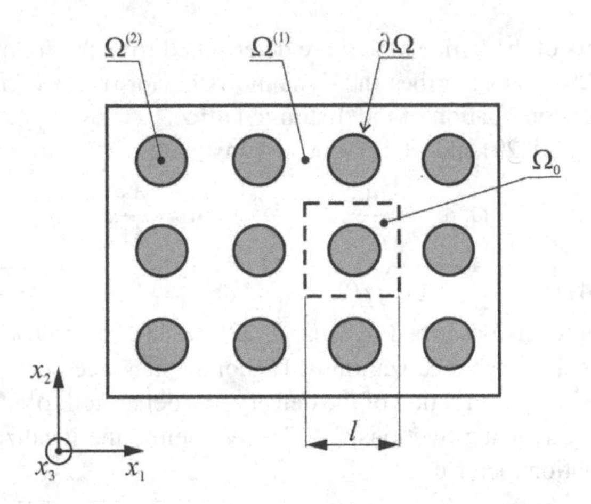

Figure 8.9 Fiber composite material [reprinted with permission from John Wiley and Sons].

From the physical point of view, Eqs. (8.28) mean the perfect bonding conditions at the interface $\partial\Omega$.

Following the Flöquet-Bloch theorem (Section 2.3), a harmonic wave propagating through a periodic structure is represented in the form:

$$u = F(\mathbf{x})\exp(i\mathbf{k}\cdot\mathbf{x})\exp(i\omega t), \tag{8.29}$$

where $F(\mathbf{x})$ is a spatially periodic function, $F(x_{1,2}) = F(x_{1,2}+l)$, $\mathbf{k}$ is the wave vector, $\mathbf{k} = k_1\mathbf{e}_1 + k_2\mathbf{e}_2$.

It should be noted that the 1D solution based on the transfer-matrix approach cannot describe the wave propagation in a 2D composite material. For the 2D problem, we have to take into account the alterability of the stress-strain state not only in the direction of the wave propagation (determined by the vector $\mathbf{k}$), but also in the orthogonal direction. The latter is caused by the heterogeneity of the medium and also influences on the wave properties.

Therefore, a 2D description of the stress-strain state must be introduced. For this aim, we use the plane-wave expansions method [266, 411]. The function $F(\mathbf{x})$ and the material properties $\mu(\mathbf{x})$, $\rho(\mathbf{x})$ are expressed into infinite Fourier series:

$$\begin{aligned}
F(\mathbf{x}) &= \sum_{n_1=-\infty}^{\infty}\sum_{n_2=-\infty}^{\infty} F_{n_1n_2}\exp\left[i\frac{2\pi}{l}(n_1x_1+n_2x_2)\right],\\
\mu(\mathbf{x}) &= \sum_{n_1=-\infty}^{\infty}\sum_{n_2=-\infty}^{\infty} \mu_{n_1n_2}\exp\left[i\frac{2\pi}{l}(n_1x_1+n_2x_2)\right],\\
\rho(\mathbf{x}) &= \sum_{n_1=-\infty}^{\infty}\sum_{n_2=-\infty}^{\infty} \rho_{n_1n_2}\exp\left[i\frac{2\pi}{l}(n_1x_1+n_2x_2)\right],
\end{aligned} \tag{8.30}$$

where

$$\mu_{k_1k_2} = \frac{1}{l^2} \iint\limits_{\Omega_0} \mu(\mathbf{x}) \exp\left[-i\frac{2\pi}{l}(k_1x_1 + k_2x_2)\right] dx_1dx_2,$$
$$\rho_{k_1k_2} = \frac{1}{l^2} \iint\limits_{\Omega_0} \rho(\mathbf{x}) \exp\left[-i\frac{2\pi}{l}(k_1x_1 + k_2x_2)\right] dx_1dx_2, \tag{8.31}$$

the operator $\iint_{\Omega_0}(\cdot)dx_1dx_2$ denotes the integration over a unit cell Ω_0 (Fig. 8.9).

Substituting ansatz (8.29) and expansions (8.30) into the wave Eq. (8.25) and collecting coefficients at the terms $\exp\left[i2\pi l^{-1}(j_1x_1 + j_2x_2)\right]$, $j_1, j_2 = 0, \pm1, \pm2, ...$, we come to an infinite system of the linear algebraic equations for the unknown coefficients $F_{n_1n_2}$:

$$\sum_{n_1=-\infty}^{\infty} \sum_{n_2=-\infty}^{\infty} F_{n_1n_2} \left\{ \mu_{\substack{j_1-n_1,\\ j_2-n_2}} \left[\left(\frac{2\pi}{l}n_1 + k_1\right)\left(\frac{2\pi}{l}j_1 + k_1\right) + \right.\right.$$
$$\left.\left. + \left(\frac{2\pi}{l}n_2 + k_2\right)\left(\frac{2\pi}{l}j_2 + k_2\right)\right] - \rho_{\substack{j_1-n_1,\\ j_2-n_2}}\omega^2 \right\} = 0. \tag{8.32}$$

System (8.32) allows a nontrivial solution if and only if the determinant of the matrix $(F_{n_1n_2})$ is zero, i.e. we have

$$\det(F_{n_1n_2}) = 0. \tag{8.33}$$

The dispersion relations for ω and $\mathbf{k}$ (8.33) are calculated approximately by the truncation of the infinite system (8.32) supposing $-j_{\max} \le j_s \le j_{\max}$. From the physical point of view such a truncation means cutting off the higher frequencies.

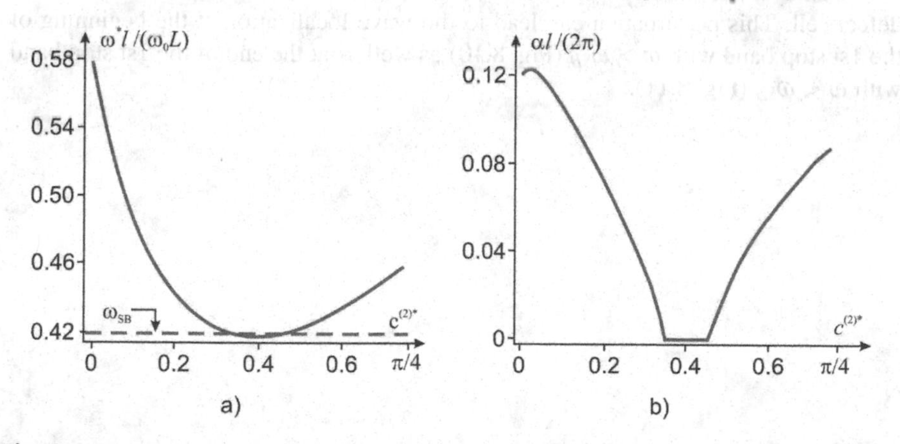

Figure 8.10 Wave localization at the beginning of the 1st stop band induced by the perturbation of the volume fraction of a fiber: a) localization frequency, b) attenuation factor [reprinted with permission from John Wiley and Sons].

Now let us suppose that composite material contain defect unit cell, which properties differ from other cells. To predict the localization effect, we apply Eq. (8.33)

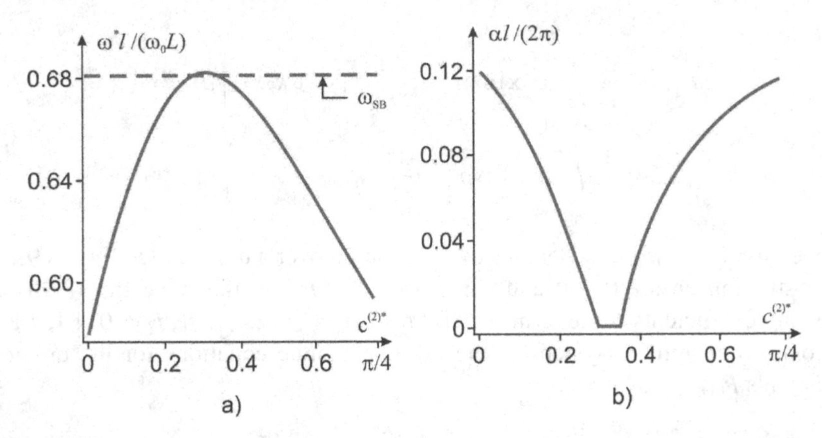

Figure 8.11 Wave localization at the end of the 1st stop band induced by the perturbation of the volume fraction of a fiber: a) localization frequency, b) attenuation factor [reprinted with permission from John Wiley and Sons].

firstly to the defect unit cell and then for the regular one. Eq. (8.33) for the defect unit cell determines the localization frequency and for the regular unit cell - the attenuation factor. Let us consider numerical results for the nickel-aluminium composite characterized by the following values of the parameters: the volume fraction of the nickel inclusion $c^{(2)} = 0.35$, the Lamé coefficients of the components are $\mu^{(1)} = 28$ GPa and $\mu^{(2)} = 75$ GPa, the mass densities of the components are $\rho^{(1)} = 2700$ kg/m^3 and $\rho^{(2)} = 8940$ kg/m^3.

We analyse the influence of perturbation of the volume fraction of a fiber in the defect cell. This perturbation can lead to the wave localization at the beginning of the 1st stop band with $\omega > \omega_{SB}$ (Fig. 8.10) as well as at the end of the 1st stop band with $\omega < \omega_{SB}$ (Fig. 8.11).

9 Nonlinear Vibrations of Viscoelastic Heterogeneous Solids of Finite Size: Internal Resonances and Modes Interactions

In this chapter, nonlinear vibrations of a layered viscoelastic solid are studied. A macroscopic wave equation taking into account the properties of the microstructure is proposed. Through use of the developed model, the interplay between the effects of nonlinearity and dissipation is analyzed. It gives possibility to predict how viscous damping influences modes coupling and to justify truncation of the original infinite system to a finite number of the leading order modes. The interaction between the effects of nonlinearity and dissipation is also investigated. We have found that in the case of a purely elastic material, an infinite number of the vibration modes may be coupled by nonlinear internal resonances. This results in periodic energy transfer between different modes and in a self-generation of higher-order modes. Hence, truncation to the modes having non-zero initial energy is prohibited.

9.1 INTRODUCTION

Real materials and structures are never perfectly elastic. In dynamical problems, part of the mechanical energy is always transformed to heat because of the internal friction. Dissipative behavior of heterogeneous solids can be determined by various factors, such as properties of the components, features of the microstructure, bonding conditions at the "matrix-inclusion" interface, and so on [62, 158, 304, 369]. In this chapter we address the issue when energy dissipation is caused by the viscoelasticity of the components, which is typical, for example, for polymer-based materials.

Viscoelastic properties of the medium can be governed by the well-known Kelvin-Voigt model (see, for example, [177]). In hydrodynamics, it corresponds to the classical behavior of a viscous gas, where shear stresses are proportional to the deformation rates and the proportionality coefficients are determined by the gas density. For solids, the Kelvin-Voigt model can be naturally deduced from a lattice-type model through passing to the continuous limit and assuming that the interaction forces between neighboring particles depend on the rate of the distance change rather than on the distance itself [304].

Palmov [343,344] studied nonlinear vibrations of semi-infinite and finite rods subjected to a dissipation. He employed the method of harmonic linearization, which gave first-order approximation coinciding with the Rayleigh-Ritz approach. Equations in slow variables were obtained and analyzed. High-frequency deformations of nonlinear rate dependent materials induced by the propagation of transient waves were considered by Varley and Rogers [445] and Seymour and Varley [404]. Mortell and Varley [323] discussed nonlinear elastic waves in finite-size bodies and studied the dynamic response of a viscoelastic rod subjected to a pulse load [322]. In the first-order approximation, the solution was developed as a superposition of two modulated waves travelling in opposite directions and not interacting with each other.

Vibrations of continuous structures can be described by dynamical systems having infinite number of degrees of freedom. Nonlinearity leads to localization of energy and its transfer from the low- to the high-frequency part of the spectrum, and vice versa. The vibration modes can then be involved in complicated interactions, resulting in internal resonances and self-generation of higher-order modes. In such a case, truncation to the modes having non-zero initial energy (which is usually applicable in linear problems) is not possible and all the resonant modes should be taken into account.

The effects of mode coupling and internal resonance have attracted the considerable attention of many authors. A number of results were obtained for vibrations of homogeneous structures and numerical [383] as well as asymptotic [24,84,85,293,331] approaches have been widely applied. However, the non-linear dynamic behavior of heterogeneous solids was studied to a significantly less extend. Only recently, one-dimensional vibrations of microstructured rods were considered by Andrianov et al. [21] in regard to a nonlinear elastic external medium and geometrical and physical nonlinearity. Heterogeneity results in dispersion of energy and, thus compensates the influence of nonlinearity. As the spatial period of the modes decreases and approaches the size of the microstructure, internal resonances are suppressed and truncation to only a few leading order modes can be justified.

It should be noted that dissipation restrains energy transfers between the vibration modes and on a large time scale modes coupling vanishes. In that sense, dispersion and dissipation acting in a nonlinear system may lead to qualitatively similar physical consequences (Zabusky and Kruskal [468] encountered this analogy analysing the FPU problem, see also [456]).

9.2 INPUT PROBLEM AND HOMOGENIZED DYNAMICAL EQUATION

Let us consider a heterogeneous solid consisting of periodically repeated viscoelastic layers $\Omega^{(1)}$ and $\Omega^{(2)}$ (Fig. 9.1). We shall study natural longitudinal vibrations in the direction x. This model can describe the properties of laminated composite materials, phononic crystals and acoustic diodes (see, for example, references [217,282,317]). Notice that layered structures are employed in band gap engineering for the design of vibration and sound control devices [216].

The mechanical behavior of each of the layers $\Omega^{(i)}$ is described by the Kelvin-Voigt model [177], which includes a purely elastic spring connected in parallel with

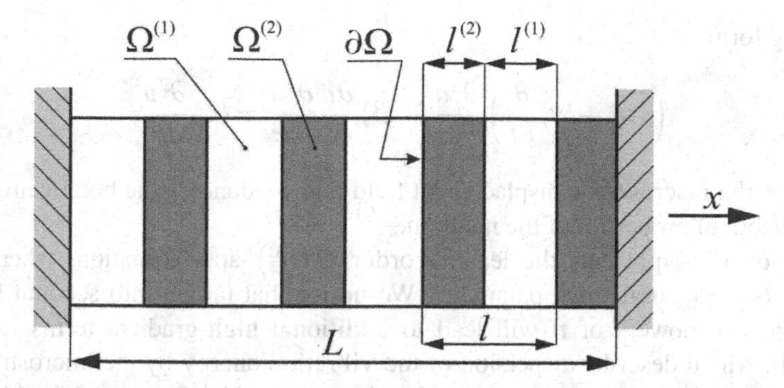

Figure 9.1 Periodically heterogeneous solid under consideration [reprinted with permission from Elsevier].

a viscous dashpot. The properties of the dashpot are assumed to be linear, while the spring exhibits nonlinear response. Geometric nonlinearity is taken into account using the Cauchy-Green strain tensor and physical nonlinearity is modeled with the help of the Murnaghan elastic potential (see Chapter 4.1). The time dependent longitudinal stress $\sigma^{(i)}$ is given by

$$\sigma^{(i)} = \alpha^{(i)}\frac{\partial u^{(i)}}{\partial x} + \frac{\beta^{(i)}}{2}\left(\frac{\partial u^{(i)}}{\partial x}\right)^2 + \nu^{(i)}\frac{\partial^2 u^{(i)}}{\partial x \partial t}, \tag{9.1}$$

where $u^{(i)}$ is the displacement; $\alpha^{(i)}$ and $\beta^{(i)}$ are the elastic coefficients; $\alpha^{(i)} = \lambda^{(i)} + 2\mu^{(i)}$, $\lambda^{(i)}$, $\mu^{(i)}$ stand for the Lamé elastic constants; $\beta^{(i)} = 3\left(\lambda^{(i)} + 2\mu^{(i)}\right) + 2\left(A^{(i)} + 3B^{(i)} + C^{(i)}\right)$, $A^{(i)}$, $B^{(i)}$, $C^{(i)}$ are the Landau elastic constants; $\nu^{(i)}$ is the viscosity.

Here and in the sequel the upper indexes (i), $i = 1,2$, refer to the different components $\Omega^{(1)}$, $\Omega^{(2)}$ of the medium.

The governing dynamical equation is as follows

$$\left(\alpha^{(i)} + \nu^{(i)}\frac{\partial}{\partial t}\right)\frac{\partial^2 u^{(i)}}{\partial x^2} + \beta^{(i)}\frac{\partial u^{(i)}}{\partial x}\frac{\partial^2 u^{(i)}}{\partial x^2} = \rho^{(i)}\frac{\partial^2 u^{(i)}}{\partial t^2}, \tag{9.2}$$

where $\rho^{(i)}$ is the mass density.

At the interface $\partial\Omega$, we assume the conditions of perfect bonding that imply continuity of the displacement and stress fields:

$$u^{(1)} = u^{(2)}, \quad \sigma^{(1)} = \sigma^{(2)} \quad \text{at} \quad \partial\Omega. \tag{9.3}$$

Let us suppose the characteristic size l of the microstructure to be essentially smaller than the spatial period L_n of a considered vibration mode. Then the input problem (9.2), (9.3) can be replaced by a homogenized dynamical equation of the following form

$$\left(\langle\alpha\rangle+\langle\nu\rangle\frac{\partial}{\partial t}\right)\frac{\partial^2 u}{\partial x^2}+\langle\beta\rangle\frac{\partial u}{\partial x}\frac{\partial^2 u}{\partial x^2}=\langle\rho\rangle\frac{\partial^2 u}{\partial t^2}, \tag{9.4}$$

where u is the macroscopic displacement field and $\langle\cdot\rangle$ denotes the homogenized (so called effective) properties of the medium.

Equation (9.4) presents the leading order $O(\eta^0)$ approximation, where $\eta = l/L \ll 1$ is the natural small parameter. We notice that taking into account higher-order terms in powers of η will lead to additional high-gradient terms in equation (9.4), which describe dispersion of the vibration energy by the microstructure. Higher-order homogenized models have been considered in Chapter 4.3. In this chapter we focus on dissipation and, for the sake of clarity, restrict the carried out analysis to the $O(\eta^0)$ approximation only.

The effective mass density ρ is given by a simple rule of mixture: $\langle\rho\rangle = c^{(1)}\rho^{(1)} + c^{(2)}\rho^{(2)}$, where $c^{(i)}$ are the volume fractions of the components, $c^{(i)} = l^{(i)}/l$.

In the first approximation, the elastic and the viscous properties can be homogenized independently of each other. This simplified procedure may provide a reasonable approach if nonlinearity and viscosity are small. The effective elastic coefficients $\langle\alpha\rangle$ and $\langle\beta\rangle$ are evaluated as follows (Chapter 4.3):

$$\langle\alpha\rangle=\frac{\alpha^{(1)}\alpha^{(2)}}{c^{(1)}\alpha^{(2)}+c^{(2)}\alpha^{(1)}}, \tag{9.5}$$

$$\langle\beta\rangle=\frac{c^{(1)}\beta^{(1)}\left(\alpha^{(2)}\right)^3+c^{(2)}\beta^{(2)}\left(\alpha^{(1)}\right)^3}{\left(c^{(1)}\alpha^{(2)}+c^{(2)}\alpha^{(1)}\right)^3}. \tag{9.6}$$

In order to determine the effective viscosity $\langle\nu\rangle$, let us apply to equations (9.2), (9.3) the Laplace transform $u_s^{(i)}(x,s)=\int_0^\infty u^{(i)}(x,t)\exp(-st)\,dt$. Assuming zero initial conditions, in the linear approximation we obtain

$$\left(\alpha^{(i)}+s\nu^{(i)}\right)\frac{\partial^2 u_s^{(i)}}{\partial x^2}=s^2\rho^{(i)}u_s^{(i)}, \tag{9.7}$$

$$u_s^{(1)}=u_s^{(2)},\quad \left(\alpha^{(1)}+s\nu^{(1)}\right)\frac{\partial u_s^{(1)}}{\partial x}=\left(\alpha^{(2)}+s\nu^{(2)}\right)\frac{\partial u_s^{(2)}}{\partial x}\quad \text{at}\quad \partial\Omega. \tag{9.8}$$

Homogenization of the problem (9.7), (9.8) gives [306]

$$\langle\alpha+s\nu\rangle\frac{\partial^2 u_s}{\partial x^2}=s^2\langle\rho\rangle u_s^{(i)}, \tag{9.9}$$

where $\alpha+s\nu=\left[c^{(1)}/\left(\alpha^{(1)}+s\nu^{(1)}\right)+c^{(2)}/\left(\alpha^{(2)}+s\nu^{(2)}\right)\right]^{-1}$.

Application of the inverse Laplace transform leads to an integro-differential equation, which complicates significantly further analysis. To avoid this, in the case of small viscosity we represent the homogenized parameter $\langle \alpha + s\nu \rangle$ as the power series expansion

$$\langle \alpha + s\nu \rangle = \frac{\alpha^{(1)}\alpha^{(2)}}{c^{(1)}\alpha^{(2)} + c^{(2)}\alpha^{(1)}} + s\frac{c^{(1)}\nu^{(1)}\left(\alpha^{(2)}\right)^2 + c^{(2)}\nu^{(2)}\left(\alpha^{(1)}\right)^2}{\left(c^{(1)}\alpha^{(2)} + c^{(2)}\alpha^{(1)}\right)^2} + O\left(s^2\right). \tag{9.10}$$

Substituting (9.10) into (9.9) and inverting the transform, one obtains the linear part of the homogenized equation (9.4), where the effective elastic coefficient $\langle \alpha \rangle$ is given by formula (9.5) and the effective viscosity $\langle \nu \rangle$ is obtained as follows

$$\langle \nu \rangle = \frac{c^{(1)}\nu^{(1)}\left(\alpha^{(2)}\right)^2 + c^{(2)}\nu^{(2)}\left(\alpha^{(1)}\right)^2}{\left(c^{(1)}\alpha^{(2)} + c^{(2)}\alpha^{(1)}\right)^2}. \tag{9.11}$$

Equation (9.4) describes the dynamic behavior of a heterogeneous solid on the macro scale associated with the spatial period of the vibration modes. We note that all the effective coefficients are determined explicitly in terms of the microscopic properties of the medium. This contrasts with many well-known phenomenological approaches like Cosserat's [129] and Biot's theories, where the material parameters are usually left undetermined.

9.3 DISCRETIZATION PROCEDURE

Let us introduce the nondimensional variables $\bar{x} = x(\pi/L)$, $\bar{t} = t(\pi/L)\sqrt{\alpha/\rho}$, $\bar{u} = u/A$, where L is the length of the entire heterogeneous solid and A is the vibration amplitude. For the simplicity, the overbars are omitted in the sequel. Equation (9.4) takes the following form:

$$\frac{\partial^2 u}{\partial x^2} + \gamma \frac{\partial^3 u}{\partial x^2 \partial t} + \varepsilon \frac{\partial u}{\partial x}\frac{\partial^2 u}{\partial x^2} = \frac{\partial^2 u}{\partial t^2}, \tag{9.12}$$

where $\gamma = \pi\langle \nu \rangle / \left(L\sqrt{\langle \alpha \rangle \langle \rho \rangle}\right)$, $\varepsilon = \pi A \langle \beta \rangle / (L\langle \alpha \rangle)$ are the non-dimensional small parameters characterizing viscosity and nonlinearity. The coefficient γ is always positive, whereas the sign of ε depends on the physical properties of the medium. For the most of materials (e.g., metals, polymers, stones) we have $\langle \beta \rangle < 0$ and $\varepsilon < 0$, which results in a soft nonlinearity. A hard nonlinearity, $\langle \beta \rangle > 0$, $\varepsilon > 0$, can be observed for elastomers and rubber-like materials.

Equation (9.12) represents an asymptotic model, which accounts the case of small viscosity and small nonlinearity. That is why the damping term is considered in a linear form and the nonlinear term is purely elastic. A refined model may include a coupled damping and nonlinear term of the order $O(\gamma\varepsilon)$. However, supposing $\gamma\varepsilon \to 0$, we can asymptotically neglect it in the present approximation.

Equation (9.12) may allow various modifications. Adding a dispersion term with the forth-order spatial derivative and neglecting the dissipative term setting $\gamma = 0$, we will obtain a version of the well-known Boussinesq equation that is widely used to study wave propagation in nonlinear heterogeneous media (see, for example, [306] and references therein). Similar models have been recently applied to describe nonlinear vibrations of finite size heterogeneous bodies [21, 456]. Employing two or three dispersion terms makes it possible to develop refined gradient models [74], which are applicable in a wider frequency range. Double- and triple-dispersion equations were used to predict the propagation of linear transient waves induced by pulse and harmonic loads [43, 126].

The governing dynamic model introduced in the particular form (9.12) is intended to study the interplay between nonlinear and dissipative effects, while the period of the vibration modes is considered to be essentially larger then the size of the microstructure. Let us consider a case of clamped-clamped edges. The boundary conditions are

$$u(0,t) = u(\pi,t) = 0. \tag{9.13}$$

In order to solve the boundary value problem (9.12), (9.13), a discretization procedure is employed. Satisfying conditions (9.13), the displacement field is represented as a Fourier sine series expansion

$$u(x,t) = q_1(t)\sin(x) + q_2(t)\sin(2x) + q_3(t)\sin(3x) + ... \tag{9.14}$$

Substituting ansatz (9.14) into equation (9.12), we come to an infinite system of nonlinear ordinary differential equations

$$\begin{aligned}
&\frac{d^2q_1}{dt^2} + \gamma\frac{dq_1}{dt} + q_1 + \varepsilon(q_1q_2 + 3q_2q_3 + 6q_3q_4 + ...) = 0,\\
&\frac{d^2q_2}{dt^2} + 4\gamma\frac{dq_2}{dt} + 4q_2 + \varepsilon\left(\frac{q_1^2}{2} + 3q_1q_3 + 8q_2q_4 + ...\right) = 0,\\
&\frac{d^2q_3}{dt^2} + 9\gamma\frac{dq_3}{dt} + 9q_3 + \varepsilon(3q_1q_2 + 6q_1q_4 + ...) = 0,\\
&\frac{d^2q_4}{dt^2} + 16\gamma\frac{dq_4}{dt} + 16q_4 + \varepsilon(6q_1q_3 + 4q_2^2 + ...) = 0,\\
&...,
\end{aligned} \tag{9.15}$$

which can be written in more general form reads

$$\frac{d^2q_n}{dt^2} + n^2\gamma\frac{dq_n}{dt} + n^2q_n + \frac{\varepsilon}{2}\sum_{m=1}^{\infty} m^2q_m\left[|m-n|q_{|m-n|} - (m+n)q_{m+n}\right] = 0, \tag{9.16}$$

where $n = 1,2,3,...$

The obtained nonlinear system is non-symmetric with respect to the odd and even vibration modes. A simple analysis reveals two fundamentally different types of the mode interactions those described by Chechin and Sakhnenko [118]. An initial excitation of the odd modes leads necessarily to the appearance of even modes, i.e., equations (9.15) do not permit the solution $q_{2n-1} \neq 0$, $q_{2n} = 0$. The even modes undergo a so called "force" excitation. On the other hand, the solution $q_{2n-1} = 0$, $q_{2n} \neq 0$ is allowed and, in general, the presence of the even modes does not give rise to the odd modes. However, vibrations by even modes appear to be unstable and a small perturbation of the initial conditions induces energy transfers to the odd modes [21]. In this case the odd modes are subjected to a "parametric" excitation.

9.4 METHOD OF MULTIPLE TIME SCALES

We seek an asymptotic solution supposing nonlinearity to be small, $|\varepsilon| \ll 1$. Let us introduce different time scales described by new independent variables $t_0 = t$, $t_1 = \varepsilon t$,..., and represent q_n as an expansion in powers of ε:

$$q_n(t) = q_{n0}(t_0,t_1) + \varepsilon q_{n1}(t_0,t_1) + O(\varepsilon^2). \tag{9.17}$$

The number of required time scales depends on the order of approximation. We will carry out expansion (9.17) to $O(\varepsilon^2)$, therefore, two time scales ("fast" t_0 and "slow" t_1) are needed. The derivatives with respect to t are represented as follows

$$\frac{d}{dt} = \frac{\partial}{\partial t_0} + \varepsilon\frac{\partial}{\partial t_1} + O(\varepsilon^2), \quad \frac{d^2}{dt^2} = \frac{\partial^2}{\partial t_0^2} + 2\varepsilon\frac{\partial^2}{\partial t_0 \partial t_1} + O(\varepsilon^2). \tag{9.18}$$

Splitting each equation of the system (9.15) in powers of ε, in the $O(\varepsilon^0)$ approximation we obtain

$$\frac{\partial^2 q_{n0}}{\partial t_0^2} + n^2\gamma\frac{\partial q_{n0}}{\partial t_0} + n^2 q_{n0} = 0. \tag{9.19}$$

The solution of equation (9.19) is

$$q_{n0} = \exp(-n^2\gamma t_0/2)[a_n(t_1)\cos(\omega_n t_0) + b_n(t_1)\sin(\omega_n t_0)], \tag{9.20}$$

where ω_n is the natural frequency of the nth mode in the linear case, $\omega_n^2 = n^2(1 - n^2\gamma^2/4)$; a_n and b_n are slow varying amplitudes.

The $O(\varepsilon^1)$ approximation of the system (9.15) gives:

$$\begin{aligned} &\frac{\partial^2 q_{n1}}{\partial t_0^2} + n^2\gamma\frac{\partial q_{n1}}{\partial t_0} + n^2 q_{n1} = \\ &-2\frac{\partial^2 q_{n0}}{\partial t_0 \partial t_1} - \frac{\varepsilon}{2}\sum_{m=1}^{\infty} m^2 q_{m0}\left[|m-n| q_{|m-n|0} - (m+n) q_{|m+n|0}\right]. \end{aligned} \tag{9.21}$$

A straightforward integration of the system (9.21) will induce secular terms in the expressions for q_{n1}. Secular terms infinitely grow in time, which is inconsistent with the physical properties of the conservative system under consideration. In order to eliminate secular terms, the coefficients of $\cos(\omega_n t_0)$ and $\sin(\omega_n t_0)$ in the r.h.s. of equations (9.21) must be equal to zero. Substituting (9.20) into (9.21) and expanding the products $q_{j0}q_{k0}$ using the standard trigonometric identities, we obtain a system of equations for a_n and b_n, which gives a possibility to investigate the interactions between different modes.

In the non-dissipative case, $\gamma = 0$, the frequencies ω_n become proportional to each other

$$\omega_j/\omega_k = j/k. \tag{9.22}$$

Thus, an infinite number of modes are coupled by internal resonances. The presence of viscosity leads to a "detuning" of relation (9.22) and, on a large time scale, prevents energy transfers between the vibration modes.

Let us suppose viscosity to be small, $\gamma \ll 1$, and represent $\cos(\omega_n t_0)$ and $\sin(\omega_n t_0)$ in expression (9.20) as follows

$$\begin{aligned} \cos(\omega_n t_0) &= \cos(nt_0) + \frac{n^3\gamma^2 t_0}{8}\sin(nt_0) + O(n^6\gamma^4 t_0^2), \\ \sin(\omega_n t_0) &= \sin(nt_0) - \frac{n^3\gamma^2 t_0}{8}\cos(nt_0) + O(n^6\gamma^4 t_0^2). \end{aligned} \tag{9.23}$$

Truncating expansions (9.23) at leading order, we hold strictly the resonance condition (9.22). The physical sense of this approximation is that viscosity is considered to be responsible for energy dissipation (i.e., for exponential attenuation of q_{n0}), but its influence on the natural frequencies is neglected and, thus, the internal resonances are not violated. The developed approach is applicable on a short time scale as $n^3\gamma^2 t_0/8 \ll 1$.

As follows from (9.20), the attenuation coefficient of the nth mode is proportional to n^2. We observe that the internal dissipation exhibits a quadratic growth with the increase of the mode number and, therefore, the lowest modes can be the most important in vibrations of real structures. Let us examine resonant interactions between modes 1 and 2. It is convenient to introduce the amplitude $r_n(t_1)$ and the phase $\varphi_n(t_1)$ as follows: $a_n = r_n\cos(\varphi_n)$, $b_n = r_n\sin(\varphi_n)$. Keeping the first two equations of system (9.21), the condition of elimination of secular terms gives:

$$\begin{aligned} \frac{dr_1}{dt_1} &= \frac{1}{4} r_1 r_2 \exp(-2\gamma t_0)\sin(\varphi_2 - 2\varphi_1), \\ r_1\frac{d\varphi_1}{dt_1} &= -\frac{1}{4} r_1 r_2 \exp(-2\gamma t_0)\cos(\varphi_2 - 2\varphi_1), \\ \frac{dr_2}{dt_1} &= -\frac{1}{16} r_1^2 \exp(\gamma t_0)\sin(\varphi_2 - 2\varphi_1), \\ r_2\frac{d\varphi_2}{dt_1} &= -\frac{1}{16} r_1^2 \exp(\gamma t_0)\cos(\varphi_2 - 2\varphi_1). \end{aligned} \tag{9.24}$$

The solution of system (9.24) in the purely elastic case ($\gamma = 0$) was analyzed by Andrianov et al. [21]. We note that vibration by the single mode 1 is not possible. If all the initial energy is accumulated in mode 1, i.e. $r_1(0) \neq 0$, $r_2(0) = 0$, then the mode 2 is necessarily excited and a periodic energy exchange between the modes takes place. On the other hand, if we start with zero initial energy the mode 1, $r_1(0) = 0$, $r_2(0) \neq 0$, then $dr_2/dt_1 = d\varphi_2/dt_1 = 0$. The amplitude r_2 and the phase φ_2 are constant in time, so the system can vibrate through mode 2 only. However, this regime is not stable in the sense that even a small perturbation of the initial conditions will induce excitation of mode 1. The described behavior is illustrated in the phase plane $(r_2, \varphi_2 - 2\varphi_1)$ (see Fig. 2 reported in [21]).

9.5 NUMERICAL SIMULATION OF THE MODES COUPLING

The distribution of energy among the vibration modes can be determined using a Lagrangian formalism. Let us introduce a generalized equation of motion for the n-th mode as follows

$$\frac{d}{dt}\left[\frac{\partial T}{\partial(dq_n/dt}\right] - \frac{\partial T}{\partial q_n} = -\frac{\partial V}{\partial q_n} - n^2\gamma\frac{dq_n}{dt}, \tag{9.25}$$

where T is the kinetic and V is the potential energy of the whole system.

Integrating equations (9.15), we obtain

$$T = \frac{1}{2}\sum_{n=1}^{\infty}\left(\frac{\partial q_n}{\partial t}\right)^2, \tag{9.26}$$

$$V = \frac{1}{2}\sum_{n=1}^{\infty}\left\{n^2 q_n^2 + \varepsilon\int\sum_{m=1}^{\infty} m^2 q_m\left[|m-n|q_{|m-n|} - (m+n)q_{m+n}\right]dq_n\right\}. \tag{9.27}$$

The term $(\partial q_n/\partial t)^2$ in expression (9.26) and the term $n^2 q_n^2$ in expression (9.27) contribute to the nth equation of system (9.15). Then, in the leading order approximation and omitting the coefficient 1/2, the total energy E_n of the nth mode can be evaluated as follows

$$E_n = \left(\frac{dq_n}{dt}\right)^2 + n^2 q_n^2 + O(\varepsilon). \tag{9.28}$$

The second term in expression (9.27) represents a correction of order $O(\varepsilon)$ to the potential energy V. It can include products $q_m q_{|m-n|}$ and $q_m q_{m+}$ and, thus, its contribution is shared between modes m, $|m-n|$ and $m+n$. Below we are restricted by the $O(\varepsilon^0)$ approximation and calculate the energy by formula (9.28).

The numerical examples are presented for the case of soft nonlinearity, which is typical for the most industrial materials, and $\varepsilon = -0.1$. The increase in the magnitude of ε makes the energy transfers faster, however, the qualitative behavior of the solution remains the same. Numerical integration is performed by the Runge-Kutta fourth-order method in the open-source CAS Maxima. Practical convergence is verified by decreasing twice the step of integration and checking that this does not affect the obtained numerical data.

Firstly, let us consider a two mode approximation and compare the developed asymptotic approach (9.24) with the direct numerical integration of the two leading equations of the nonlinear system (9.15). The initial energy is assumed to be localized in mode 1. System (9.24) has a singularity at $r_2 = 0$; to avoid this the initial conditions are taken as follows

$$r_2 = q_2 = 0.01, \qquad r_1 = q_1 = \sqrt{1-4r_1^2} = 0.99979998,$$
$$\frac{dr_1}{dt} = \frac{dq_1}{dt} = \frac{dr_2}{dt} = \frac{dq_1}{dt} = 0 \quad \text{at} \quad t = 0. \tag{9.29}$$

Fig. 9.2 displays a periodic energy exchange between modes 1 and 2. The asymptotic approximation of the method of multiple time scales is formally applicable at a relatively short time scale $t \sim O(\varepsilon^{-1})$. However, it is numerically useful far beyond their nominal range of validity. Next, let us examine interactions between the

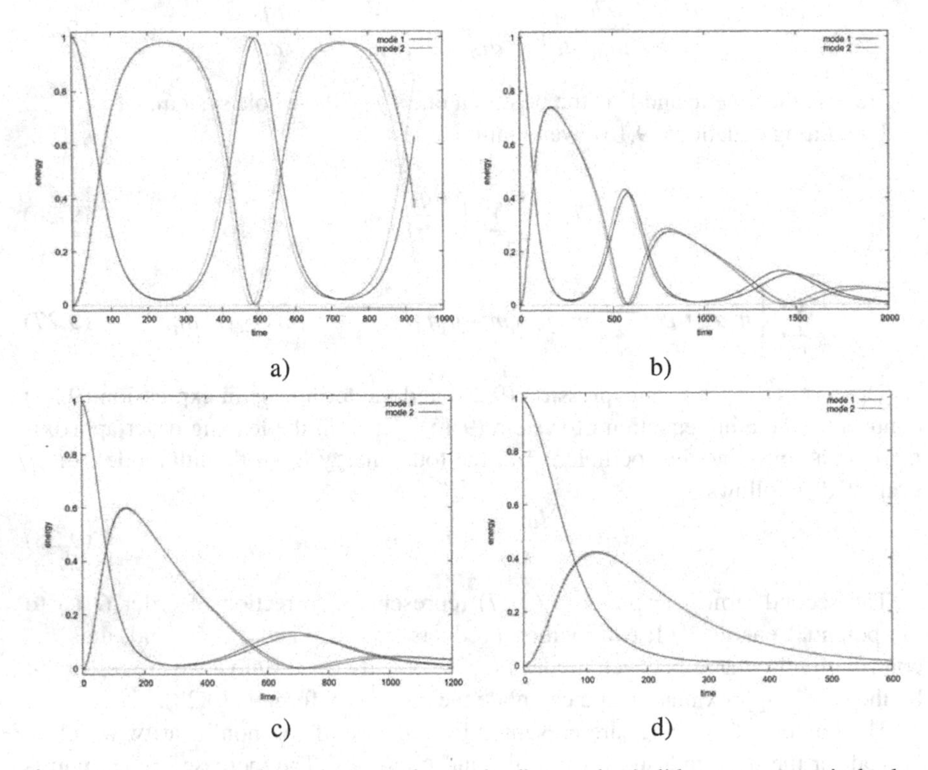

Figure 9.2 Energy exchange between the two leading modes (solid curves - numerical solution, dotted curves - asymptotic solution): (a) $\gamma = 0$ (non-dissipative case), (b) $\gamma = 0.0005$, (c) $\gamma = 0.001$, (d) $\gamma = 0.002$ [reprinted with permission from Elsevier].

six leading modes. The initial conditions correspond to the case when all the input energy is localized in the mode 1:

$$q_1 = 1, \quad q_k = 0, \quad k = 2, ..., 6, \quad \frac{dq_n}{dt} = 0, \quad n = 1, ..., 6 \quad \text{at} \quad t = 0. \tag{9.30}$$

Fig. 9.3 presents the results of the numerical integration of system (9.15) truncated with $n = 1, ..., 6$. We observe that the increase in dissipation suppresses energy

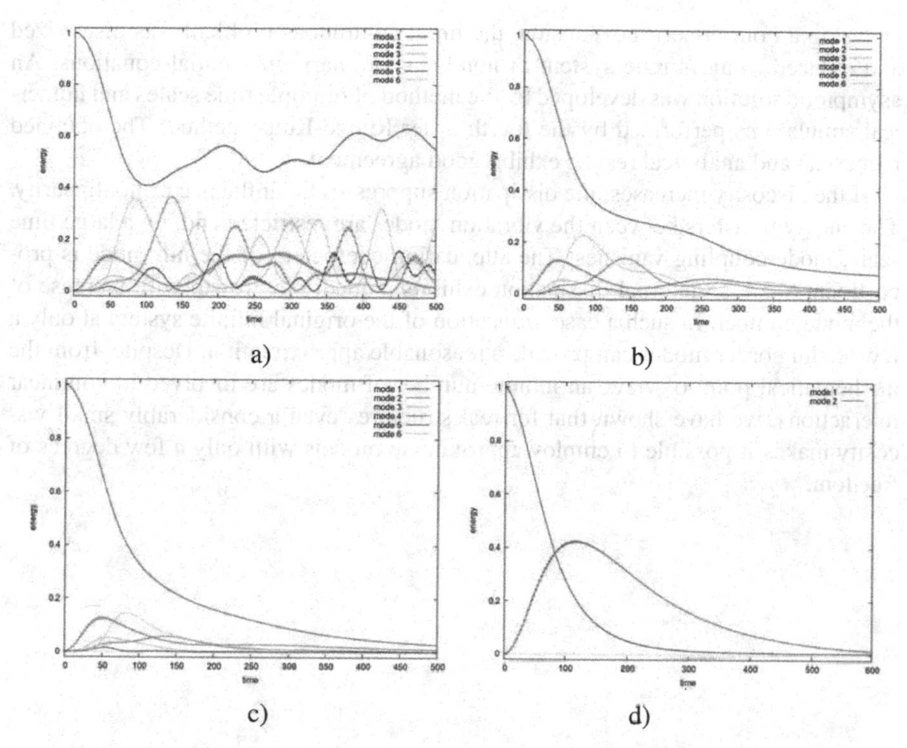

Figure 9.3 Energy transfers between the six leading modes: (a) $\gamma = 0$ (non-dissipative case), (b) $\gamma = 0.0005$, (c) $\gamma = 0.001$, (d) $\gamma = 0.002$ [reprinted with permission from Elsevier].

transfers and makes it possible to truncate the original infinite system to only a few leading order modes.

9.6 CONCLUDING REMARKS

We studied natural vibrations of a heterogeneous solid consisting of periodically repeated layers of two different materials. The viscoelastic properties of the components were represented by the Kelvin-Voigt model. Geometric nonlinearity was introduced by the Cauchy-Green strain tensor and physical nonlinearity was modeled employing the Murnaghan elastic potential. The characteristic size of the microstructure (e.g., length of the unit cell) was supposed to be essentially smaller than the scale of the macroscopic problem (e.g., spatial periods of the vibration modes). A homogenized dynamical equation was proposed to describe the dynamic properties of the medium on the macro level.

An important contribution of the chapter is that the coefficients of the homogenized equation were determined explicitly basing on rigorous theoretical origins. The developed macroscopic model encapsulate information about the properties of the microstructure.

Using a Fourier series expansion, the input continuous problem was discretized and reduced to an infinite system of nonlinear ordinary differential equations. An asymptotic solution was developed by the method of multiple time scales and numerical simulations performed by the fourth-order Runge-Kutta method. The obtained numerical and analytical results exhibit good agreement.

If the viscosity increases, the dissipation suppresses the influence of nonlinearity. The energy transfers between the vibration modes are restricted and, on a large time scale, mode coupling vanishes. The attenuation coefficient of the nth mode is proportional to n^2, so internal dissipation exhibits a quadratic growth with increase of the mode number. In such a case, truncation of the original infinite system at only a few leading order modes can provide a reasonable approximation. Despite, from the mathematical point of view, an infinite number of modes are involved in nonlinear interactions, we have shown that for real structures even a considerably small viscosity makes it possible to employ approximate models with only a few degrees of freedom.

10 Nonlocal, Gradient and Local Models of Elastic Media: 1D Case

This chapter deals with local/nonlocal and gradient one-dimensional models of elastic media. First, modeling of a lattice of elastically couple particles is considered. Second, the classical continuous approximations of a discrete problem is illustrated and discussed. Next, splashes, envelope continualization and intermediate continuous models are described and investigated. Benefits of the use of the Padé approximations for the construction of continuous models are outlined. Normal modes expansion and theories of elasticity with couple-stresses are revisited. Then, the correspondence between functions of discrete arguments and continuous systems as well as relations between the kernels of integro-differential equations of the discrete and continuous systems are outlined. Moreover, the following topics are discussed: dispersive wave propagation, Green function, double-dispersive equations, Toda lattices, discrete kinks, continualization of the Fermi-Pasta-Ulam lattice, anticontinuum limit and 2D lattices. Finally, the problems regarding molecular dynamlcs vs. continualization, continualization vs. discretization, as well as opened problems are addressed.

10.1 INTRODUCTION

Over the last few years, a number of new materials and designs have appeared that require more sophisticated research than the capabilities of a classical continuous media. There are, in particular, ultra dispersive and nanocrystalline materials [137, 195, 355]. The nanocrystalline material is represented by a regular or quasi-regular lattice of granules, fullerenes, nanotubes, carbon nanostructures, etc. and possesses internal degrees of freedom [374]. This situation is also usual for various problems of nanomechanics [217], because correct description of the mechanical properties of nanomaterials required taking into account size effects. Models based on a classical continuous media cannot govern high frequency vibrations, material behavior in the vicinities of cracks and on the fronts of destruction waves, and during phase transitions [340]. It is not possible to reach bifurcation points, i.e., thresholds of lattice stability under catastrophic deformations [6]. Wave dispersion in granular materials [354, 375] represents an important example of microstructure effects too. Microstructural effects are essential for correct description of softening phenomena [356] in damage mechanics [41, 115, 136] and in the theory of plasticity [174, 384]. Do not forget also problems of biomechanics and molecular biology. As it is mentioned in [237], "From the behavior of calcium waves in living cells to the discontinuous propagation of action potentials in the heart or chains of neurons and from chains of chemical reactors or arrays of Josephson junctions to optical waveguides, dislocations and the DNA double strand, the relevant models of physical reality are inherently discrete".

Acoustic (or mechanical) metamaterials, i.e. cellular periodic structures, only in the long wavelength range behave like continuous materials. The study of manifestations of discreteness, dispersion, dissipation and nonlinearity of acoustic waves in metamaterials is not only of theoretical interest. It is also about the prospects of practical applications of metamaterials [137, 183, 476], for example, the creation of super-absorbers of sound.

Discrete lattice-type models are widely used to describe vibrations in crystals, in foams, in cellular structures and bone tissues. Some novel applications include modeling of polymer molecules, atomic lattices (e.g., graphene). Discrete models can also appear in engineering, e.g., for simulating lightweight truss structures with attached masses.

In contrast to continuous media, discrete lattices sometimes allow evaluation of exact dispersion relations in a closed analytical form, which is convenient for further analysis. That is why they can be considered as benchmark models to highlight the main effects of the wave propagation in heterogeneous structures.

The effects mentioned may be analyzed within the frame of discrete models, using molecular dynamics [255], quasicontinuum analysis [426, 455] or other numerical approaches. However, a sought result is difficult to obtain using high tech computers in an economical way. For example, modern practical problems are still intractable for molecular dynamics, even if the most powerful computing facilities are used.

Situation can be described by Dirac's words [138]: "The physical laws necessary for the mathematical theory of a large part of physics and whole of chemistry are thus completely known, and the difficulty is only that the exact application of these laws leads to equations much too complicated to be solvable. Therefore, it becomes desirable that approximate practical methods of applying quantum mechanics should be developed, which can lead to an explanation of the main features of complex atomic systems without too much computations".

Hence, refinement of the existing theories of continuous media for the purpose of more realistic predictions of mechanical properties seems to be the actual goal. The most challenging problems in multiscale analysis is that of finding continuous models for atomistic lattice starts from the first principles. In statistical physics, these question as already addressed 100 years ago. Most important is the question how to obtain irreversible thermodynamics as a macroscopic limit from microscopic models that are reversible. Recently, a certain progress has been outlined in solving these problems [288, 471], but it is still far from the final conclusions.

The purpose of this chapter is much more modest than the global tasks described above. On simple models, we consider various ways of obtaining models of intermediate complexity between classical continuous and discrete ones. Here the following strategies exist.

(i) *Phenomenological approach.* Additional terms are added to the energy functional or to the constitutive relation. The structure and character of these terms are postulated in references [8, 204, 307]. Phenomenological approach is useful in the applied sciences when it is necessary to solve problems of practical significance. However, the range of applicability of these approaches is usually unknown, and the basic pecularities of behavior of discrete systems remain unclear.

(ii) *Statistical approach.* Starting from an continuous inhomogeneous theories it is aimed to obtain refined (e.g. couple-stress) theories of homogeneous continuum [73]. We know only two work in which this idea is brought to concrete results [291, 292]. In these papers polycrystal are considered as a micro-inhomogeneous elastic medium, which consists of homogeneous anisotropic crystallites with random orientation. It is shown that the elastic energy density of the polycrystalline medium depends non-locally on the non-uniform field of average deformations. In the case of smooth fields of average deformations it could be considered that the energy density locally depends not only on the values of deformations, but also on the values of their derivatives. This dependence in the isotropic case is characterized by five gradient modules, in addition to the two elastic moduli of the classical theory of elasticity. For the first time, explicit expressions are obtained for the gradient modules of polycrystalline materials.

(iii) *Homogenization and continualization approaches* [20, 28, 39, 40, 43, 80, 81, 94, 236–238, 257–259, 353, 382]. They are based on various approximations of nonlocal (discrete, pseudodifferential, integral) operators by the local (differential) ones.

This part of book is devoted to the continualization approaches and their features.

10.2 A CHAIN OF ELASTICALLY COUPLED MASSES

In this Section we follow paper [329]. We study a chain of $n+2$ material points with the same masses m, located in equilibrium states in the points of the axis x with coordinates jh $(j = 0, 1, ..., n, n+1)$ and suspended by elastic couplings of stiffness c (Fig. 10.1).

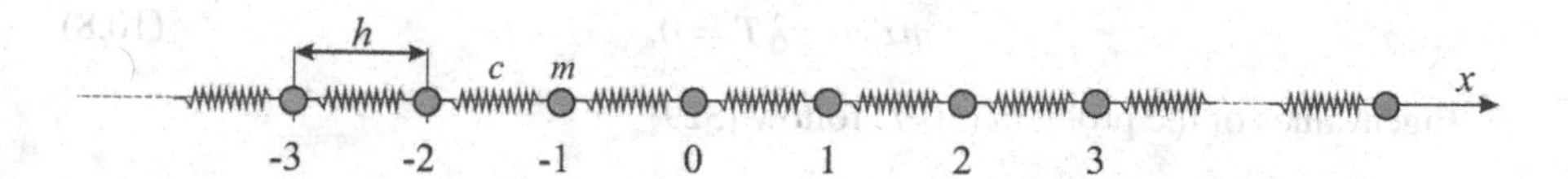

Figure 10.1 A chain of elastically coupled masses.

Owing to the Hooke's law the elastic force acting on the j-th mass is as follows

$$\sigma_j(t) = c[y_{j+1}(t) - y_j(t)] - c[y_j(t) - y_{j-1}(t)] = c[y_{j-1}(t) - 2y_j(t) + y_{j+1}(t)], \quad j = 1, 2, ..., n, \tag{10.1}$$

where $y_j(t)$ is the displacement of the j-th material point from its static equilibrium position.

Applying the second Newton's law one gets the following system of ODEs governing chain dynamics

$$my_{jtt}(t) = c[y_{j-1}(t) - 2y_j(t) + y_{j+1}(t)], \quad j = 1, 2, ..., n. \tag{10.2}$$

This system of ODEs is called Lagrange lattice, because Lagrange studied it in his "Analytical Mechanics" [267]. Sometimes one can read term "Newton lattice" [98], referring to Principia (livre II, proposition XLIX), however, this section of Principia contains only some general discussions about the speed of sound. In fact, the ODEs (10.2) were appeared only in Lagrange book.

System (10.2) can be recast into the following form:

$$m\sigma_{jtt}(t) = c(\sigma_{j+1} - 2\sigma_j + \sigma_{j-1}), \quad j = 1, ..., n. \tag{10.3}$$

Let the chain ends be fixed

$$y_0(t) = y_{n+1}(t) = 0. \tag{10.4}$$

In general, the initial conditions have the following form:

$$y_j(t) = \varphi_j^{(0)}, \quad y_{jt}(t) = \varphi_j^{(1)} \quad \text{at} \quad t = 0. \tag{10.5}$$

As it has been shown in [329], for any solution of the BVP (10.2), (10.4), (10.5) the total energy is constant. Besides, the solutions mentioned so far are not asymptotically stable due to the Lyapunov stability definition.

A solution to the BVP (10.2), (10.4), (10.5) can be expressed by elementary functions applying the discrete variant of the method of variables separation. For this purpose normal vibrations are constructed in the form

$$y_j(t) = C_j T(t), \quad j = 1, \ldots, n, \tag{10.6}$$

where constants C_j are defined via solution of the following eigenvalue problem

$$-\lambda C_j = C_{j+1} - 2C_j + C_{j-1}, \quad j = 1, \ldots, n, \quad C_0 = C_{n+1} = 0. \tag{10.7}$$

Function $T(t)$ satisfies the following equation:

$$mT_{tt} + c\lambda T = 0. \tag{10.8}$$

Eigenvalues of the problem (10.7) follow [329]

$$\lambda_k = 4\sin^2 \frac{k\pi}{2(n+1)}, \quad k = 1, 2, \ldots, n. \tag{10.9}$$

A solution to Eq. (10.8) has the form $T = A\exp(i\omega t)$. Hence, Eqs. (10.8), (10.9) yield the following Lagrange formula for frequencies ω_k of discrete system (10.3):

$$\omega_k = 2\sqrt{\frac{c}{m}} \sin \frac{k\pi}{2(n+1)}, \quad k = 1, 2, \ldots, n. \tag{10.10}$$

Since all values λ_k are distinct, then all eigenvalues are different. Therefore, each of the eigenvalues is associated with one eigenvector $\mathbf{C}_k(C_1^{(k)}, C_2^{(k)}, ..., C_n^{(k)})$ of the form:

$$\mathbf{C}_k = \operatorname{cosec}\frac{k\pi}{n+1}\left(\sin\frac{k\pi}{n+1}, \sin\frac{2k\pi}{n+1}, ..., \sin\frac{nk\pi}{n+1}\right), \quad k = 1, 2, \ldots, n. \quad (10.11)$$

Eigenvectors are mutually orthogonal, whereas

$$|\mathbf{C}_k|^2 = \frac{n+1}{2}\operatorname{cosec}^2\frac{k\pi}{n+1}, \qquad k = 1, 2, \ldots, n. \quad (10.12)$$

Each of the eigenfrequencies (10.10) is associated with a normal vibration

$$y_j^{(k)}(t) = C_j^{(k)}[A_k\cos(\omega_k t) + B_k\sin(\omega_k t)], \quad k = 1,\ 2, \ldots, n. \quad (10.13)$$

A general solution of the BVP (10.3)–(10.5) is obtained as a result of superposition of normal vibrations

$$y_j(t) = \sum_{k=1}^{n} C_j^{(k)}[A_k\cos(\omega_k t) + B_k\sin(\omega_k t)], \quad j = 1, \ldots, n. \quad (10.14)$$

Let us study now the problem of chain movement under action of a unit constant force on the point number zero. Motion of such system is governed by Eq. (10.3) with the following boundary and initial conditions

$$\sigma_0(t) = 1, \qquad \sigma_{n+1}(t) = 0, \quad (10.15)$$

$$\sigma_j(t) = 0, \sigma_{jt}(t) = 0 \quad \text{at} \quad t = 0. \quad (10.16)$$

In what follows the initial BVP with non-homogeneous BCs (10.3), (10.15), (10.16) will be reduced to that of homogeneous BCs for Eq. (10.3) with non-homogeneous initial conditions, and then the method of superposition of normal vibrations can be applied. The formulas of normal forms obtained so far can be applied in a similar way with exchange of $y_j(t)$ for $\sigma_j(t)$. As result, the following exact solution to the BVP (10.3), (10.15), (10.16) is obtained [289, 329]:

$$\sigma_j = \frac{1}{n+1}\sum_{k=1}^{n}\sin\frac{\pi kj}{n+1}ctg\frac{\pi k}{2(n+1)}[1-\cos(\omega_k t)], j = 1, 2, \ldots, n. \quad (10.17)$$

10.3 CLASSICAL CONTINUOUS APPROXIMATIONS

For large values of n usually continuous approximation of discrete problem is applied. Let us introduce the continuous coordinate x scaled in such a way that $x = nh$ at the nodes of the lattice, where h is the distance between the particles. Then classical continuous approximation for discrete model described by the Eqs. (10.3), (10.15), (10.16) takes the form

$$m\sigma_{tt}(x,t) = ch^2\sigma_{xx}(x,t), \quad (10.18)$$

$$\sigma(0,t) = 1, \tag{10.19}$$

$$\sigma(l,t) = 0, \tag{10.20}$$

$$\sigma(x,0) = \sigma_t(x,0) = 0, \tag{10.21}$$

where $l = (n+1)h$, $\sigma(x,t) = chy_x(x,t)$.

Let us analyse BC (10.19). Using expression for force

$$\sigma_0(t) = c[y_1(t) - y_0(t)] = ch\left[\frac{\partial}{\partial x} + \frac{h}{2}\frac{\partial^2}{\partial x^2} + \frac{h^2}{6}\frac{\partial^3}{\partial x^3} + \cdots\right] y(x,t) \tag{10.22}$$

and leaving the first term on the right-hand side of equation (10.22), we obtain the BC (10.19).

BVP (10.18)–(10.22) can be used, for example, for modeling of stresses in couplings of the rolling stocks [261].

Having in hand a solution to continuous BVP (10.18)–(10.22), one obtains an approximate solution of a discrete problem due to the following formulas

$$\sigma_j(t) = \sigma(jh,t), \qquad j = 0,1,...,n,n+1. \tag{10.23}$$

Formally, the approximation described so far can be obtained in the following way. Let us denote the difference operator occurring in Eq. (10.3) as D, i.e.

$$m\sigma_{jtt}(t) = cD\sigma(t). \tag{10.24}$$

Applying the translation operator $\exp(h\partial/\partial x)$, one gets [319]

$$D = \exp\left(h\frac{\partial}{\partial x}\right) + \exp\left(-h\frac{\partial}{\partial x}\right) - 2 = -4\sin^2\left(-\frac{ih}{2}\frac{\partial}{\partial x}\right). \tag{10.25}$$

Let us explain Eq. (10.25) in more details. The Maclaurin formula for infinitely differentiable function $F(x)$ has the following form:

$$F(x+1) = \left[1 + \frac{\partial}{\partial x} + \frac{1}{2!}\frac{\partial^2}{\partial x^2} + \cdots\right] F(x) = \exp\left(\frac{\partial}{\partial x}\right) F(x). \tag{10.26}$$

Observe that $\exp(h\partial/\partial x)$ belongs to the so called pseudo-differential operators. Using relations (10.24)–(10.26), we cast Eq. (10.3) into pseudo-differential equation of the following form:

$$m\frac{\partial^2\sigma}{\partial t^2} + 4c\sin^2\left(-\frac{ih}{2}\frac{\partial}{\partial x}\right)\sigma = 0. \tag{10.27}$$

On the other hand, splitting the pseudo-differential operator into the Maclaurin series yields

$$\begin{gathered}\sin^2\left(-\frac{ih}{2}\frac{\partial}{\partial x}\right) = -\frac{1}{2}\sum_{k=1}\frac{h^{2k}}{(2k)!}\frac{\partial^{2k}}{\partial x^{2k}} = \\ -\frac{h^2}{4}\frac{\partial^2}{\partial x^2}\left(1 + \frac{h^2}{12}\frac{\partial^2}{\partial x^2} + \frac{h^4}{360}\frac{\partial^4}{\partial x^4} + \frac{h^6}{10080}\frac{\partial^6}{\partial x^6} + \ldots\right).\end{gathered} \tag{10.28}$$

Keeping in right-hand of the Eq. (10.28) only the first term, one obtains a continuous approximation (10.18). Note that an application of the Maclaurin series implies that displacements of the neighborhood masses differ slightly from each other. From a physical point of view, it means that we study vibrations of the chain with a few masses located on the spatial period (Fig. 10.2), i.e. the long-wave approximation takes place. This is so called in-phase or acoustical mode.

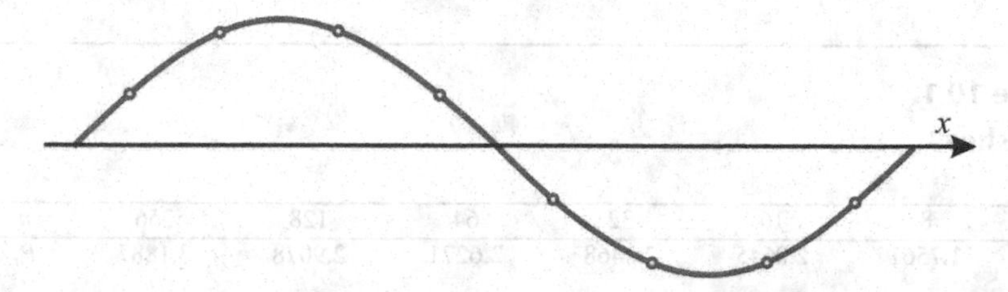

Figure 10.2 Spatial form of oscillation at a fixed time $t = const$ (points - discrete system, curve - continuous system) for acoustic mode.

Note that the vertical axis in Fig. 10.2 represents the displacement in the x direction, since the problem is 1D.

Continuous system (10.18) possesses the following discrete infinite spectrum

$$\alpha_k = \pi\sqrt{\frac{c}{m}}\frac{k}{n+1}, \qquad k = 1, 2, ... \tag{10.29}$$

Relations (10.29) relatively good approximate low frequencies of discrete system (10.10), whereas the n-th frequency α_k of a continuous system strongly differs from the corresponding n-th frequency ω_k of a discrete system. Accuracy of continuous approximations can be improved, what will be discussed further.

Replacing the finite spectrum of the discrete system (10.10) with the first n terms of the infinite spectrum (10.29) can be called the Debye approximation, since he used such an approximation in his theory the specific heat (heat capacity) in a solid.

10.4 "SPLASHES"

One can obtain an exact solution to the BVP (10.18)–(10.22), using the d'Alembert method matched with operational calculus [261, 289, 329]:

$$\sigma(x,t) = H\left(nh \arcsin\left|\sin\left(\frac{\pi}{2n}\sqrt{\frac{c}{m}}t\right)\right| - x\right), \tag{10.30}$$

where $H(...)$ is the Heaviside function.

From Eq. (10.30) one obtains the following estimation:

$$|\sigma(x,t)| \leq 1. \tag{10.31}$$

It was believed that estimation (10.31) can be applied also to a discrete system [222]. However, analytical as well as numerical investigations [165–168, 260, 261] indicated a need to distinguish between global and local characteristics of a discrete system. In other words, classical continuous approximation can be used only for the low frequency part of the discrete system spectrum [303]. Numerical investigations show that for given masses in a discrete chain quantity the $P_j = |\sigma_j(t)|$ may exceed the values of unity in certain time instants (Table 10.1 reported in [165]).

Table 10.1
Splashes

n	8	16	32	64	128	256	$n \to \infty$
P_n	1.7561	2.0645	2.3468	2.6271	2.9078	3.1887	$P_n \to \infty$

Observe that splash amplitude does not depend on the parameter m/c. In addition, the amplitude of the chain vibrations increases with increase of n, whereas its total energy does not depend on n. However, this is not a paradox. Namely, amplitude of vibrations has an order of sum of quantities $\sigma_j(t)$, whereas its potential energy order is represented by a sum of squares of the quantities mentioned [329].

On the other hand, a vibration amplitude of a mass with a fixed number is bounded for $n \to \infty$, but amplitude of vibrations of a mass with a certain number increasing with increase of n tends to infinity for $n \to \infty$ following $\ln n$ [329]. "In the language formulas! of mechanics what we just said means that when analyzing the so-called "local properties" of a one-dimensional continuous medium, one cannot treat the medium as the limiting case of a linear chain of point masses, obtained when the number of points increases without limit" [261].

Earlier the same effect of continualization was predicted by Ulam, who wrote [439]: "The simplest problems involving an actual infinity of particles in distributions of matter appear already in classical mechanics.

Strictly speaking, one has to consider a true infinity in the distribution of matter in all problems of the physics of continua. In the classical treatment, as usually given in textbooks of hydrodynamics and field theory, this is, however, not really essential, and in most theories serves merely as a convenient limiting model of *finite* systems enabling one to use the algorithms of the calculus. The usual introduction of the continuum leaves much to be discussed and examined critically. The derivation of the equations of motion for fluids, for example, runs somewhat as follows. One images a very large number N of particles, say with equal masses constituting a net approximating the continuum, which is to be studied. The forces between these particles are assumed to be given, and one writes Lagrange equations for the motion of N particles. The finite system of ordinary differential equations becomes in the limit $N = \infty$ one or several *partial* differential equations. The Newtonian laws of conservation of energy and momentum are seemingly correctly formulated for the limiting case of the continuum. There appears at once, however, at least possible objection to the unrestricted validity of this formulation. For the very fact that the limiting equations imply tacitly the continuity and differentiability of the functions describing the motion of the continuum seems to impose various *constraints* on the possible motions of the approximating finite systems. Indeed, at any stage of the limiting process, it is quite conceivable for two neighboring particles to be moving in opposite directions with a relative velocity which does not need to tend to zero as N becomes infinite, whereas the continuity imposed on the solution of the limiting continuum excludes such a situation. There are, therefore, constraints on the class of possible motions which are not explicitly recognized. This means that a viscosity or other type of constraints must be introduced initially, singling out "smooth" motions from the totality of all possible ones. In some cases, therefore, the usual differential equations of hydrodynamics may constitute a misleading description of the physical process".

The explanation of the illusory contradiction between discrete and continuous models is quite simple. The exact solution to the discrete problem (10.17) contains both slow and fast spatialy changing harmonics. The solution of continuous system (10.30) accurately describes only the slow components of a solution. Thus, the correct continuous model of the discrete system must take into account also high harmonics.

10.5 ENVELOPE CONTINUALIZATION

A classical continuous approximation gave good description of a low part of the vibration spectrum of a finite chain of masses. In what follows we study now another limiting case, antiphase vibration, when $\sigma_k = -\sigma_{k+1}$ (Fig. 10.3). In this case one obtain the following ODE:

$$m\Omega_{tt} + 4c\Omega = 0. \tag{10.32}$$

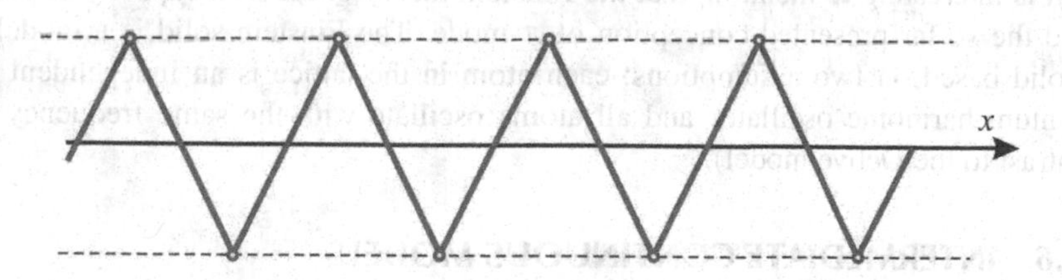

Figure 10.3 Antiphase vibrations of a mass chain (optical or π-mode).

This is so-called optical or π-vibrational mode. In the case of vibrations close to the π-mode, the "envelope continualization" or "semidiscrete approximation" is usually used [247, 250, 251, 382] (Fig. 10.4). Namely, first we use "staggering transformation" [164]

$$\sigma_k = (-1)^k \Omega_k, \tag{10.33}$$

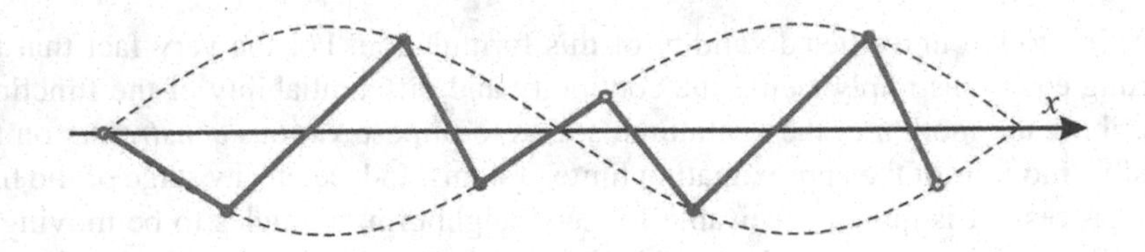

Figure 10.4 Envelope continualization (semidiscrete approach).

and then the Eq. (10.3) is reduced to the following form:

$$m\Omega_{ktt} + c(4\Omega_k + \Omega_{k-1} - 2\Omega_k + \Omega_{k+1}) = 0. \quad (10.34)$$

Then, the following relations are applied:

$$\Omega_{k-1}(t) - 2\Omega_k(t) + \Omega_{k+1}(t) = -4\sin^2\left(-\frac{ih}{2}\frac{\partial}{\partial x}\right)\Omega(x,t) =$$
$$\left(h^2\frac{\partial^2}{\partial x^2} + \frac{h^4}{12}\frac{\partial^4}{\partial x^4} + \frac{h^6}{360}\frac{\partial^6}{\partial x^6} + \cdots\right)\Omega(x,t), \quad k = 0,1,2,...,n,n+1. \quad (10.35)$$

Using Eqs. (10.36), (10.33) and considering h^2 as a small parameter, one gets (we take zeroth and first-order approximations only)

$$m\Omega_{tt} + 4c\Omega + ch^2\Omega_{xx} = 0. \quad (10.36)$$

Appropriate BCs for Eq. (10.36) for the case of natural oscillations are:

$$\Omega = 0 \quad \text{at} \quad x = 0, l. \quad (10.37)$$

It is interesting to mention, that the Einstein theory of the heat capacity of solids used the so far presented conception of π-mode. The Einstein solid is a model of a solid based on two assumptions: each atom in the lattice is an independent 3D quantum harmonic oscillator and all atoms oscillate with the same frequency (in contrast to the Debye model).

10.6 INTERMEDIATE CONTINUOUS MODELS

In what follows we are going to construct improved continuous approximations. Local modeling of such systems (nonlocal theories of elasticity) requires gradient formulation. The integral formulation may be reduced to a gradient form by truncating the series expansion of the nonlocal kernel in the dual space [362]. In what follows we apply the gradient formulation approach.

If, in the series (10.28), we keep three first terms, the following model is obtained

$$m\frac{\partial^2\sigma}{\partial t^2} = ch^2\left(\frac{\partial^2}{\partial x^2} + \frac{h^2}{12}\frac{\partial^4}{\partial x^4} + \frac{h^4}{360}\frac{\partial^6}{\partial x^6}\right)\sigma. \quad (10.38)$$

Equation (10.38) includes the fourth- and sixth-order spatial derivatives and, consequently, the auxiliary BCs are required. This is a typical difficulty that arises when higher-order models, derived originally for infinite media, are applied to bounded domains [258, 259]. To overcome this difficulty, several approaches are possible. The first, originating from [231], can be called phenomenological. In [231] authors used exact expression for frequencies spectrum and proposed to introduce auxiliary BCs in a way that guarantee the most accurate frequencies approximations. As result they obtained the following BCs:

$$\sigma = \sigma_{xx} = \sigma_{xxxx} = 0 \quad \text{at} \quad x = 0, l. \tag{10.39}$$

The same approach was used in [28] for the pulse load problem. The analytical solution was compared with a numerical simulation of the discrete lattice. Three types of higher-order auxiliary BCs were examined and BC gave the best agreement of continuous and discrete solutions was choosen.

Second approach based on the variational principle [166]. In this case the auxiliary boundary conditions treats as the natural boundary conditions for the suitable variational formulation of problem. Proposed in [166] BCs coincide with (10.39).

It can also be noted that the setting of boundary conditions during continualization is quite similar to the introduction of additional points in the finite difference method. There is some arbitrariness here, and the success of a choise is evaluated a posteriori. Specifically, we can arbitrarily choose values in auxiliary points $k < 0$ $(k > n+1)$ [17]. In particular, if we want to keep the translation symmetry and choose $y_{-1} = -y_1(t)$, etc., the BCs (10.39) are obtained.

Comparison of n-th frequency of a continuous system (10.38), (10.39) with that of a discrete system exhibits essential accuracy improvement (coefficient 2.1 instead of 2 in an exact solution yields an error of $\sim 5\%$).

In a general case, keeping in series (10.28) N terms, one gets equations of the so-called intermediate continuous models [166]

$$m\frac{\partial^2 \sigma}{\partial t^2} = 2c\sum_{k=1}^{N} \frac{h^{2k}}{(2k)!}\frac{\partial^{2k}\sigma}{\partial x^{2k}}. \tag{10.40}$$

BCs for Eq. (10.40) have the following form:

$$\frac{\partial^{2k}\sigma}{\partial x^{2k}} = 0 \quad \text{for} \quad x = 0, l, \quad k = 0, 1, \ldots, N-1. \tag{10.41}$$

From Eq. (10.40) one obtains

$$\alpha_k^2 = 2\frac{c}{m}\sum_{k=1}^{N} \frac{h^{2k}}{(2k)!}\left(\frac{k\pi}{l}\right)^{2k}. \tag{10.42}$$

By reflectional symmetry and periodicity it suffices to consider the wave numbers in the interval Brillouin first zone. Solution is then stable if and only if r.h.p. of (10.42) is positive for all k. This condition is the far reaching generalization of the Legendre-Hadamard condition of strong ellipticity in continuum elasticity [435]. One can see that BVP (10.40),(10.41) is stable for odd N. In this case Eq. (10.40) is of hyperbolic type [166].

Application of intermediate continuous models allows take into account the above mentioned splashes effect.

10.7 USING OF PADE APPROXIMATIONS

The construction of intermediate continuous models is mainly based on the splitting of a nonlocal difference operator into Maclaurin series. However, very often application of Padé approximations (PA) is more effective for approximation [47]. Collins [125] proposed to construct continuous models using PA. Then this approximation was widely used by Rosenau [385–389]. Further developments of this approach were done in [438, 453, 454]. Sometimes these continuous models are called quasicontinuum approximations.

If only two terms are left in the series (10.28), then the PA can be cast into the following form:

$$\frac{\partial^2}{\partial x^2}+\frac{h^2}{12}\frac{\partial^4}{\partial x^4}\approx\frac{\frac{\partial^2}{\partial x^2}}{1-\frac{h^2}{12}\frac{\partial^2}{\partial x^2}}. \tag{10.43}$$

For justification of this procedure Fourier or Laplace transforms can be used. The corresponding quasicontinuum model reads

$$m\left(1-\frac{h^2}{12}\frac{\partial^2}{\partial x^2}\right)\sigma_{tt}-ch^2\sigma_{xx}=0. \tag{10.44}$$

The BCs for Eq. (10.44) have the form

$$\sigma=0 \quad \text{for} \quad x=0,l. \tag{10.45}$$

Therefore, using Eq. (10.44) we need not to formulate additional unphysical BCs. The error regarding estimation of n-th frequency in comparison to that of a discrete chain is of $\sim 16.5\%$. However, Eq. (10.45) is of lower order in comparison to Eq. (10.38).

In the theory of elasticity equation of the type (10.45) is usually called Love Eq. [190] (but as Love mentioned [285], this equation was obtained earlier by Rayleigh [379]). Term σ_{xxtt} can be treated as the lateral inertia. Eq. (10.45) has hyperbolic type. Kaplunov et al. [230] refer to these type theories as theories with modified inertia.

Passage to Eq. (10.45) can be treated as regularization procedure for Eq. (6.3) with $N=2$ which is conditionally stable. The model governed by Eq. (10.45) is unconditionally stable and propagating waves cannot transfer energy quicker than the velocity c. However, the model governed by Eq. (10.45) predicts that short waves transfer elastic energy with almost zero speed [314], that is not correct from the physical standpoint.

Equation (10.38) can be using PA for two last terms in r.h.p. transformed to the following form

$$m\left(1-\frac{h^2}{30}\frac{\partial^2}{\partial x^2}\right)\frac{\partial^2\sigma}{\partial t^2}=ch^2\left(\frac{\partial^2}{\partial x^2}+\frac{h^2}{20}\frac{\partial^4}{\partial x^4}\right)\sigma. \tag{10.46}$$

In the theory of elasticity equation of the type (10.46) is usually called Bishop Eq. [79]. Eq. (10.46) with BCs

$$\sigma=\sigma_x=0 \quad \text{or} \quad \sigma=\sigma_{xx}=0 \quad \text{at} \quad x=0,l \tag{10.47}$$

is conditionally stable.

It is worth noting that Eq. (10.37) describing semidiscrete approach also can be regularized using PA as follows [162–164]:

$$m\left(1-\frac{h^2}{4}\frac{\partial^2}{\partial x^2}\right)\Omega_{tt}+c\Omega=0. \tag{10.48}$$

Finally, having in hand both long and short waves asymptotics, one may also apply two-point PA [11, 12]. In what follows we construct two-point PA using the first term of the series (10.28) as a one of limiting cases. We suppose

$$\begin{gathered}\sin^2\left(-\frac{ih}{2}\frac{\partial}{\partial x}\right)=-\frac{1}{2}\sum_{k=1}\frac{h^{2k}}{(2k)!}\frac{\partial^{2k}}{\partial x^{2k}}=\\ -\frac{h^2}{4}\frac{\partial^2}{\partial x^2}\left(1+\frac{h^2}{12}\frac{\partial^2}{\partial x^2}+\frac{h^4}{360}\frac{\partial^4}{\partial x^4}+\dots\right)\approx\frac{-\frac{h^2}{4}\frac{\partial^2}{\partial x^2}}{1-a^2\frac{\partial^2}{\partial x^2}}.\end{gathered} \tag{10.49}$$

The improved continuous approximation is governed by the following equation

$$m\left(1-\alpha^2h^2\frac{\partial^2}{\partial x^2}\right)\sigma_{tt}-ch^2\sigma_{xx}=0. \tag{10.50}$$

Now we require the n-th frequency of a continuous and discrete system to coincide (compare (10.10) and (10.29))

$$\alpha_n\approx 2\sqrt{c/m}. \tag{10.51}$$

Using condition (10.51) one gets

$$\alpha^2=0.25-\pi^{-2}. \tag{10.52}$$

Highest error in estimation of the eigenfrequencies appears for $k=[0.5(n+1)]$ and consists of 3%.

Observe that approximate models described by Eq. (10.46) with values of a^2 slightly differs from (10.52) are already known. Eringen [155–157] using a correspondence between the dispersion curves of the continuous and discrete models

[87,98] obtained $\alpha^2 = 0.1521$. This value is very close to that proposed in [42,315] on the basis of a some physical hypothesis "dynamically consistent model".

Interesting, that Mindlin and Herrmann used very similar to two-point PA idea for construction their well-known equation for longitudinal waves in rod [319].

Dispersion relation (10.50) does not satisfy condition $d\alpha_k/dk = 0$ at the end of first Brillouin zone [98]. That is why in [156,273] so-called bi-Helmholtz type equation was proposed. In [273], this equation takes the following form

$$m\left(1-\alpha^2h^2\frac{\partial^2}{\partial x^2}+\alpha_1^2h^4\frac{\partial^4}{\partial x^4}\right)\sigma_{tt}-ch^2\sigma_{xx}=0, \tag{10.53}$$

where $\alpha_1 = 1/\pi$.

BCs associated with this equation are

$$\sigma=\sigma_{xx}=0 \quad \text{at} \quad x=0,l \quad \text{or} \quad \sigma=\sigma_x=0 \quad \text{at} \quad x=0,l. \tag{10.54}$$

Unfortunately, this equation is unstable. This is the price one has to pay is that the group velocity is off at the end of the Brillouin zone [273].

Let us also analyse lattice with inertially linked masses (Fig. 10.5).

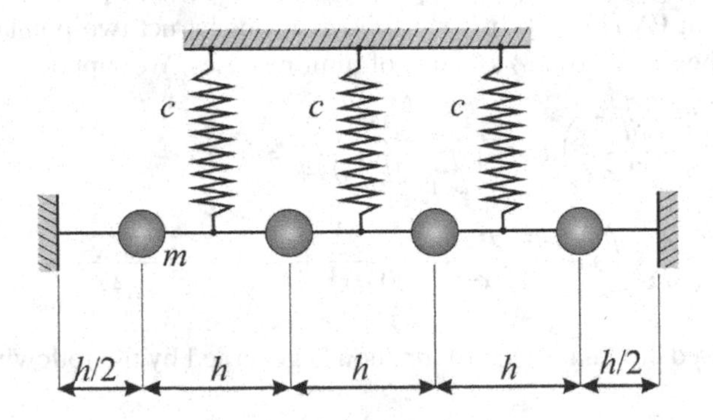

Figure 10.5 A chain of inertially coupled masses.

Governing equations can be written as follows [271]

$$4cy_j(t)=m\frac{d^2}{dt^2}\left[y_{j-1}(t)-2y_j(t)+y_{j+1}(t)\right], \quad j=1,2,....,n. \tag{10.55}$$

Frequencies ω_k of discrete system (10.55) are

$$\omega_k=\sqrt{\frac{c}{m}}\sin^{-1}\frac{k\pi}{2(n+1)}, \qquad k=1,2,\ \ldots,n. \tag{10.56}$$

Continuous approximation for this model is as follows

$$4cy(x,t)=mh^2\frac{d^2}{dx^2}y_{tt}(x,t), \tag{10.57}$$

and describes highest frequencies.

Antiphase mode of the following forms

$$cy(x,t) = -my_{tt}(x,t), \tag{10.58}$$

presents the lowest frequency.

Quasicontinuous models are governed by the following PDEs

$$c\left(1 - \frac{h^2}{12}\frac{d^2}{dx^2}\right)y(x,t) = mh^2\frac{d^2}{dx^2}y_{tt}(x,t), \tag{10.59}$$

$$c\left(1 - \alpha^2 h^2 \frac{d^2}{dx^2}\right)y(x,t) = mh^2\frac{d^2}{dx^2}y_{tt}(x,t). \tag{10.60}$$

BCs for Eqs. (10.57),(10.59) and (10.60) take the form

$$y(0,t) = y(l,t) = 0. \tag{10.61}$$

10.8 NORMAL MODES EXPANSION

In the previous sections, we studied the effectiveness of various continualization approaches for calculating the natural oscillations of discrete system. The possibility of a satisfactory approximation of natural frequencies over the entire range of their variation is shown. However, the study of free or forced oscillations requires normal modes expansion. We consider as an example the discrete problem (10.3), (10.15), (10.16), which possesses the exact solution (10.17).

Let us use for continuous approximation Eqs. (10.50),(10.52) with BCs (10.19) and ICs (10.22) and the following ansatz

$$\sigma(x,t) = 1 - \frac{x}{l} + u(x,l). \tag{10.62}$$

Function $u(x,t)$ is defined by the following BVP:

$$m\left(1 - \alpha^2 h^2 \frac{\partial}{\partial x^2}\right)u_{tt} - ch^2 u_{xx} = 0, \tag{10.63}$$

$$u(0,t) = u(l,t) = 0, \tag{10.64}$$

$$u(x,0) = -1 + \frac{x}{l}, \qquad u_t(x,0) = 0. \tag{10.65}$$

Solution to the BVP (10.63)–(10.65) can be obtained via Fourier method

$$\sigma = 1 - \frac{x}{l} - \frac{2}{\pi}\sum_{k=1}^{\infty}\frac{1}{k}\sin\left(\frac{k\pi x}{l}\right)\cos(\beta_k t), \tag{10.66}$$

where β_k are defined by $\beta_k = \frac{l}{\sqrt{l^2+\alpha^2\pi^2h^2k^2}}\alpha_k$.

For the approximation of the chain motion we keep in an infinite sum only n first harmonics

$$\sigma = 1 - \frac{x}{l} - \frac{2}{\pi}\sum_{k=1}^{n}\frac{1}{k}\sin\left(\frac{k\pi x}{l}\right)\cos(\beta_k t). \tag{10.67}$$

Let us compare solutions (10.17) and (10.67). Function $1 - x/l$ serves as a good approximation of truncated series

$$\frac{1}{n+1}\sum_{k=1}^{n}\sin\frac{\pi kj}{n+1}cot\frac{\pi k}{2(n+1)}. \tag{10.68}$$

Frequencies β_k stand for the good approximation for frequencies ω_k. But the coefficients in the truncated series (10.67) and (10.68) differ from each other. Projections onto normal modes of discrete and continuous systems,

$$\frac{1}{n+1}cot\frac{\pi k}{2(n+1)} \tag{10.69}$$

and $\frac{2}{\pi k}$ respectively, strongly differ from each other for $k \gg 1$. This is because during projection into normal modes for the discrete system one must use summation over k from 0 to n, whereas for the continuous system - integration with regard to x from 0 to 1. The problem occurring so far can be overcome using the Euler-Maclaurin formulas [2]

$$\sum_{k=0}^{n+1} f(k) = \int_0^{n+1} f(x)dx + \frac{1}{2}[f(0) + f(n+1)] + \\ \sum_{j=1}^{\infty}\frac{(-1)^{j+1}}{j+1}B_j\left[\frac{d^j f(n+1)}{dx^j} - \frac{d^j f(0)}{dx^j}\right], \tag{10.70}$$

where B_i are the Bernoulli numbers [2], having the following values: $B_0 = 1$, $B_1 = -1/2$, $B_2 = 1/6$, $B_3 = 0$, ...

In addition, the following formulas hold [377]:

$$B_n = -\frac{1}{n+1}\sum_{k=1}^{n} C_{n+1}^{k+1}B_{n-k}, \tag{10.71}$$

$$\sum_{k=0}^{n+1}\sin^2\frac{k\pi j}{n+1} = \frac{n+1}{2}, \tag{10.72}$$

$$\sum_{k=0}^{n+1}\left(1 - \frac{j}{n+1}\right)\sin\frac{k\pi j}{n+1} = \frac{1}{n+1}ctg\frac{\pi j}{2(n+1)}. \tag{10.73}$$

The corresponding integrals read

$$\int_0^{n+1}\sin^2\frac{\pi jx}{n+1}dx = \frac{n+1}{2}, \tag{10.74}$$

$$\int_0^{n+1}\left(1-\frac{x}{l}\right)\sin\frac{\pi jx}{n+1}dx=\frac{2}{\pi j}. \qquad (10.75)$$

Observe that the values of sum (10.72) and integral (10.74) coincide. Using the Euler-Maclaurin formula one obtains

$$\sum_{k=0}^{n+1}\left(1-\frac{j}{n+1}\right)\sin\frac{k\pi j}{n+1}=\int_0^{n+1}\left(1-\frac{x}{l}\right)\sin\frac{\pi jx}{n+1}dx+$$
$$\frac{1}{2}\left[\sin 0+\sin(j\pi)\right]-\frac{j\pi}{6(n+1)}\cos 0+\ldots=\frac{2}{\pi j}\left[1-\frac{\pi^2 j^2}{12(n+1)^2}\right]. \qquad (10.76)$$

Owing to formula (10.76), one can construct a simple expression relatively good approximating sum (10.73) for arbitrary j values from $j=1$ to $j=n$. For this purpose, we change second term of the right hand side of Eq. (10.76) in the following way:

$$\sum_{k=0}^{n+1}\left(1-\frac{j}{n+1}\right)\sin\frac{k\pi j}{n+1}\approx\frac{2}{\pi j}\left[1-\frac{j^2}{(n+1)^2}\right]. \qquad (10.77)$$

10.9 THEORIES OF ELASTICITY WITH COUPLE-STRESSES

Since the release of the classic monograph by Cosserat's brothers [129] and Le Roux paper [274], activity has continued in the direction of generalizations of the theory of elasticity, which expand the possibilities for describing the effects accompanying elastic deformations of materials with an internal structure [257–259]. A good description of these theories can be found in comprehensive reviews including [308, 446]. It was expected that the results of such work will lead to a significant change in the values of the concentration coefficients and the types of stress singularities [318, 434]. As a problem that impedes the use of generalized theories for solving real problems, there is a lack of quantitative data on the physical constants of these theories [308, 446]. As it is mentioned in [446], "Generalized elasticity theories even for isotropic materials contain many additional constants that are difficult or impossible to determine experimentally".

Theoretically, gradient moduli were previously determined [291] for a micro-inhomogeneous medium composed of homogeneous and isotropic grains, whose Lamé's constants are random variables. Recently, gradient modules have been defined for two-phase composites in the case of a low concentration of inclusions in a homogeneous matrix [46, 78]. It is shown that it is possible to estimate the values of the gradient modules and the corresponding characteristic lengths for the crystal lattice of metals by the methods of the density functional theory [406]. However, at present, the problem of determining the constants of the theory of elasticity with couple-stress by directly continualization of a real atomic lattice is far from solving. In this regard, the work [292] is of great interest. In this paper closed-form expressions for five physical constants of the Toupin-Mindlin theory of isotropic gradient elasticity for polycrystalline materials were obtained. A polycrystal is considered as a micro-inhomogeneous elastic medium, which consists of homogeneous anisotropic crystallites with random orientation. It is shown that the elastic energy density of the polycrystalline medium depends nonlocally on the nonuniform field of average deformations. In the case of smooth fields of average deformations it could be considered that the energy density locally depends not only on the values of deformations, but also on the values of their derivatives. This dependence in the isotropic case is characterized by five gradient modules, in addition to the two elastic moduli of the classical theory of elasticity. For the first time, explicit expressions are obtained for the gradient modules of polycrystalline materials.

Although a single sample of a polycrystalline body is described by the local classical theory of elasticity of micro-inhomogeneous bodies, the averaged macroscopic properties of a representative ensemble of polycrystalline samples turn out to be homogeneous in a macroscopic sense. Therefore, the transition from the consideration of a particular sample to the description of the averaged macroscopic properties of an ensemble of samples can be called homogenization. In this case, it turns out that macroscopic properties are described by a nonlocal theory of elasticity. It is necessary to distinguish the homogenization of a continuous microinhomogeneous medium from the continualization of a discrete elastic medium, such as a crystal lattice [257–259]. Formally, both approaches lead to theories with a nonlocal dependence of the density of elastic energy on the displacement field, but physically these are completely different theories. But from the physical point of view, the result of the lattice continualization differs significantly from the homogenization of a continuous microinhomogeneous medium. First, nonlocal effects for the lattice are include into the theory from the very beginning due the long-range interatomic forces. Secondly, the characteristic lengths of nonlocal lattice effects correspond in order of magnitude to interatomic distances, i.e. their influence is many orders of magnitude less than in the case of polycrystals. And thirdly, a correct description of the lattice behavior requires consideration of its anisotropy, which further complicates the already cumbersome description of nonlocal effects. For example, the local Toupin-Mindlin gradient theory obtained as a result of homogenization requires the introduction of five additional physical constants. But after the continualization of even a simple cubic lattice, eleven additional physical constants appear [45, 199].

10.10 CORRESPONDENCE BETWEEN FUNCTIONS OF DISCRETE ARGUMENTS AND APPROXIMATING ANALYTICAL FUNCTIONS

For continualization we must construct a function $u(x,t)$ representing a continuous approximation of the function of discrete argument $u_j(t)$

$$u(jh,t) = u_j(t), \tag{10.78}$$

where h is the distance between points of discrete function.

Observe that it is defined with an accuracy to any arbitrary function, which equals zero in nodal points $x = jh$. For this reason, from a set of interpolating functions one may choose the smoothest function owing to filtration of fast oscillating terms. As it is shown in [258], the interpolating function is determined uniquely, if one requires its Fourier image

$$\bar{u}(q,t) = \int_{-\infty}^{\infty} u(x,t)e^{-iqx}dx; u(x,t) = \frac{1}{2\pi}\int_{-\infty}^{\infty} \bar{u}(q,t)e^{iqx}dx \tag{10.79}$$

to differ from zero only on the segment $-\pi/h \leq q \leq \pi/h$. From this condition one obtains

$$u(x,t) = \sum_{k=1}^{n} u_k(t)\frac{\sin\frac{\pi(x-kh)}{h}}{\pi(x-kh)}. \tag{10.80}$$

Function

$$\operatorname{sinc}(x) = \frac{\sin(\pi x)}{\pi x} \tag{10.81}$$

is familia normalized sinc function, often called Whittaker-Kotelnikov-Shannon (WKS) interpolating function. The value of this function at the removable singularity at zero is understood to be the limit value 1. The sinc function is then analytic everywhere and hence an entire function. Also this function has interesting properties:

$$\int_{-\infty}^{\infty} \frac{\sin \pi x}{\pi x}dx = 1, \tag{10.82}$$

$$\lim_{h\to 0} \frac{\sin\frac{\pi x}{h}}{\pi x} = \delta(x). \tag{10.83}$$

We note that the concept of concentrated force in the theory of elasticity with couple-stresses must be changed in comparison with classical theory of elasticity. Namely, delta-function must be replaced to the sinc function due size effect caused by internal structure of material. This statement is illustrated by identity (10.83).

10.11 THE KERNELS OF INTEGRO-DIFFERENTIAL EQUATIONS OF THE DISCRETE AND CONTINUOUS SYSTEMS

Let us consider the chain of material points with the equal masses m, located in equilibrium states in the points of the axis x with coordinate's $jh, j = 0, \pm 1, \pm 2, ...$ and suspended by elastic couplings with stiffness c. The governing ODEs are as follows

$$m\frac{d^2 y_j(t)}{dt^2} = c(y_{j+1} - 2y_j + y_{j-1}), \quad j = 1,2,3,... \tag{10.84}$$

where $y_j(t)$ is the displacement of the j-th material point from its static equilibrium position.

Instead of the dimensional coordinates x, y, t, we introduce dimensionless coordinates $\xi = x/h$, $\tau = t\sqrt{c/m}, z_j = y_j/h$ and rewrite (10.84) as

$$\frac{d^2 z_j}{d\tau^2} = z_{j+1} - 2z_j + z_{j-1}. \tag{10.85}$$

Differential-difference Eq. (10.85) can be reduced to the following integro-differential equation [182, 400]:

$$\frac{\partial^2 z(x,\tau)}{\partial \tau^2} = -\int_{-\infty}^{\infty} \Phi_d(\xi - \xi_1) y(\xi_1, \tau) d\xi_1, \tag{10.86}$$

$$\Phi_d(\xi) = \frac{2}{\pi} \int_{-\pi}^{\pi} \sin^2\left(\frac{q}{2}\right) \cos(q\xi) dq = \frac{2}{\pi\xi} \frac{(2\xi^2 - 1)}{(\xi^2 - 1)} \sin(\pi\xi). \tag{10.87}$$

Equations (10.86), (10.87) describe the media with nonlocal interactions. The classical continuous approximation is:

$$\frac{\partial^2 z}{\partial \tau^2} = \frac{\partial^2 z}{\partial \xi^2}. \tag{10.88}$$

Thus, replacement of the discrete media with the continuous one leads to two reasons for error. The first one is connected with replacement of the integral with finite limits of integration on the integral with infinite limits of integration

$$\Phi(\xi) = \frac{4}{\pi} \int_{0}^{\infty} \sin^2\left(\frac{q}{2}\right) \cos(q\xi) dq. \tag{10.89}$$

Such a replacement from a physical standpoint is caused due the transition from the discrete to the continuous media.

The second reason for error type is connected with approximation of the operator $4\sin^2(q/2)$. The general approach to improvement of approximation (10.88) consists in approximation of the integral operator from the r.h.p. of Eq. (10.87) with higher-order derivatives in the assumption that nonlocal property is weak. For example, replacing integration limits in integral (10.87) on infinite ones and using truncated Maclaurin series

$$4\sin^2\left(\frac{q}{2}\right) \approx q^2 - \frac{1}{12}q^4 + \frac{1}{360}q^6 \tag{10.90}$$

we obtain intermediate continuous model (Chapter 10.6):

$$\frac{\partial^2 z}{\partial \tau^2} = \left(\frac{\partial^2}{\partial \xi^2} + \frac{1}{12}\frac{\partial^4}{\partial \xi^4} + \frac{1}{360}\frac{\partial^6}{\partial \xi^6}\right) z. \tag{10.91}$$

One of the ways for improvement of the local approximation can consist in more precise approximation of the operator $\sin^2\left(\frac{q}{2}\right)$. Using for this aim PA one obtains (Chapter 10.7)

$$4\sin^2\left(\frac{q}{2}\right) \approx q^2 / \left(1 + \frac{1}{12}q^2\right). \tag{10.92}$$

Using Eq. (10.92) one obtains the following model:

$$\left(1 - \frac{1}{12}\frac{\partial^2}{\partial \xi^2}\right) \frac{\partial^2 z}{\partial \tau^2} = \frac{\partial^2 z}{\partial \xi^2}. \tag{10.93}$$

Two-point PA gives more precise approximation (Chapter 10.7)

$$4\sin^2\left(\frac{q}{2}\right) \approx \frac{q^2}{1+\alpha^2 q^2}, \qquad \alpha^2 = 0.25 - \pi^{-2}. \tag{10.94}$$

Using Eq. (10.94) one obtains the following approximation:

$$\left(1-\alpha^2 \frac{\partial^2}{\partial \xi^2}\right)\frac{\partial^2 z}{\partial \tau^2} = \frac{\partial^2 z}{\partial \xi^2}. \tag{10.95}$$

We investigate the accuracy of the approximation of the kernel of an integro-differential equation (10.86), which describes the deformation of a discrete media. Integral (10.87) has the following asymptotics

$$\Phi_d(\xi) \to \frac{4}{\pi}\frac{\sin(\pi\xi)}{\xi} \approx 1.2732\frac{\sin(\pi\xi)}{\xi} \quad \text{at} \quad \xi \to \infty, \tag{10.96}$$

$$\Phi_d(\xi) \to 2 \quad \text{at} \quad \xi \to 0, \tag{10.97}$$

$$\Phi_d(\xi) \to -1 \quad \text{at} \quad \xi \to 1. \tag{10.98}$$

The kernels of integro-differential equations of classical continuous approximation (10.90) $\Phi_c(\xi)$ and intermediate continuous approximation (10.91) $\Phi_{ci}(\xi)$ have the following expressions and asymptotics:

$$\Phi_c(\xi) = \frac{1}{\pi}\int_0^\pi q^2 \cos(q\xi)\,dq = \frac{(\pi^2\xi^2-2)\sin(\pi\xi) + 2\pi\xi\cos(\pi\xi)}{\pi\xi^3}, \tag{10.99}$$

$$\Phi_c(\xi) \to \pi\frac{\sin(\pi\xi)}{\xi} \quad \text{at} \quad \xi \to \infty, \tag{10.100}$$

$$\Phi_c(\xi) \to \frac{\pi^2}{3} \approx 3.2899 \quad \text{at} \quad \xi \to 0, \tag{10.101}$$

$$\Phi_c(\xi) \to -2 \quad \text{at} \quad \xi \to 1, \tag{10.102}$$

$$\begin{gathered}\Phi_{ci}(\xi) = \frac{1}{\pi}\int_0^\pi \left(q^2 - \frac{q^4}{12} + \frac{q^6}{360}\right)\cos(q\xi)\,dq = \\ \frac{\left(1-\frac{\pi^2}{12}+\frac{\pi^4}{360}\right)\pi\sin(\pi\xi)}{\xi} + \frac{\left(2-\frac{\pi^2}{3}+\frac{\pi^4}{60}\right)\cos(\pi\xi)}{\xi^2} - \\ \frac{\left(2-\pi^2+\frac{\pi^4}{12}\right)\sin(\pi\xi)}{\pi\xi^3} + \frac{\left(2-\frac{\pi^2}{3}\right)\cos(\pi\xi)}{\xi^4} - \\ \frac{(2-\pi^2)\sin(\pi\xi)}{\pi\xi^5} + \frac{2\cos(\pi\xi)}{\xi^6} - \frac{2\sin(\pi\xi)}{\pi\xi^7},\end{gathered} \tag{10.103}$$

$$\Phi_{ci}(\xi) \to \frac{\left(1-\frac{\pi^2}{12}+\frac{\pi^4}{360}\right)\pi\sin(\pi\xi)}{\xi} \approx 1.4077\frac{\sin(\pi\xi)}{\xi} \quad \text{at} \quad \xi\to\infty, \tag{10.104}$$

$$\Phi_{ci}(\xi) \to \frac{\pi^2}{3}\left(1-\frac{\pi^2}{20}+\frac{\pi^4}{840}\right) \approx 2.0479 \quad \text{at} \quad \xi\to 0, \tag{10.105}$$

$$\Phi_{ci}(\xi) \to -6+\frac{2\pi^2}{3}-\frac{\pi^4}{60} \approx -1.0437 \quad \text{at} \quad \xi\to 1. \tag{10.106}$$

The kernels of integro-differential equations continuous approximation (10.93) $\Phi_{qc}(\xi)$ and improved continuous approximation (10.95) $\Phi_{qci}(\xi)$ can be written as follows

$$\Phi_{qc}(\xi) = \Phi_0\left(b^2,\xi\right) \quad \text{for} \quad b^2 = 1/3, \tag{10.107}$$

$$\Phi_{qci}(\xi) = \Phi_0\left(b^2,\xi\right) \quad \text{for} \quad b^2 = 1-4/\pi^{-2}, \tag{10.108}$$

where

$$\Phi_0\left(b^2,\xi\right) = \frac{8}{\pi}\int\limits_0^{\pi/2}\frac{q^2\cos(2q\xi)}{1+b^2q^2}dq. \tag{10.109}$$

Evaluating the integral (10.109) one obtains

$$\begin{aligned}\Phi_0\left(b^2,\xi\right) = \frac{4\sin\xi\pi}{\pi b^2\xi}+\frac{8}{\pi b^3}\left[\operatorname{Re}Si\left(\pi\xi+i\frac{2\xi}{b}\right)\sinh\left(\frac{2\xi}{b}\right)+\right.\\ \left.\left(\operatorname{Im}Ci\left(\pi\xi+i\frac{2\xi}{b}\right)-\operatorname{Im}Ci\left(i\frac{2\xi}{b}\right)\right)\cosh\left(\frac{2\xi}{b}\right)\right],\end{aligned} \tag{10.110}$$

where $Si(\cdot)$ and $\mathrm{Ci}(\cdot)$ are the familiar sine and cosine integrals, $Ci(y) = \gamma + \ln y + \int_0^y t^{-1}(\cos t - 1)\,dt$, $\mathrm{Si}(y) = \int_0^y t^{-1}\sin t\,dt$, γ is the Euler constant, $\gamma = 0.5772156649...$ ([2], Chapter 5).

Asymptotics of kernels of continuous approximation (10.93) $\Phi_{qc}(\xi)$ and improved continuous approximation (10.95) $\Phi_{qci}(\xi)$ are

$$\Phi_{qc}(\xi) \to \frac{12\pi}{12+\pi^2}\frac{\sin(\pi\xi)}{\xi} \approx 1.7238\frac{\sin(\pi\xi)}{\xi} \quad \text{at} \quad \xi\to\infty, \tag{10.111}$$

$$\Phi_{qc}(\xi) \to 4\frac{3\pi-6\sqrt{3}arctg\left(\frac{\pi}{2\sqrt{3}}\right)}{\pi} \approx 2.2532 \quad \text{at} \quad \xi\to 0, \tag{10.112}$$

$$\Phi_{qc}(\xi) \to -1.2123 \quad \text{at} \quad \xi\to 1, \tag{10.113}$$

$$\Phi_{qci}(\xi) \to \frac{4}{\pi}\frac{\sin(\pi\xi)}{\xi} \quad \text{at} \quad \xi\to\infty, \tag{10.114}$$

$$\Phi_{qci}(\xi) \to \frac{4\pi^2\left(\sqrt{\pi^2-4}-2arctg\left(\frac{\sqrt{\pi^2-4}}{2}\right)\right)}{\sqrt{\left(\pi^2-4\right)^3}} \approx 1.8360 \quad \text{at} \quad \xi\to 0, \tag{10.115}$$

$$\Phi_{qci}(\xi) \to -0.92036 \quad \text{at} \quad \xi \to 1. \tag{10.116}$$

For a good continuous approximation is the most important as a more accurate approximation at $\xi \to \infty$ the kernel of a integro-differential equation [258]. The best approximation gives the continuous approximation (10.95), which provides an accurate approximation up to the order $O(\xi^{-2})$.

10.12 DISPERSIVE WAVE PROPAGATION

Let us analyse dispersive wave propagation in the following mass-spring lattice

$$mY_{j\tau\tau}(\tau) = c[Y_{j-1}(\tau) - 2Y_j(\tau) + Y_{j+1}(\tau)]; j = 0, \pm 1, \pm 2, \pm 3, ..., \tag{10.117}$$

where $Y_j(\tau)$ is the displacement of the j-th material point from its static equilibrium position, τ is the time.

Let us introduce new variables $y_j(t) = Y_j(\tau)/h$, $t = \tau\sqrt{c/m}$, then Eq. (10.117) can be transformed to the following one:

$$y_{jtt}(t) = y_{j-1}(t) - 2y_j(t) + y_{j+1}(t); j = 0, \pm 1, \pm 2, \pm 3, ... \tag{10.118}$$

Suppose that at $t = 0$ all masses in the lattice are at rest with the sole exception of the mass numbered zero, which has been displaced by 1

$$y_j(0) = \delta_{j0}, y_{jt}(0) = 0, \tag{10.119}$$

where δ_{ji} is the Kronecker delta.

Solution of initial value problem (IVP) (10.4) was firstly obtained by Schrödinger [400] and can be written as follows [182]

$$y_m(t) = J_{2m}(2t) = \frac{h}{2\pi} \int_{-\pi/h}^{\pi/h} \cos\left[2t \sin\left(\frac{qh}{2}\right)\right] \cos(mhq)\, dq, \tag{10.120}$$

where $J_{2m}(2t)$ is the Bessel function of the first kind with integer order 2m, m denotes the m-th node.

Let us analyse continuous approximations. Let us introduce the continuous coordinate x scaled in such a way that $x = jh$ at the nodes of the lattice. We assume that $y_m(t)$ is discrete approximation to a continuous function $u(mh,t)$. We have

$$y_m(t) = u(mh,t). \tag{10.121}$$

Then initial-boundary value problem for continuous function $u(x,t)$ can be written as follows

$$u_{tt} - h^2 u_{xx} = 0, \tag{10.122}$$

Conditions (10.119) are transformed using Whittaker-Kotelnikov-Shannon interpolating function $\text{sinc}(x) = \sin(x)/x$ (see Chapter 10.10)

$$u(x,0) = \frac{h\sin(\pi x/h)}{\pi x}, \qquad u_t(x,0) = 0. \tag{10.123}$$

The solution of initial-boundary value problem (10.122), (10.123) may be expressed as follows

$$u(x,t) = \frac{h}{2\pi}\int_{-\pi/h}^{\pi/h} \cos(qht)\cos(qx)\,dq = \frac{1}{2\pi}\left[\frac{\sin(\pi(t-x/h))}{t-x/h} + \frac{\sin(\pi(t+x/h))}{t+x/h}\right]. \tag{10.124}$$

The continuous approximation qualitatively correctly describes the behavior of the discrete lattice far from the wave front. The construction of a continuous model that is completely adequate to the discrete case requires additional research.

10.13 GREEN'S FUNCTION

Dynamic Green's functions for nonlocal (10.3), local (10.18) and improved local (10.50) models are

$$G(x) = -\frac{1}{2\sqrt{mch}\omega}\sin\left[2\left(\arcsin\left(\frac{\omega}{2h}\sqrt{\frac{m}{c}}\right)\right)|x|\right] = -\frac{1}{ch}\cos\left[\left(\arcsin\left(\frac{\omega}{2h}\sqrt{\frac{m}{c}}\right)\right)|x|\right], \tag{10.125}$$

$$G(x) = -\frac{1}{2\sqrt{mch}\omega}\sin\left(\omega\sqrt{\frac{m}{c}}\frac{|x|}{h}\right), \tag{10.126}$$

$$G(x) = -\frac{1}{2\sqrt{mch}\omega}\sin\left(\frac{\omega}{h\sqrt{1-\alpha^2 m\omega^2 h^2/c}}\sqrt{\frac{m}{c}}|x|\right). \tag{10.127}$$

Kunin [258, 259] proposes to construct continuous approximation of the Green's functions of a discrete chain using by the first roots of the characteristic equation. Note that approximation (10.127) gives also satisfactory results.

Very detail analysis of Green's function for continuous approximations can be found in [116, 117].

10.14 DOUBLE- AND TRIPLE-DISPERSIVE EQUATIONS

Let us consider an infinite lattice consisting of identical particles of the mass m connected by massless springs of rigidity c. The equation of motion reads

$$m\frac{d^2u_n}{dT^2} + c\,(2u_n - u_{n+1} - u_{n-1}) = 0, \tag{10.128}$$

where u is the displacement; n is the index number of the particle, $n = 0, \pm 1, \pm 2, \ldots$; T is the time variable.

The time-harmonic wave is represented by the expression

$$u = A\exp(-ikn)\exp(i\Omega T), \tag{10.129}$$

where A is the amplitude; k is the non-dimensional wave number and Ω is the frequency.

Substituting (10.129) into (10.128), we obtain the dispersion relation as follows

$$\omega^2 = 4\sin^2(k/2), \tag{10.130}$$

where ω is the non-dimensional frequency, $\omega^2 = \Omega^2 m/c$.

A simple analysis of formula (10.130) shows that wave propagation is allowed in the frequency range $0 < \omega < 2$. At the long-wave limit, as $k \to 0$, the frequency vanishes, $\omega \to 0$, so no vibrations occur and the motion is simply a rigid body translation. The opposite limit, $k \to \pi$, describes a standing wave with zero group velocity and non-zero frequency, $\omega \to 2$. This regime can be considered as a "hidden" or "trapped" mode, in the sense that no energy is transmitted on macro scale, but on micro scale the lattice exhibits antyphase oscillations. As $\omega > 2$, the wave number k becomes complex. Then the spectrum of the lattice exhibits a band gap and the signal decays exponentially with an attenuation coefficient equal to the imaginary part of the wave number.

In the limiting cases discussed above, the dispersion relation (10.130) can be expressed by asymptotic relations

$$\omega^2 \sim k^2 \quad \text{at} \quad k \to 0, \tag{10.131}$$

$$\omega^2 \sim 4 - (\pi - k)^2 \quad \text{at} \quad k \to \pi. \tag{10.132}$$

Matching expressions (10.131) and (10.132) with the help of two-point PA (Chapter 10.7), we obtain

$$\omega^2 \approx \frac{a_1 k^2 + a_2 k^4}{1 + a_3 k^2 + a_4 k^4}. \tag{10.133}$$

Here the coefficients a_1, ... ,a_4 are determined in such a way that the leading terms of the power series expansions of PA (10.133) at $k \to 0$ and $k \to \pi$ must coincide with formulas (10.131) and (10.132) up to $O(k^2)$, $O\left[(\pi - k)^2\right]$ accordingly. Using this condition, we derive

$$a_1 = 1, \quad a_2 = \frac{64 - \pi^4}{\pi^6}, \quad a_3 = \frac{\pi^2 - 8}{4\pi^2}, \quad a_4 = \frac{64 + 4\pi^2 - \pi^4}{4\pi^6}. \tag{10.134}$$

In Fig. 10.6, formula (10.133) is compared with the exact dispersion relation (10.130). We note that the numerical results are essentially indistinguishable.

For any real k, the denominator of the derived PA does not equal zero. Therefore, expression (10.133) can be rewritten as follows

$$\omega^2 - a_1 k^2 - a_2 k^4 + a_3 \omega^2 k^2 + a_4 \omega^2 k^4 = 0. \tag{10.135}$$

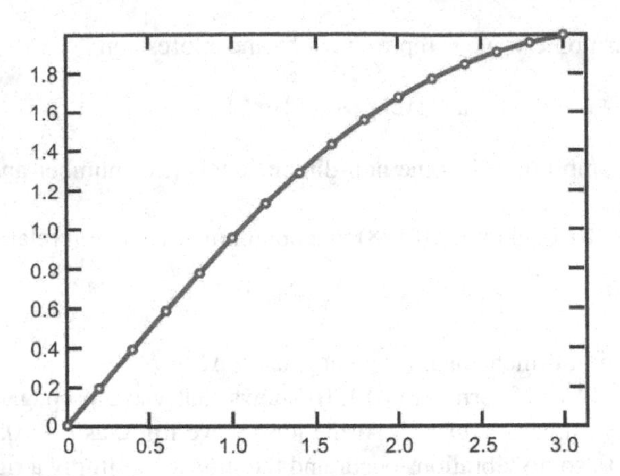

Figure 10.6 Dispersion relation of the lattice. Solid curves - exact solution (10.130), circles - PA (10.133).

Let us introduce the continuous coordinate X scaled in such a way that $X = nl$ at the nodes of the lattice, where l is the distance between the particles.

Using non-dimensional variables, the time-harmonic solution (10.129) reads

$$u = A\exp\left(-ikx\right)\exp\left(i\omega t\right), \tag{10.136}$$

where $x = X/l$, $t = T(c/m)^{1/2}$, $k = 2\pi l/L$, L is the wave length.

Then one can easily show

$$\begin{gathered}\frac{\partial^2 u}{\partial x^2} = -k^2 u, \quad \frac{\partial^4 u}{\partial x^4} = k^4 u, \quad \frac{\partial^2 u}{\partial t^2} = -\omega^2 u,\\ \frac{\partial^4 u}{\partial x^2 \partial t^2} = \omega^2 k^2 u, \quad \frac{\partial^6 u}{\partial x^4 \partial t^2} = -\omega^2 k^4 u.\end{gathered} \tag{10.137}$$

The dispersion equation (10.135) allows us to obtain a PDE that describes wave propagation for all wave numbers $0 \leq k \leq \pi$:

$$a_1 \frac{\partial^2 u}{\partial x^2} - \frac{\partial^2 u}{\partial t^2} - a_2 \frac{\partial^4 u}{\partial x^4} + a_3 \frac{\partial^4 u}{\partial x^2 \partial t^2} - a_4 \frac{\partial^6 u}{\partial x^4 \partial t^2} = 0. \tag{10.138}$$

Equation (10.138) includes three dispersive terms and may be considered as a generalization of double-dispersive equations, which were employed by many authors to simulate elastic waves in waveguides with a free lateral surface (see, for example, [369] and references therein). In the theory of waves in structured solids, double- and triple-dispersion equations can be obtained by imposing some additional internal degrees of freedom on the system (see, for example, a review [74]). It should be noted that such gradient models include a number of phenomenological parameters, which for real materials remain usually unknown. In contrary, the approach proposed in this section allows us to evaluate all the coefficients of Eq. (10.138) theoretically basing on the information about the internal structure of the medium and its properties. The developed macroscopic model is able to describe the long-wave case and, at the same time, it is valid in a high-frequency domain in the vicinity of the stop band threshold.

10.15 TODA LATTICE

Let nonlinear springs in 1D problem have potential energy $U(r_n)$, where $r_n = y_{n+1} - y_n$. The chain dynamics is governed by the following infinite system of PDEs [401]:

$$\frac{\partial^2 r_n}{\partial t^2} = \frac{\partial U(r_{n+1})}{\partial r_{n+1}} - 2\frac{\partial U(r_n)}{\partial r_n} + \frac{\partial U(r_{n-1})}{\partial r_{n-1}}, \tag{10.139}$$

where $r_n = y_{n+1} - y_n$.

Assume that solution to the (10.139) is a travelling wave, $r_n = R(z)$, $z = n - vt$. Then Eq. (10.139) is cast into the following form:

$$v^2 R''(z) = U'(R(z-1)) - 2U'(R(z)) + U'(R(z+1)), \tag{10.140}$$

where $(...)' = \frac{d(...)}{dz}$.

Introducing notation $\Phi(z) = U'(R(z))$ and applying the Fourier transformation to Eq. (10.140) one gets

$$\bar{R}(q) = \int_{-\infty}^{\infty} R(z)\exp(-iqz)dz; \bar{\Phi}(q) = \int_{-\infty}^{\infty} \Phi(z)\exp(-iqz)dz. \tag{10.141}$$

From (10.140) one obtains

$$v^2 \bar{R}(q) = f(q)\bar{\Phi}(q), \tag{10.142}$$

where $f(q) = 4\frac{\sin^2(q/2)}{q^2}$.

Term $f(q)$ can be approximated by PA or two-point PA in the following way

$$f(q) = \frac{1}{1+q^2/12} \quad \text{or} \quad f(q) = \frac{1}{1+\alpha^2 q^2}, \tag{10.143}$$

where $\alpha^2 = 0.25 - \pi^{-2}$.

In result one obtains equation of continuous approximation

$$v^2(1 - \frac{1}{12}\frac{d^2}{dz^2})R(z) = U'(R(z)) \tag{10.144}$$

or

$$v^2\left(1 - a^2\frac{d^2}{dz^2}\right)R(z) = U'(R(z)). \tag{10.145}$$

We show also a simultaneous application of PA with a perturbation procedure using an example of the Toda lattice [401]:

$$m\frac{d^2 y_n}{dt^2} = a[\exp(b(y_{n-1} - y_n)) - \exp(b(y_n - y_{n+1}))]. \tag{10.146}$$

System of ODEs (10.146) can be rewritten as pseudo-differential equation:

$$m\frac{\partial^2 y}{\partial t^2} - 4a\left[\sinh\left(\frac{1}{2}\frac{\partial}{\partial x}\right)\right]^2 \exp(-by) = 0. \tag{10.147}$$

Equation (10.147) can be approximately factorized [258]

$$\left[\sqrt{\frac{m}{ab}}\frac{\partial}{\partial t} - 2\sinh\left(\frac{1}{2}\frac{\partial}{\partial x}\right) + \frac{b}{2}y\frac{\partial}{\partial x}\right]\left[\sqrt{\frac{m}{ab}}\frac{\partial}{\partial t} + 2\sinh\left(\frac{1}{2}\frac{\partial}{\partial x}\right) - \frac{b}{2}y\frac{\partial}{\partial x}\right]y = 0. \tag{10.148}$$

For wave spreading in right direction one obtains

$$\left[\sqrt{\frac{m}{ah}}\frac{\partial}{\partial t} + 2\sinh\left(\frac{1}{2}\frac{\partial}{\partial x}\right) - \frac{h}{2}y\frac{\partial}{\partial x}\right]y = 0. \tag{10.149}$$

Formally expanding function $\sinh\left(\frac{1}{2}\frac{\partial}{\partial x}\right)$ in a Maclaurin series up to the terms of the third order one obtains from (10.149)

$$\sqrt{\frac{m}{ab}}\frac{\partial y}{\partial t} + \frac{\partial y}{\partial x} + \frac{1}{24}\frac{\partial^3 y}{\partial x^3} - \frac{b}{2}y\frac{\partial y}{\partial x} = 0. \tag{10.150}$$

Using variables transform $t_1 = \sqrt{abt/m}, y_1 = 0.5yb,$ the Kortewegde Vries Eq. [258, 401] of the following form is obtained:

$$\frac{\partial y_1}{\partial t_1} + \frac{\partial y_1}{\partial x} + \frac{1}{24}\frac{\partial^3 y_1}{\partial x^3} - y_1\frac{\partial y_1}{\partial x} = 0. \tag{10.151}$$

Let us construct regularized long-wave equation, described in references [1, 72, 357]. Using PA one gets

$$1 + \frac{1}{24}\frac{\partial^2}{\partial x^2} \sim \left(1 - \frac{1}{24}\frac{\partial^2}{\partial x^2}\right)^{-1}. \tag{10.152}$$

Then, using (10.152) one obtains from Eq. (10.151) (up to the highest terms) the following regularized long-wave equation:

$$\left(1 - \frac{1}{24}\frac{\partial^2}{\partial x^2}\right)\frac{\partial y_1}{\partial t_1} + \frac{\partial y_1}{\partial x} - y_1\frac{\partial y_1}{\partial x} = 0. \tag{10.153}$$

On the other hand, there are nonlinear lattices with a special type nonlinearities allowing to achieve exact solutions (Toda, Ablowitz-Ladik, Langmuir, Calogero, etc. [72, 135, 139, 184, 305, 357]) in the case of infinite lattices or in the case of boundary conditions. In many cases non-integrable systems like a discrete variant of sine-Gordon equation possesses soliton-like solution.

Occurrence of exact solutions of a discrete system allows applying the following approach: fast changeable solution part is constructed using a discrete model, whereas a continuous approximation is applied for slow components. The latter approach is used by Maslov and Omel'yanov [305], who analyzed Toda lattice with variable coefficients

$$m_k \frac{d^2 y_k}{dt^2} = -a_k[\exp b_k(y_k - y_{k-1}) - 1] + \\ a_{k+1}[\exp(b_{k+1}(y_{k+1} - y_k) - 1], \quad k = 0, \pm 1, \pm 2, \ldots \tag{10.154}$$

They construct soliton solutions in the following way: rapidly changing part of soliton is constructed using Toda lattice with constant coefficients, and for slowly part of solution continuous approximation is used. Then these solutions are matched.

10.16 DISCRETE KINKS

We consider the nonlinear Klein-Gordon equation

$$C_1 \frac{\partial^2 z}{\partial y^2} - \frac{\partial^2 z}{\partial t^2} - \frac{\partial V(z)}{\partial z} = 0. \tag{10.155}$$

Here $V(z)$ is the potential having at least two degenerate minimums.

The travelling wave solution of Eq. (10.155) describes the following equation:

$$C \frac{\partial^2 z}{\partial x^2} - \frac{\partial V(z)}{\partial z} = 0, \tag{10.156}$$

where $x = y - ct; z(x) \equiv z(y - ct)$; $C = C_1 - c^2$.

The reason for the transition from the PDE (10.155) to the ODE (10.156) is the invariance of the equations to Lorentz transformations.

Discrete analog of Eq. (10.156) is given by

$$-C(z_{n-1} - 2z_n + z_{n+1}) + \left.\frac{\partial V(z)}{\partial z}\right|_{z=z_n} = 0, \tag{10.157}$$

where $z_n = z(n)$; without loss of generality the mesh width is taken as 1.

The estimation of Peierls-Nabarro barrier value for the kinks in the Eq. (10.157) represents a problem of great importance. The following approach is proposed in [173,208]. The energy is estimated based on a discrete model and the discrete values of variables are replaced with the values calculated within the framework of the continuous model. By virtue of the fact that the approximation accuracy is small, in [172] it was proposed perturbation approach basing on the averaging. As a result for the Φ^4 model with $V(z) = \frac{1}{4}(z^2 - 1)^2$ potential, the following equation is derived

$$C \frac{d^2 z}{dx^2} - z(z^2 - 1)\left(1 + \frac{1}{12C} - \frac{1}{4C} z^2\right) = 0, \tag{10.158}$$

and for sine-Gordon equation with $V(z) = 1 - \cos(z)$ potential the improved equation takes the form of double sine-Gordon equation

$$C \frac{d^2 z}{dx^2} - \sin(z) + \frac{1}{24C} \sin(2z) = 0. \tag{10.159}$$

Solution of Eq. (10.158) is given by [172]

$$z(x) = \frac{\tanh\left(\sqrt{1-\frac{1}{6C}}\frac{x}{\sqrt{2C}}\right)}{1-\frac{1}{6C}\operatorname{sech}^2\left(\sqrt{1-\frac{1}{6C}}\frac{x}{\sqrt{2C}}\right)}, \tag{10.160}$$

for Eq. (10.159) it is:

$$z(x) = 2\pi - 2\arctan\left[\sqrt{1-\frac{1}{12C}}\operatorname{cosech}\left(\sqrt{1-\frac{1}{12C}}\frac{x}{\sqrt{C}}\right)\right], \quad x \geq 0, \tag{10.161}$$

$$z(x) = -2\arctan\left[\sqrt{1-\frac{1}{12C}}\operatorname{cosech}\left(\sqrt{1-\frac{1}{12C}}\frac{x}{\sqrt{C}}\right)\right], \quad x \leq 0. \tag{10.162}$$

The continuous approximation for the discrete model (10.157) yields

$$C\frac{d^2z}{dx^2} - \left(1 - \frac{1}{12}\frac{d^2}{dx^2}\right)\frac{\partial V(z)}{\partial z} = 0. \tag{10.163}$$

For the Φ^4 model one obtains

$$C\frac{d^2z}{dx^2}\left(1 - \frac{1}{12C} + \frac{1}{4C}z^2\right) - z(z^2-1) + \frac{1}{2}z\left(\frac{dz}{dx}\right)^2 = 0. \tag{10.164}$$

If $C >> 1$, the coefficient $1 - \frac{1}{12C} + \frac{1}{4C}z^2$ can be transformed using PA:

$$1 - \frac{1}{12C} + \frac{1}{4C}z^2 \sim \left(1 + \frac{1}{12C} - \frac{1}{4C}z^2\right)^{-1}. \tag{10.165}$$

Using the Eq. (10.165) and omitting the higher-order terms, the Eq. (10.164) is replaced by the following one

$$C\frac{\partial^2 z}{\partial x^2} - z(z^2-1)\left(1 + \frac{1}{12C} - \frac{1}{4C}z^2\right) + \left\{\frac{1}{2}z\left(\frac{\partial z}{\partial x}\right)^2\right\} = 0. \tag{10.166}$$

Let us estimate an influence of the braced terms. In the neighborhood of zero the linear terms have a dominant role. Using the solution (10.160), we find

$$\left(\frac{\partial z}{\partial x}\right)^2 \sim \frac{1}{2C\operatorname{sech}^4\left(\frac{x}{\sqrt{2C}}\right)}. \tag{10.167}$$

Therefore the braced term can be neglected and Eq. (10.166) is the same as Eq. (10.163) obtained in [355] by distinctly other means. In the similar way for the sine-Gordon equation the continuous approximation can be reduced to Eq. (10.164).

The equations obtained on the basis of the continuous approximation are:

$$C\frac{d^2z}{dx^2} - z(z^2-1)\left(1+\frac{\alpha^2}{C} - \frac{3\alpha^2}{C}z^2\right) = 0, \tag{10.168}$$

$$C\frac{d^2z}{dx^2} - \sin(z) + \frac{\alpha^2}{2C}\sin(2z) = 0. \tag{10.169}$$

Solution of Eq. (10.168) is

$$z(x) = \frac{\tanh\left(\sqrt{1-\frac{2\alpha^2}{C}}\frac{x}{\sqrt{2C}}\right)}{1-\frac{2\alpha^2}{C}\operatorname{sech}^2\left(\sqrt{1-\frac{2\alpha^2}{C}}\frac{x}{\sqrt{2C}}\right)}, \tag{10.170}$$

whereas for Eq. (10.169) we have

$$z(x) = 2\pi - 2\arctan\left[\sqrt{1-\frac{\alpha^2}{C}}\operatorname{cosech}\left(\sqrt{1-\frac{\alpha^2}{C}}\frac{x}{\sqrt{C}}\right)\right], \quad x \geq 0, \tag{10.171}$$

$$z(x) = -2\arctan\left[\sqrt{1-\frac{\alpha^2}{C}}\operatorname{cosech}\left(\sqrt{1-\frac{\alpha^2}{C}}\frac{x}{\sqrt{C}}\right)\right], \quad x \leq 0. \tag{10.172}$$

Solutions (10.170), (10.172) make possible to obtaln the values for the Peierls-Nabarro barrier with the accuracy better than the accuracy received when using the solutlons (10.160)- (10.162).

10.17 CONTINUALIZATION OF β-FPU LATTICE

We study vibrations of a monatomic lattice consisting of identical particles connected by cubically nonlinear elastic springs (Fig. 10.7). Neighbouring particles are depicted by different colors illustrating that they can move with different amplitudes, although the masses of the particles are the same. The length of the lattice is L and the distance between the particles is l. Every particle is denoted by a number n, $n = 0,1,2,\ldots,N$, where $N = L/l$. The total number of the particles is $N+1$.

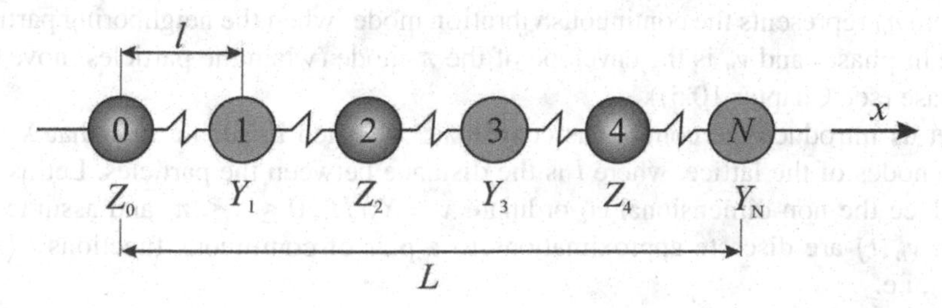

Figure 10.7 The lattice under consideration.

We seek a continuous model capable to describe both acoustic and optical forms of motion. Let us employ different notations Y and Z for the displacements of black and white particles accordingly. Vasiliev et al. [447] called this approach as a two-field model (see also [371]). It is also consistent with the continualization procedures proposed by Zabusky and Deem [467]. Below we follow paper [27].

The governing equations of motion read

$$\begin{aligned} & m\frac{d^2Y_n}{dT^2} + c_1(2Y_n - Z_{n-1} - Z_{n+1}) + \\ & c_3(Y_n - Z_{n-1})^3 + c_3(Y_n - Z_{n+1})^3 = 0, \\ & m\frac{d^2Z_{n+1}}{dT^2} + c_1(2Z_{n+1} - Y_n - Y_{n+2}) + \\ & c_3(Z_{n+1} - Y_n)^3 + c_3(Z_{n+1} - Y_{n+2})^3 = 0. \end{aligned} \tag{10.173}$$

Here T is time, m is the mass of the particle, c_1 and c_3 are the elastic coefficients of the spring.

Systems of this type are called β-Fermi-Pasta-Ulam (β-FPU) lattice [456].

In a dimensionless form, equations (10.173) can be written as follows

$$\begin{aligned} & \frac{d^2y_n}{dt^2} + (2y_n - z_{n-1} - z_{n+1}) + \varepsilon(y_n - z_{n-1})^3 + \varepsilon(y_n - z_{n+1})^3 = 0, \\ & \frac{d^2z_{n+1}}{dt^2} + (2z_{n+1} - y_n - y_{n+2}) + \varepsilon(z_{n+1} - y_n)^3 + \varepsilon(z_{n+1} - y_{n+2})^3 = 0, \end{aligned} \tag{10.174}$$

where $t = T\sqrt{c_1/m}$; $y_n = Y_n/A$; $z_n = Z_n/A$; A is the vibration amplitude; $\varepsilon = A^2c_3/c_1$. We suppose that neighboring particles do not collide each other, so $|A| < l/2$.

In order to separate low- and high-frequency motions of the lattice, let us introduce the acoustic u_n and optical v_n fields as follows

$$\begin{aligned} & u_{n-1} = \frac{y_n + z_{n-1}}{2}, \quad u_n = \frac{y_n + z_{n+1}}{2}, \quad u_{n+1} = \frac{y_{n+2} + z_{n+1}}{2}, \\ & v_{n-1} = \frac{y_n - z_{n-1}}{2}, \quad v_n = \frac{y_n - z_{n+1}}{2}, \quad v_{n+1} = \frac{y_{n+2} - z_{n+1}}{2}. \end{aligned} \tag{10.175}$$

Here u_n represents the continuous vibration mode (when the neighboring particles move in-phase) and v_n is the envelope of the π-mode (when the particles move out-of-phase (see Chapter 10.5)).

Let us introduce the continuous coordinate X scaled in such a way that $X = nl$ at the nodes of the lattice, where l is the distance between the particles. Let us also introduce the non-dimensional co-ordinate $x = X\pi/L$, $0 \le x \le \pi$, and assume that $u_n(t)$, $v_n(t)$ are discrete approximations to a pair of continuous functions $u(x,t)$, $v(x,t)$, i.e.

$$u_n(t) = u(x,t)|_{x=n\eta}, \qquad v_n(t) = v(x,t)|_{x=n\eta}, \tag{10.176}$$

where $\eta = \pi l/L = \pi/N$.

We seek a long-wave solution, whose spatial period (or the period of the envelope) is much larger then the distance between the particles and $\eta \ll 1$. Using a Taylor series expansion, one can write

$$u_{n\pm 1}(t) = u(x\pm\eta,t) = \sum_{i=0}^{\infty} \frac{(\pm\eta)^i}{i!}\frac{\partial^i u(x,t)}{\partial x^i},$$
$$v_{n\pm 1}(t) = v(x\pm\eta,t) = \sum_{i=0}^{\infty} \frac{(\pm\eta)^i}{i!}\frac{\partial^i v(x,t)}{\partial x^i}. \tag{10.177}$$

Adding together and subtracting from each other equations (10.174), and taking into account expressions (10.175)–(10.177), we obtain the macroscopic dynamic equations for u and v. Within an error $O(\eta^6)$ they read

$$\frac{\partial^2 u}{\partial t^2} - \eta^2\frac{\partial^2 u}{\partial x^2} - 12\varepsilon\eta^2 v\left(v\frac{\partial^2 u}{\partial x^2} + 2\frac{\partial u}{\partial x}\frac{\partial v}{\partial x}\right) - \frac{1}{12}\eta^4\frac{\partial^4 u}{\partial x^4}$$
$$\varepsilon\eta^4\left\{v^2\frac{\partial^4 u}{\partial x^4} + 3\frac{\partial^2 u}{\partial x^2}\left[\left(\frac{\partial u}{\partial x}\right)^2 + \left(\frac{\partial v}{\partial x}\right)^2\right] + 4v\left(\frac{\partial u}{\partial x}\frac{\partial^3 v}{\partial x^3} + \frac{\partial^3 u}{\partial x^3}\frac{\partial v}{\partial x}\right) + \right. \tag{10.178}$$
$$\left. 6v\frac{\partial^2 u}{\partial x^2}\frac{\partial^2 v}{\partial x^2} + 6\frac{\partial u}{\partial x}\frac{\partial v}{\partial x}\frac{\partial^2 v}{\partial x^2}\right\} = 0,$$

$$\frac{\partial^2 v}{\partial t^2} + 4v + 16\varepsilon v^3 + \eta^2\frac{\partial^2 v}{\partial x^2} + 12\varepsilon\eta^2 v\left[\left(\frac{\partial u}{\partial x}\right)^2 + \left(\frac{\partial v}{\partial x}\right)^2 + v\frac{\partial^2 v}{\partial x^2}\right] +$$
$$\frac{1}{12}\eta^4\frac{\partial^4 v}{\partial x^4} + \varepsilon\eta^4\left\{v^2\frac{\partial^4 v}{\partial x^4} + 3\frac{\partial^2 v}{\partial x^2}\left[\left(\frac{\partial u}{\partial x}\right)^2 + \left(\frac{\partial v}{\partial x}\right)^2\right] + \right.$$
$$\left. 3v\left[\left(\frac{\partial^2 u}{\partial x^2}\right)^2 + \left(\frac{\partial^2 v}{\partial x^2}\right)^2\right] + 4v\left[\frac{\partial u}{\partial x}\frac{\partial^3 u}{\partial x^3} + \frac{\partial v}{\partial x}\frac{\partial^3 v}{\partial x^3}\right] + 6\frac{\partial u}{\partial x}\frac{\partial^2 u}{\partial x^2}\frac{\partial v}{\partial x}\right\} = 0. \tag{10.179}$$

It should be noted that acoustic and optical motions are coupled through the nonlinear terms in (10.178), (10.179). The variables η and ε are natural small parameters characterizing, accordingly, the rate of heterogeneity and the rate of non-linearity; $\eta \ll 1$, $|\varepsilon| \ll 1$.

In a general case, the asymptotic behavior of the model (10.178), (10.173) can depend on a scaling relation between the small parameters η, ε. Imposing $\varepsilon \sim \eta^\alpha$, one can obtain different asymptotic solutions for different α.

Equations (10.178), (10.173) widely used for infinite lattices [371,467]. Their use to describe the dynamics of a finite chain requires further investigation, because, one way or another, the system has to be discretized. It may be more appropriate to use the normal modes of linear oscillations of a discrete system from the very beginning, without going over to a continuous representation.

10.18 ACOUSTIC BRANCH OF α-FPU LATTICE

Classical continualization of α-FPU lattice [456] or homogenization of a periodically heterogeneous rod [26] leads to the same PDE

$$E_1 \frac{\partial^2 u}{\partial x^2} + E_2 \frac{\partial u}{\partial x}\frac{\partial^2 u}{\partial x^2} + E_3 \frac{\partial^4 u}{\partial x^4} = \rho \frac{\partial^2 u}{\partial t^2}. \tag{10.180}$$

Coefficients of Eq. (10.180) have different values for rod and lattice, but in both cases they take into account nonlinearity, dispersion and dissipation effects. It should be noted that dissipation restrains energy transfers between the vibration modes and on a large time scale modes coupling vanishes. In that sense, dispersion and dissipation acting in a nonlinear system may lead to qualitatively similar physical consequences (Zabusky and Kruskal encountered this analogy analysing the FPU problem, see [456]).

Therefore, the main conclusions made in [21] for the problem of nonlinear vibrations of a composite rod of finite length can be carried over to the case of acoustic branch of α - FPU lattice.

If the distance between particles relatively small in comparison to the amplitude of the vibrations, the effect of internal resonance takes place. It results in periodic energy transfers between different modes and in a modulation of their amplitudes. The resonant modes are coupled in the main approximation, so the truncation to the modes having non-zero initial energy is not possible.

If the distance between particles increases, the dispersion supresses the influence of nonlinearity. The energy transfers to higher-order modes are restricted, so truncation to only a few leading modes can provide a reasonable approach. The further increase of dispersion eliminates the energy exchange between the modes and the internal resonances become negligible.

10.19 ANTI-CONTINUUM LIMIT

Consider the discrete nonlinear Schrödinger equation in the form

$$i\sigma_{jt}(t) = \varepsilon(\sigma_{j+1} - 2\sigma_j + \sigma_{j-1}) + |\sigma_j|^{2p}\sigma_j, \tag{10.181}$$

$p \in N$ and ε define the coupling constant.

The anti-continuum limit [44] corresponds to $\varepsilon = 0$, in which case the Eq. (10.181) becomes an infinite system of uncoupled ODEs.

10.20 2D LATTICE

In order to analyze the 2D case, we use 9-cell square lattice (Fig. 10.8) [316, 361]. The central particle is supposed to interact with eight neighbors in the lattice. The mass centers of four of them are on horizontal and vertical lines, while the mass centers of the other four neighboring particles lie along diagonals. Interactions between the neighboring particles are modeled by elastic springs of three types.

Horizontal and vertical springs with rigidity c_1 define interaction forces of extension/compression of the material. Governing equations of motion are [361]:

$$\begin{aligned}
m\ddot{u}_{m,n} = {} & c_1(u_{m,n-1} - 2u_{m,n} + u_{m,n+1}) + c_2(u_{m-1,n} - 2u_{m,n} + u_{m+1,n}) + \\
& 0.5c_0(u_{m-1,n-1} + u_{m+1,n-1} + u_{m-1,n+1} + u_{m+1,n+1} + \\
& v_{m-1,n-1} - v_{m+1,n-1} - v_{m-1,n+1} + v_{m+1,n+1} - 4u_{m,n}), \\
m\ddot{v}_{m,n} = {} & c_1(v_{m,n-1} - 2v_{m,n} + v_{m,n+1}) + c_2(v_{m-1,n} - 2v_{m,n} + v_{m+1,n}) + \\
& 0.5c_0(v_{m-1,n-1} + v_{m+1,n-1} + v_{m-1,n+1} + v_{m+1,n+1} + \\
& u_{m-1,n-1} - u_{m+1,n-1} - u_{m-1,n+1} + u_{m+1,n+1} - 4v_{m,n}).
\end{aligned} \tag{10.182}$$

Here $u_{i.j}, v_{i.j}$ is the displacement vector for a particle situated at point (x_i, y_j), $x_i = ih, y_j = jh$, c_0, c_1, c_2 are the stiffnesses of diagonal longitudinal, axial longitudinal and axial shear springs, respectively.

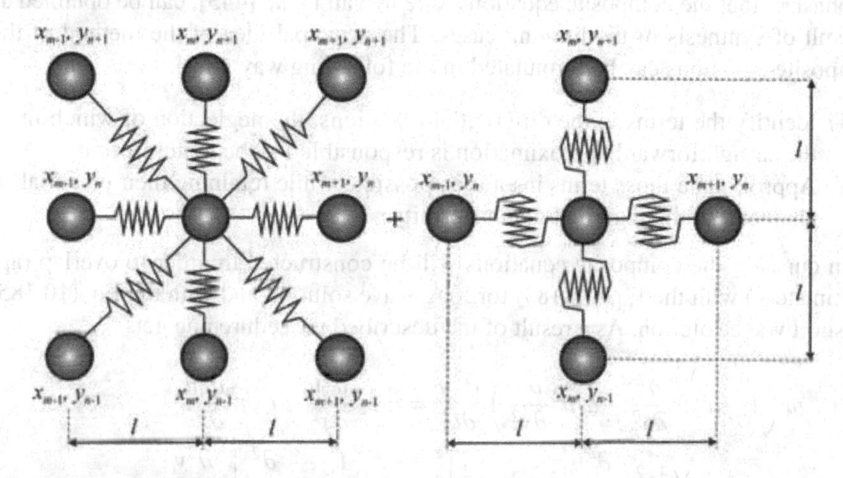

Figure 10.8 2D lattice.

The standard continualization procedure for Eq. (10.182) involves introducing a continuous displacement field $u(x,y), v(x,y)$ such that $u(x_m, y_n) = u_{m,n}; v(x_m, y_n) = v_{m,n}$, and expanding $u_{m\pm1,n\pm1}, v_{m\pm1,n\pm1}$ into Maclaurin series around $u_{m,n}, v_{m,n}$. Second-order continuous theory in respect to the small parameter h is [361]:

$$\begin{aligned}
m\frac{\partial^2 u}{\partial t^2} &= c_1 h^2 \frac{\partial^2 u}{\partial x^2} + c_2 h^2 \frac{\partial^2 u}{\partial y^2} + 2c_0 h^2 \frac{\partial^2 v}{\partial x \partial y}, \\
m\frac{\partial^2 v}{\partial t^2} &= c_1 h^2 \frac{\partial^2 v}{\partial x^2} + c_2 h^2 \frac{\partial^2 v}{\partial y^2} + 2c_0 h^2 \frac{\partial^2 u}{\partial x \partial y}.
\end{aligned} \tag{10.183}$$

Naturally, one can construct equations of higher order, but, as it is shown in [361], it leads to the principal difficulties.

Let us construct semidiscrete approximation.

Using staggered transformations

$$u_k = (-1)^k u, v_k = (-1)^k v, \tag{10.184}$$

one obtains:

$$\begin{aligned} m\frac{\partial^2 u}{\partial t^2} &= 4\left(c_1\frac{\partial^2 u}{\partial x^2} + c_2\frac{\partial^2 u}{\partial y^2}\right), \\ m\frac{\partial^2 v}{\partial t^2} &= 4\left(c_1\frac{\partial^2 v}{\partial x^2} + c_2\frac{\partial^2 v}{\partial y^2}\right). \end{aligned} \tag{10.185}$$

As in 1D case (see Eq. (10.32)), Eq. (10.185) do not contain parameter h^2. The existence of the continuous approximations (10.183) and (10.185) give a possibility to construct the composite equations, which is uniformly suitable in the whole interval of the frequencies and the oscillation forms of the 2D lattice of masses. Let us emphasize, that the composite equations, due to Van Dyke [443], can be obtained as a result of synthesis of the limiting cases. The principal idea of the method of the composite equations can be formulated in the following way [443]:

(i) Identify the terms in the differential equations, the neglection of which in the straightforward approximation is responsible for the nonuniformity.
(ii) Approximate those terms insofar as possible while retaining their essential character in the region of nonuniformity.

In our case the composite equations will be constructed in order to overlap (approximately) with the Eq. (10.183) for long wave solution and with the Eq. (10.185) for short wave solution. As a result of the described procedure one gets

$$\begin{aligned} &m\left(1 - a^2h^2\frac{\partial^2}{\partial x^2} - a^2h^2\frac{\partial^2}{\partial y^2}\right)\frac{\partial^2 u}{\partial t^2} = c_1h^2\frac{\partial^2 u}{\partial x^2} + c_2h^2\frac{\partial^2 u}{\partial y^2} - \\ &(c_1+c_2)h^4\gamma^2\frac{\partial^4 u}{\partial x^2\partial y^2} + 2c_0h^2\left[1 + \frac{1}{4(c_1+c_2)}\frac{\partial^2}{\partial t^2}\right]\frac{\partial^2 v}{\partial x\partial y}, \\ &m\left(1 - a^2h^2\frac{\partial^2}{\partial x^2} - a^2h^2\frac{\partial^2}{\partial y^2}\right)\frac{\partial^2 v}{\partial t^2} = c_1h^2\frac{\partial^2 v}{\partial x^2} + c_2h^2\frac{\partial^2 v}{\partial y^2} - \\ &(c_1+c_2)h^4\gamma^2\frac{\partial^4 v}{\partial x^2\partial y^2} + 2c_0h^2\left[1 + \frac{1}{4(c_1+c_2)}\frac{\partial^2}{\partial t^2}\right]\frac{\partial^2 u}{\partial x\partial y}. \end{aligned} \tag{10.186}$$

Here $a^2 = 0.25 - \pi^{-2}$, $\gamma^2 = (4 - \pi^2 + 8\pi^2\alpha^2)/(4\pi^2)$.

For 1D case from Eq. (10.186) one obtains Eq. (10.50) for u and the same equation for v. For small variability in spatial and time variables Eq. (10.186) can be approximated by Eq. (10.183), for large variability in spatial and time variables Eqs. (10.186) can be approximated by Eq. (10.185).

For further results see [18, 25, 32].

10.21 MOLECULAR DYNAMICS SIMULATIONS AND CONTINUALIZATION: HANDSHAKE

Molecular dynamics (MD) simulations have become prominent as a tool for elucidating complex physical phenomena, such as solid fracture and plasticity. However, the length and time scales probed using MD are still fairly limited. To overcome this problem, it is possible to use MD only in localized regions in which the atomic-scale dynamics is important, and a continuous simulation method (for example, FEM) everywhere else [452]. Then the problem of coupling MD and continuum mechanics simulations appears. There have been a large number of contributions on the subject, see, e.g., [65, 66] and references therein for a small representative example. From a physical point of view, the problem is related to coupling between either, nonlocal and local continuum mechanics [287], or discrete and continuum models (atomistic-to-continuum coupling) [58,346,360,463], or, possibly, local and global models [64]. As a rule, the proposed approach relies on the so-called bridging (overlap) domains ("gluing zones", "handshake region"), see, e.g., [212]. As noted in [66], these methods are closely related to the overlapping Schwarz methods [422] stemming from the classical alternating Schwarz algorithm. The main idea of these methods can be described as follows: part of the system in the vicinity of the defect or boundary is considered within the framework of discrete media, whereas the rest is assumed to be homogenized.

Within the overlap domain, the discrete and homogenized solutions should match in some sense. For example, the discrete solutions may be interpolated (or continuous solutions could be discretized). "In this transition region, approximations are made such as treating finite element nodes as atoms, or vice versa, to accommodate the incompatibility between a nonlocal atomistic description and a local finite element description" [346]. Linear interpolation was used in [58] for addressing this issue, however, other types of interpolation could also be applied. Gluing of local and global solutions may also rely on energy method [66]; another option is the Lagrange multiplier method or augmented Lagrangian method [463].

An alternative approach involves formulation of certain artificial boundary conditions on the interface between discrete and continuous domains. As mentioned in [151], numerical simulations of crystal defects are necessarily restricted to finite computational domains, supplying artificial boundary conditions that emulate the effect of embedding the defect in an effectively infinite crystalline environment.

Let us finally discuss problems closely connected with brittle fracture of elastic solids [416–421]. Observe that a continuous model, which does not include material structure, is not suitable since any material crack occurs on the material structure level. This is a reason why some of the essential properties of damages of the classical continuous model theory are not exhibited. For instance, in a discrete medium the propagated waves transport part of the energy from the elastic body into the crack edges. This is why application of nonlocal theories during explosion process modeling (for instance for the composite materials) looks very promising. As it is mentioned in [256], "The continuous development of advanced materials ensures that a "one size fits all" approach will no longer serve the engineering community in terms of predicting and preventing fatigue failures and reducing their associated costs".

10.22 CONTINUALIZATION AND DISCRETIZATION

The connection between these operations is deeply nontrivial, which is shown in detail in the Section 11. Here we dwell on several issues directly related to the approaches described above. In the theory of finite differences the method of differential approximation is widely used [407]. It may be described as follows. Let us compare local operator

$$-h^2\frac{\partial^2}{\partial x^2} \tag{10.187}$$

and nonlocal operator

$$\sin^2\left(-\frac{ih}{2}\frac{\partial}{\partial x}\right). \tag{10.188}$$

In previous sections we approximated nonlocal operator by infinite number of local ones:

$$4\sin^2\left(-\frac{ih}{2}\frac{\partial}{\partial x}\right)=-h^2\frac{\partial^2}{\partial x^2}+\frac{h^4}{12}\frac{\partial^4}{\partial x^4}-\frac{h^6}{360}\frac{\partial^6}{\partial x^6}+\dots \tag{10.189}$$

Using (10.189) one obtains

$$-h^2\frac{\partial^2}{\partial x^2}=4\sin^2\left(-\frac{ih}{2}\frac{\partial}{\partial x}\right)-\frac{h^4}{12}\frac{\partial^4}{\partial x^4}+\frac{h^6}{360}\frac{\partial^6}{\partial x^6}-\dots \tag{10.190}$$

Using the same procedure for local operators $\frac{\partial^4}{\partial x^4}, \frac{\partial^6}{\partial x^6},\dots$, one obtains nonlocal difference scheme [302]. As result local operator (10.187) can be exactly approximated by infinite numbers of nonlocal ones [429]

$$-h^2\frac{\partial^2 y(x,t)}{\partial x^2}=\sum_{k=-\infty,k\neq 0}^{\infty}\frac{(-1)^k}{k^2}y_{n-k}(t)+\frac{\pi^2}{3}y_n(t). \tag{10.191}$$

10.23 POSSIBLE GENERALIZATION AND APPLICATIONS AND OPEN PROBLEMS

Demands of engineering and biology [134] require further study of lattices of complex structure: hexagonal, zigzag, kiral, etc. [110, 126, 250, 294, 296, 297, 373], as well as lattices with interaction not only nearest neighbors [249].

The creation of new metamaterials necessitates the study of mass-in-mass systems and lattices with internal resonators [213, 478]. May be, three-point or multy-point Padé approximations will be useful for this aim.

An interesting objects for study are lattices with large deformations [95, 335, 341, 342]. An example of complex non-smooth nonlinearity is Hertzian contacts describing elastic interactions among grains of a monatomic lattice in the absence of any precompression (i.e., in the so-called sonic-vacuum regime) [335]. We can also mention investigation of discrete breathers and nonlinear normal modes [119].

Surface stress [154] and microstructural instabilities [244] may also be the subject of further study.

Note that discrete systems can be described equations with microinhomogeneous coefficients. Homogenization of this system [253] can be used before continualization.

Finally, the most challenging problem is a study of the real interaction of atoms in a chain based on the Mie, Lennard - Johnes or more realistic potentials. The most promising here is combination of MD with quantum mechanics approaches.

11 Regular and Chaotic Dynamics Based on Continualization and Discretization

In this chapter, continualization and discretization algorithms preserving various properties of original difference-differential equations (DDE) and ordinary differential equations (ODE) are proposed and analyzed. Logistic ODE serves as an example of a deterministic system and logistic DDE is a classic example of a simple system with very complicated (chaotic) behavior. Here we present examples of deterministic discretization and chaotic continualization. Continualization procedure is based on the Padé approximations. In order to correctly characterize the dynamics of the obtained ODE, we have measured such characteristic parameters of chaotic dynamical systems as the Lyapunov exponents and the Lyapunov dimensions.

11.1 INTRODUCTION

Differential and difference equations are the main tools for mathematical modeling of physical, economic, environmental, and social processes [309,397,405]. In order to study differential equations, the entire arsenal of Calculus and Functional Analysis is used. Difference equations are the standard objects for numerical analysis. The relationship between differential and difference equations is nontrivial, and in addition discretization or continualization algorithms often may change fundamentally the nature of their solution. Therefore, the study of approaches that allow these operations to preserve the basic properties of the original systems stands for the hot topic of research. In order to give more light to the considered problem, we choose and analyze the simplest possible examples, i.e., logistic differential and difference equations.

Verhulst belongs to the first who studied the following ODE

$$\frac{dN}{dt} = rN\left(1 - \frac{N}{K}\right) \tag{11.1}$$

and named it logistic equation [449,450].

Later investigators proposed variations of the Verhulst equation and they considered the difference logistic equation yielded by logistic models. The first application of ODE (11.1) was connected with population problems, and more generally with the problems arised in ecology. If the Verhulst model is used for describing change in population size N over time t, then in equation (11.1) r is the so called Malthusian parameter (rate of maximum population growth) and K is the parameter responsible for carrying capacity (i.e., the maximum sustainable population). Equation (11.1) is widely used in problems of ecology, economics, chemistry, medicine, pharmacology, and epidemiology [211, 309, 321, 397, 405, 424]. As a rule, this model is oversimplified for quantitative estimations, but reflects the key qualitative features of processes under consideration.

Equation (11.1) can be reduced to the form

$$\frac{dx}{dt} = rx(1-x), \tag{11.2}$$

where $x = N/K$.

We introduce the following initial condition

$$x(0) = a. \tag{11.3}$$

The Cauchy problem (11.2), (11.3) has the exact solution

$$x = \frac{a}{a+(1-a)e^{-rt}}. \tag{11.4}$$

The discrete logistic equation can be written as follows

$$x_{n+1} = Rx_n(1-x_n), \tag{11.5}$$

where parameter R characterizes the rate of reproduction (growth) of the population. We have $R = rh$, and parameter h defines the time between consecutive measurements.

It should be mentioned that the nonlinear difference equation (11.5) exhibits period doubling bifurcation leading to chaos [211, 321].

In what follows we will analyze a slightly different discrete logistic equation

$$x_{n+1} - x_n = Rx_n(1-x_n). \tag{11.6}$$

This equation has two equilibrium positions

$$x_n = 0, \tag{11.7}$$

$$x_n = 1. \tag{11.8}$$

Difference equation (11.6) is obtained from ODE (11.2) using a forward difference scheme for the derivative with the step of discretization h.

Ordinary difference equation ($O\Delta E$) (11.6) is close enough in behavior to solutions of the following equation [309, 321]

$$x_{n+1} = x_n \exp[R(1-x_n)]. \tag{11.9}$$

On the other hand, $O\Delta E$ (11.9) is close enough in behavior to solutions of equation (11.6) for $x_n \approx 1$. Initial condition for $O\Delta E$s (11.5), (11.6) or (11.9) is taken as follows

$$x_0 = a. \tag{11.10}$$

The discrete Cauchy problems regarding (11.5) and (11.10); (11.6) and (11.10), or (11.9) and (11.10) for sufficiently large values of the parameter R exhibits the complicated chaotic behavior of the system [211, 309, 321, 397, 405, 424]. For $O\Delta E$ (11.6), as it is shown in [405] for $R = 2.3$, the solution starts to oscillate periodically around the value $x = 1$. This solution is stable as long as $R < \sqrt{6} \approx 2.449$. For $R = 2.500$ the process comes to steady periodic oscillations with a period of four. It can be mentioned that the chaotic threshold for $O\Delta E$ (11.9) takes the value of 2.6824 [309, 321].

After description of the objects of our research, we proceed to the formulation of its goals. Both continuous and discrete logistic equations have been extensively investigated. The results of these studies are described in a large number of research including books and review papers [113, 211, 309, 321, 397, 424]. Our study focuses on comparison of the results obtained through the discretization and continualization of nonlinear ODE and $O\Delta E$, and we employed logistic equations as convenient and simple examples to study the mentioned problem.

The following questions arise while investigating the mentioned problems.

First problem: is it possible to discretize ODE (11.2) in such a way that the resulting $O\Delta E$ has only regular solutions? This practically important issue has been studied resonably well [68, 205, 234, 359], hence we address this issue briefly.

On the other hand, many researchers pointed out that discrete logistic models are more adequate to describe the essence of the physical, economic or biological processes because they include chaotic regimes [405]. In this regard, our second problem is: does a continualization of the original $O\Delta E$ exist that the resulting ODE possesses a chaotic solution?

It is difficult to expect that standard continualization, based on the Taylor series, will provide the desired result. However, it can be achieved based on the use of Padé approximants (Chapter 10).

11.2 INTEGRABLE $O\Delta E$

As it is mentioned in [113], non-invertible maps, such as the logistic map, may display chaos. Therefore, it is of interest to find the transformation of the original discrete logistic equation into a form leading to deterministic solutions.

Following the ideas of reference [205], we rewrite $O\Delta E$ (11.6) into the following form

$$x_{n+1} - x_n = Rx_n - Rx_{n+1}x_n. \tag{11.11}$$

This presentation makes it possible to express x_{n+1} not as polynomial, but as a rational function of x_n. Equation (11.11) with initial condition (11.10) has the exact solution of the form

$$x_n = \frac{a}{a + (1-a)(1+R)^{-n}}. \tag{11.12}$$

Thus, representation (11.11) allows one to obtain a difference scheme without chaotic behavior.

11.3 CONTINUALIZATION WITH PADE APPROXIMANTS

Let us try to construct a continuous model, describing the chaotic behavior like that exhibited by the original $O\Delta E$. As it is mentioned in [113], for generating chaotic behavior a nonlinear ODE must have an order $x \geq 3$.

To construct a logistic-like ODE with chaotic behavior, additional modifications were included introduction of a piecewise constant argument, delay, and fractional derivative [9,221,327,403,430]. Here, we use only continualization based on Maclaurin expansion and Padé approximations.

In order to carry out the continualization of $O\Delta E$ (11.6), let us introduce the continuous coordinate x scaled in such a way that $x_n = x(nh)$. Assuming that $x(t)$ is a slightly changing function, we employ the Maclaurin expansion at the vivinity of equilibrium position (11.7)

$$x_{n+1} - x_n = hx_t + \frac{h^2}{2}x_{tt} + \frac{h^3}{6}x_{ttt} + \dots \tag{11.13}$$

The third-order equation yielded by using the above described procedure takes the following form

$$h^3 x_{ttt} + 3h^2 x_{tt} + 6hx_t - 6Rx + 6Rx^2 = 0, \tag{11.14}$$

and it describes completely regular deterministic trajectories.

Let us consider fluctuations around the second equilibrium position of equation (11.6), $x_n = 1$. Changing the variables

$$x_n = 1 + y_n, \qquad |y_n| << 1, \tag{11.15}$$

one obtains

$$y_{n+1} - y_n = -Ry_n(1 + y_n). \tag{11.16}$$

The initial condition for this difference equation is

$$y_0 = \alpha, \qquad 0 < \alpha << 1. \tag{11.17}$$

Suppose $y(t)$ is a slightly changing function, then the Maclaurin expansion yields

$$y_{n+1} - y_n = hy_t + \frac{h^2}{2}y_{tt} + \frac{h^3}{6}y_{ttt} + \dots \tag{11.18}$$

The fifth order ODE based on this expansion can be written as follows

$$h\frac{d}{dt}\left[1 + \frac{h}{2}\frac{d}{dt} + \frac{h^2}{6}\frac{d^2}{dt^2} + \frac{h^3}{24}\frac{d^3}{dt^3} + \frac{h^4}{120}\frac{d^4}{dt^4}\right]y = -Ry(1+y). \tag{11.19}$$

We transform the differential operator in square brackets into the diagonal Padé approximation:

$$1+\frac{h}{2}\frac{d}{dt}+\frac{h^2}{6}\frac{d^2}{dt^2}+\frac{h^3}{24}\frac{d^3}{dt^3}+\frac{h^4}{120}\frac{d^4}{dt^4}\approx\frac{1}{3}\frac{h^2\frac{d^2}{dt^2}+6h\frac{d}{dt}+60}{h^2\frac{d^2}{dt^2}-8h\frac{d}{dt}+20}. \tag{11.20}$$

Then

$$\left(h^3\frac{d^3}{dt^3}+6h^2\frac{d^2}{dt^2}+60h\frac{d}{dt}\right)y=-3R\left(h^2\frac{d^2}{dt^2}-8h\frac{d}{dt}+20\right)y(1+y), \tag{11.21}$$

and after routine transformations we get

$$\begin{gathered}h^3\frac{d^3y}{dt^3}+3h^2(2+R(1+2y))\frac{d^2y}{dt^2}+12h(5-2R)\frac{dy}{dt}-\\ 48hRy\frac{dy}{dt}+6Rh^2\left(\frac{dy}{dt}\right)^2+60Ry(y+1)=0.\end{gathered} \tag{11.22}$$

Let us formulate initial conditions for the third-order ODE (11.22). Based on the initial condition of the original difference equation (11.17), one gets

$$y(0)=\alpha. \tag{11.23}$$

We choose additional initial conditions for equation (11.22) in the following form

$$\begin{gathered}\text{at}\quad t=0:\quad y=\alpha,\qquad \frac{dy}{dt}\approx\frac{y_1-y_0}{h}=-\frac{R\alpha(1+\alpha)}{h},\\ \frac{d^2y}{dt^2}\approx\frac{y_2-2y_1+y_0}{h^2}=\frac{R^2\alpha(1+\alpha)(\alpha+(1+\alpha)(1-R\alpha))}{h^2}.\end{gathered} \tag{11.24}$$

The initial value problem (11.22), (11.24) can be transformed using dimensionless time $T=t/h$ to the following form

$$\begin{gathered}\frac{d^3y}{dT^3}+3(2+R(1+2y))\frac{d^2y}{dT^2}+12(5-2R)\frac{dy}{dT}-\\ 48Ry\frac{dy}{dT}+6R\left(\frac{dy}{dT}\right)^2+60Ry(y+1)=0,\end{gathered} \tag{11.25}$$

$$\begin{gathered}\text{at}\quad T=0:\quad y=\alpha,\qquad \frac{dy}{dT}=-R\alpha(1+\alpha),\\ \frac{d^2y}{dT^2}=R^2\alpha(1+\alpha)(\alpha+(1+\alpha)(1-R\alpha)).\end{gathered} \tag{11.26}$$

11.4 NUMERICAL RESULTS

Numerical integration of Cauchy problem (11.25), (11.26) is carried out using the Adams-Bashforth-Moulton method (a predictor-corrector method).

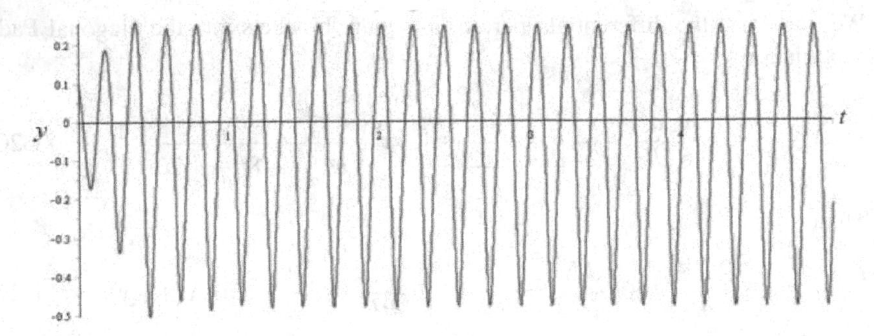

Figure 11.1 Numerical solution of Cauchy problem (11.25), (11.26) for $R = 2.5$ shows periodic motion.

The presented numerical results can be divided into three classes: periodic oscillations, periodic oscillations with subharmonics, and chaotic oscillations. For $2.5 \leq R \leq 2.88$ one obtains periodic oscillations (Figs. 11.1-11.3).

Fig. 11.3 reports phase trajectory, Poincaré section (Fig. 11.3a), and trajectories in a 3D space (Fig. 11.3b). The small changing at initial conditions for the function y leads to a small change of solution (Fig. 11.3c).

For the values $2.5 \leq R \leq 2.88$, a small change at the initial conditions does not lead to a radical change in the behavior of the system. For $2.89 \leq R \leq 3.0$ appear subharmonics components, which one may observe in a periodic solution (Figs 11.4-11.6).

In this case, under small changes of the initial conditions, the nature of oscillations of the system undergoes significant changes (Fig. 11.6c).

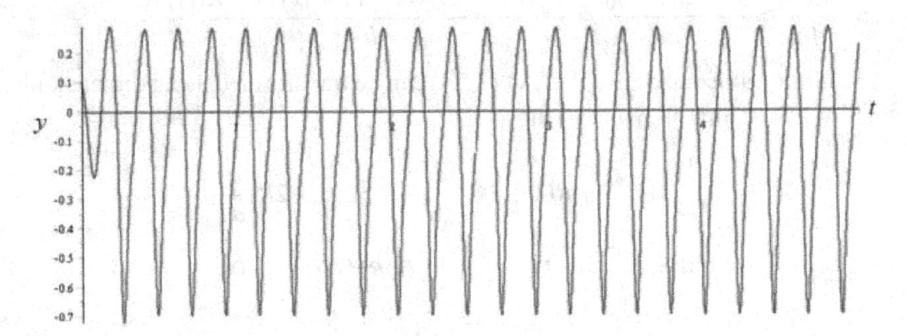

Figure 11.2 With increasing of parameter R ($R = 2.86$ for this figure) numerical solution of Cauchy problem (11.25), (11.26) shows periodic motion, slightly different from that shown in Fig. 11.1.

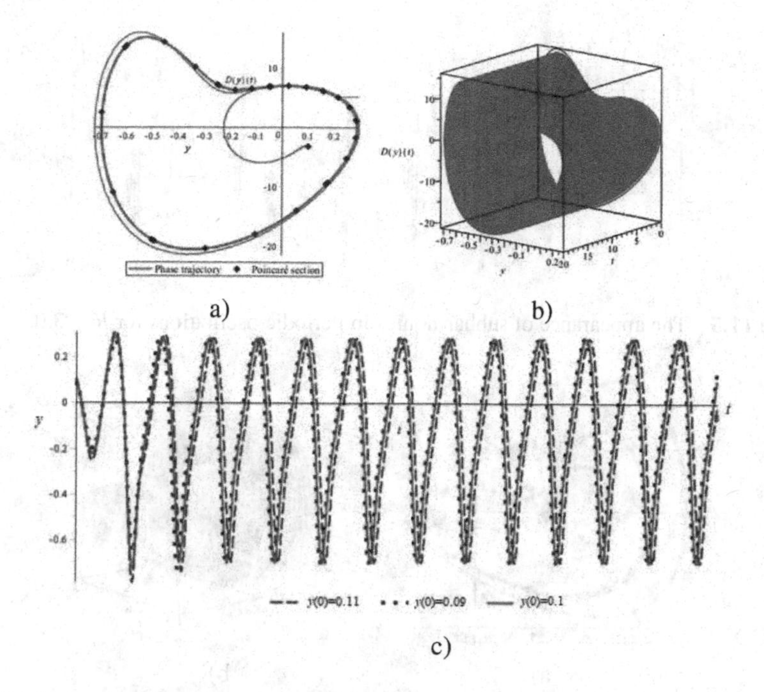

Figure 11.3 Numerical solution of Cauchy problem (11.25), (11.26) for $R = 2.86$.

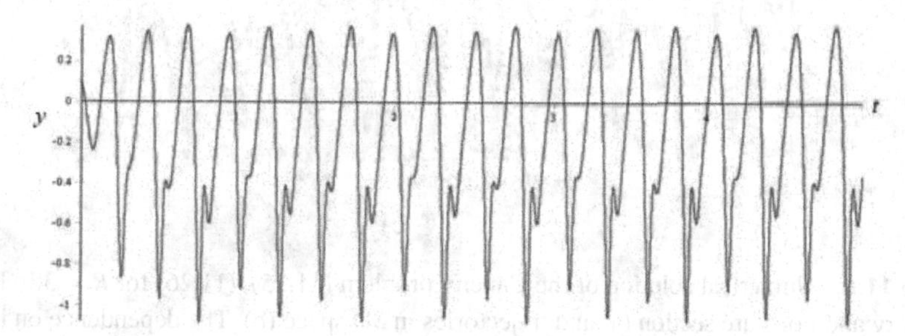

Figure 11.4 Numerical solution of Cauchy problem (11.25), (11.26) for $R = 2.95$ shows the appearance of subharmonic components in periodic oscillations.

For $R > 3.0$ the behavior of the system becomes chaotic (Figs 11.7-11.9).

With the chaotic behavior of the system, small changes at the initial conditions yields significant changes in the output oscillations (Fig. 11.9c).

At the same time, the structure of the phase trajectories does not depend on the initial conditions (Figs 11.10 and 11.11). In order for achieving the correct characterization of the dynamics and confirmations of the chaotic behavior of the system, the Lyapunov exponents and Lyapunov dimensions for different values of R are calculated. If the system has at least one positive Lyapunov exponent, then it is chaotic [113, 321, 424]. Figs 11.12 and 11.13 show the time progress of Lyapunov

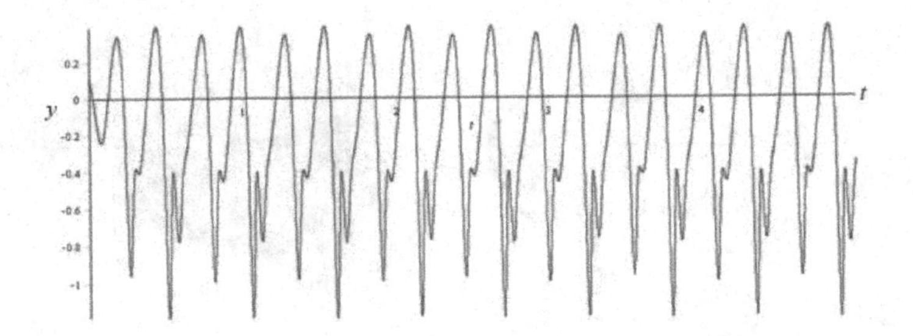

Figure 11.5 The appearance of subharmonics in periodic oscillations for $R = 3.0$.

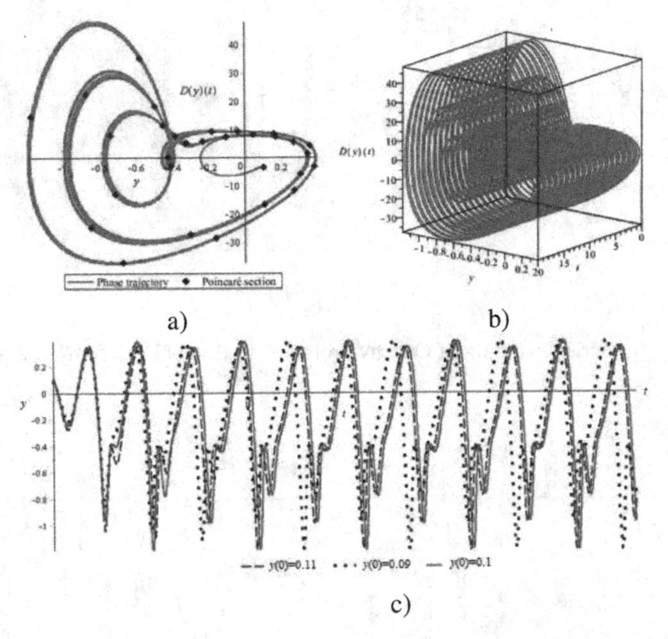

Figure 11.6 Numerical solution of the Cauchy problem (11.25), (11.26) for $R = 3.0$. Phase trajectory and Poincaré section (a) and trajectories in 3D space (b). The dependence on initial conditions is small (see (c)).

exponents for $R = 2.5$ and $R = 3.0$, respectively. Since all Lyapunov exponents are negative, the system is not chaotic.

At $R > 3.0$, the largest Lyapunov exponent becomes positive, which indicates the appearance of chaos in the system (Figs 11.14 and 11.15).

The obtained values of Lyapunov exponents confirm the earlier conclusions about the absence or presence of chaos in the system.

The Lyapunov dimension D_L can be calculated using the formula:

$$D_L = j + \sum_{i=1}^{j} \frac{\lambda_i}{|\lambda_{j+1}|}, \tag{11.27}$$

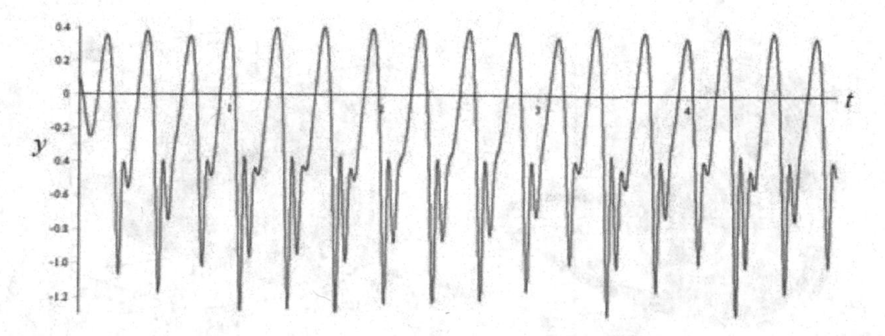

Figure 11.7 Chaotic oscillations for $R = 3.05$.

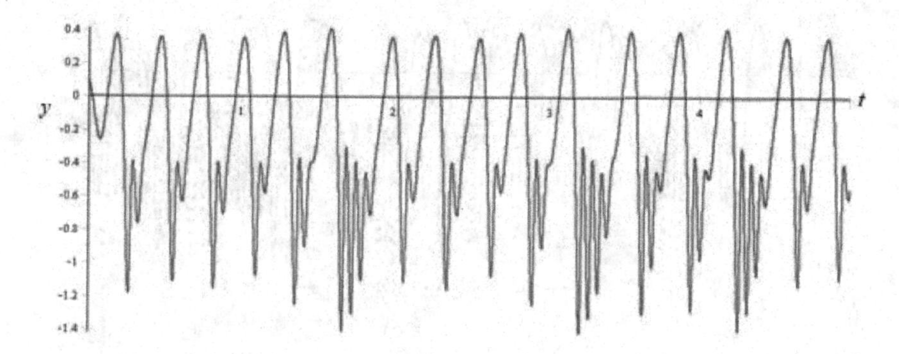

Figure 11.8 Chaotic oscillations for $R = 3.1$.

where j is defined from the following conditions

$$\sum_{i=1}^{j} \lambda_i > 0 \quad \text{and} \quad \sum_{i=1}^{j+1} \lambda_i < 0. \tag{11.28}$$

For $R = 3.05$ and $R = 3.10$ one obtains $D_L(3.05) = 2.0102$ and $D_L(3.10) = 2.0316$, which are consistent with that of a third-order chaotic system [179].

In this section, we looked at some toy problems, the solutions of which lead to interesting conclusions. Logistic ODE serves as example of deterministic system and discrete logistic equation is a classic example of simple system with chaotic behavior. Standart discretization of logistic ODE leads to the difference equation with chaotic behavior. Classic continualization transform discrete logistic equation to the deterministic ODE. However, using fairly simple techniques of nonstandart discretization and enhanced generalized continualization (see Chapter 10), it is possible to construct deterministic OΔE and ODE with chaotic behavior. This expands the possibilities of mathematical modeling of real processes.

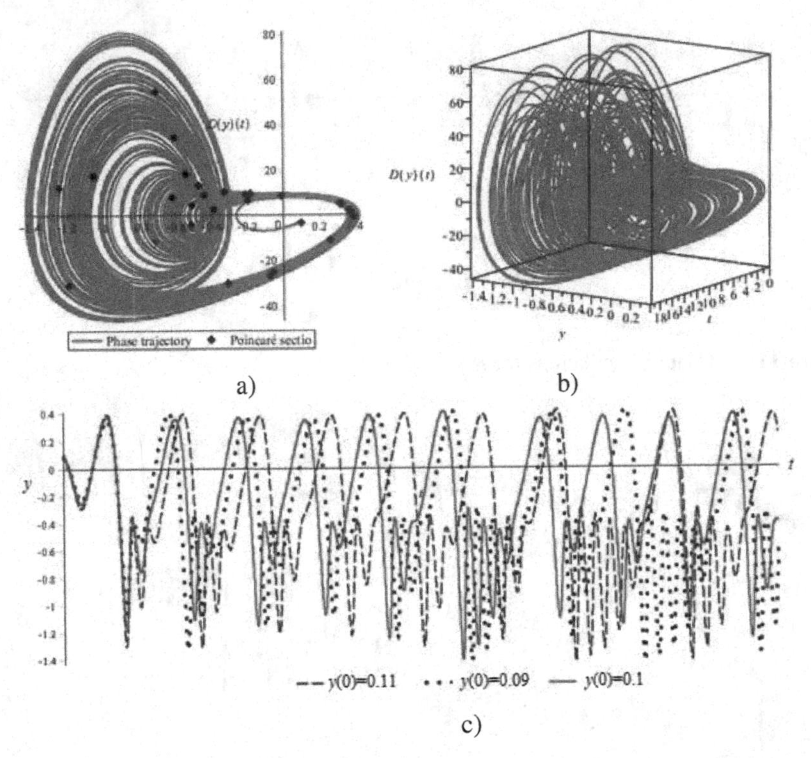

Figure 11.9 Numerical solution of the Cauchy problem (equations (11.25) and (11.26)) for $R = 3.1$. Phase trajectory and Poincaré section (a) and trajectories in 3D space (b). A very small change in initial conditions created a significantly different outcome (see (c)).

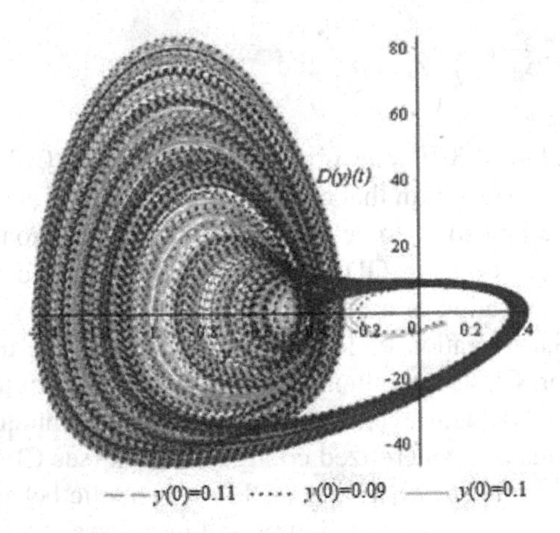

Figure 11.10 Plane phase portrait for $R = 3.1$.

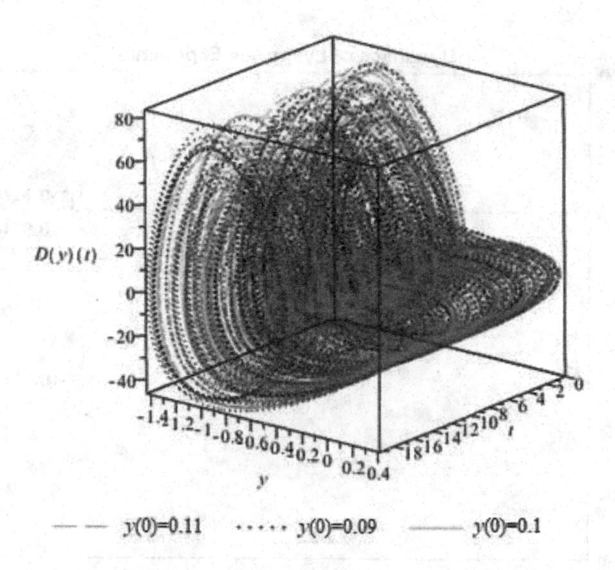

Figure 11.11 3D phase portrait for $R = 3.1$.

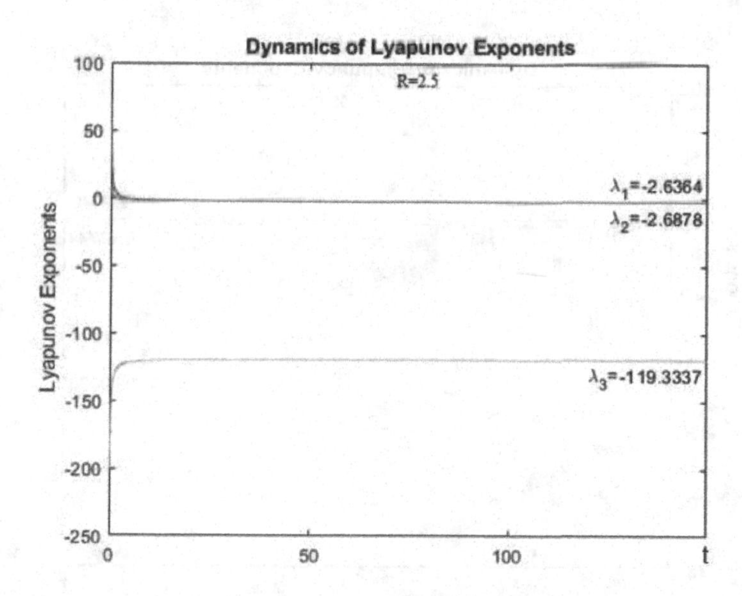

Figure 11.12 Dynamics of Lyapunov exponents for $R = 2.5$.

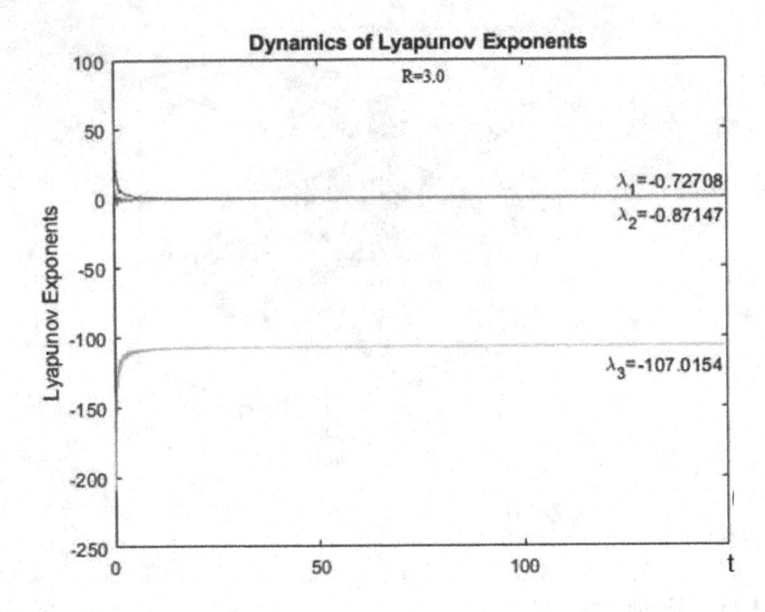

Figure 11.13 Dynamics of Lyapunov exponents for $R = 3.0$.

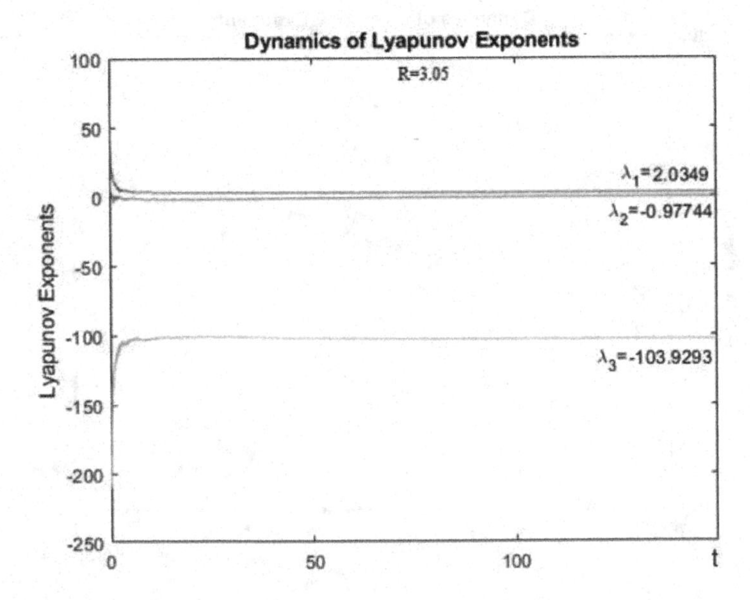

Figure 11.14 Dynamics of Lyapunov exponents for $R = 3.05$.

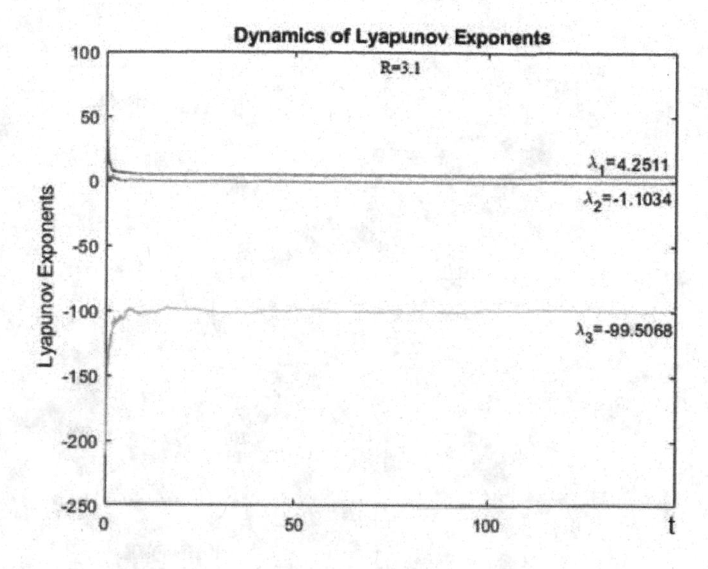

Figure 11.15 Dynamics of Lyapunov exponents for $R = 3.1$.

References

1. Ablowitz, M.J. and Segur, H. 1981. *Solitons and the Inverse Scattering Transform*. Philadelphia: SIAM.
2. Abramowitz, M. and Stegun, I.A. (eds.) 1965. *Handbook of Mathematical Functions, with Formulas, Graphs, and Mathematical Tables*. New York: Dover Publications.
3. Abramyan, A.K., Bessonov, N.M., Indeitsev, D.A., Mochalova, Yu.A. and Semenov, B.N. 2011. Influence of oscillation localization on film detachment from a substrate. *Vestnik St. Petersburg Univ.: Math.* **44**(1):5–12 .
4. Achenbach, J.D. and Herrmann, G. 1968. Dispersion of free harmonic waves in fibre-reinforced composites. *AIAA J.* **6**:1832–1836.
5. Achenbach, J.D. and Zhu, H. 1989. Effect of interfacial zone on mechanical behavior and failure of fibre-reinforced composites. *J. Mech. Phys. Sol.* **7**:381–393.
6. Aero, E., Fradkov, A., Andrievsky, B. and Vakulenko, S. 2006. Dynamics and control of oscillations in a complex crystalline lattice. *Phys. Lett. A* **353**(1):24–29.
7. Ahsani, S., Boukadia, R., Droz, C., Claeys, C., Deckers, E. and Desmet W. 2020. Diffusion based homogenization method for 1D wave propagation. *Mech. Sys. Sig. Proc.*, **136**:106515.
8. Aifantis, E.C. 1999. Gradient deformation models at nano, micro, and macro scales. *ASME J. Eng. Mater. Techn.* **121**:189–202.
9. Akhmet, M., Altintan, D. and Ergenc, T. 2010. Chaos of the logistic equation with piecewise constant argument. *arXiv*:1006.4753.
10. Allaire, G., Lamacz, A. and Rauch, J. 2018. Crime pays; homogenized wave equations for long times. *Mathematics* arXiv:1803.09455 [math.AP].
11. Andrianov, I.V. 1991. Continuous approximation of higher-frequency oscillation of a chain. *Doklady AN Ukr. SSR Ser. A* **2**:13–15 (in Russian).
12. Andrianov, I.V. 2002. The special feature of limiting transition from a discrete elastic media to a continuous one. *J. Appl. Math. Mech.* **66**(2):261–265.
13. Andrianov, I.V. 2021. Mathematical models in pure and applied mathematics. In: *Nonlinear Dynamics of Discrete and Continuous Systems*, 15–29. Advanced Structured Materials 139. Abramian A.K., Andrianov I.V., Gaiko V.A. (eds.).Cham: Springer Nature.
14. Andrianov, I.V. and Awrejcewicz, J. 2000. Numbers or understanding: analytical and numerical methods in the theory of plates and shells. *Facta Univ., series Mech., Aut. Contr. Robot., University of Nis* **2**(10):1319–1327.
15. Andrianov, I.V. and Awrejcewicz, J. 2003. Homo analyticus or homo computicus? *Facta Univ., series Mech., Aut. Contr. Robot., University of Nis* **3**(13):765–770.
16. Andrianov, I.V. and Awrejcewicz, J. 2001. New trends in asymptotic approaches: summation and interpolation methods. *Appl. Mech. Rev.* **54**(1):69–92.
17. Andrianov, I.V. and Awrejcewicz, J. 2005. Continuous models for chain of inertially linked masses. *Eur. J. Mech. A/Sol.* **24**(3):532–536.
18. Andrianov, I.V. and Awrejcewicz, J. 2008. Continuous models for 2D discrete media valid for higher-frequency domain. *Comp. & Struct.* **86**:140–144.
19. Andrianov, I.V., Awrejcewicz, J. and Danishevskyy, V.V. 2018. *Asymptotical Mechanics of Composites. Modelling Composites without FEM*. Berlin: Springer Nature.

20. Andrianov, I.V., Awrejcewicz, J. and Weichert, D. 2010. Improved continuous models for discrete media. *Math. Probl. Eng.* ID 986242.
21. Andrianov, I.V., Awrejcewicz, J., Danishevskyy, V.V. and Markert, B. 2017. Influence of geometric and physical nonlinearities on the internal resonances of a finite continuous rod with a microstructure. *J. Sound Vib.* **386**:359–371.
22. Andrianov, I.V., Bolshakov, V.I., Danishevskyy, V.V. and Weichert, D. 2008. Higher-order asymptotic homogenization and wave propagation in periodic composite materials. *Proc. R. Soc. A* **464**:1181–1201.
23. Andrianov, I.V., Danishevskyy, V.V. and Kalamkarov, A.L. 2013. Vibration localization in one-dimensional linear and nonlinear lattice: discrete and continuum models. *Nonl. Dyn.* **72**:37–48.
24. Andrianov, I.V., Danishevskyy, V.V., and Markert, B. 2015. Nonlinear vibrations and mode interactions for a continuous rod with microstructure. *J. Sound Vib.* **351**:268–281.
25. Andrianov, I.V., Danishevskyy, V.V. and Rogerson, G. 2018. Elastic waves in periodically heterogeneous two-dimensional media: locally periodic and anti-periodic modes. *Proc. Roy. Soc. A* **474**:2215.
26. Andrianov, I.V., Danishevskyy, V.V. and Rogerson, G. 2018. Internal resonances and modes interactions in non-linear vibrations of viscoelastic heterogeneous solids. *J. Sound Vib.* **433**:55–64.
27. Andrianov, I.V., Danishevskyy, V.V. and Rogerson, G. 2020. Vibrations of nonlinear elastic lattices: low- and high-frequency dynamic models, internal resonances and modes coupling. *Proc. R. Soc. A* **476**:20190532.
28. Andrianov, I.V., Danishevskyy, V.V., Kaplunov, J. and Markert, B. 2019. Wide frequency higher-order dynamic model for transient waves in a lattice. In: *Problems of Nonlinear Mechanics and Physics of Materials, Advanced Structured Materials* 94, Andrianov, I.V., Manevich, A.I., Mikhlin, Yu.V., Gendelman, O.V. (eds.) Springer Nature, pp. 3–12.
29. Andrianov, I.V., Danishevskyy, V.V., Ryzhkov, O.I. and Weichert, D. 2013. Dynamic homogenization and wave propagation in a nonlinear 1D composite material. *Wave Motion* **50**:271–281.
30. Andrianov, I.V., Danishevskyy, V.V., Topol, H. and Luyt, A.S 2018. Shear wave propagation in layered composites with degraded matrices at locations of imperfect bonding. *Wave Motion* **78**:9–31.
31. Andrianov, I.V., Danishevskyy, V.V., Topol, H. and Rogerson, G. 2016. Propagation of Floquet-Bloch shear waves in viscoelastic composites: analysis and comparison of interface/interphase models for imperfect bonding. *Acta Mech.* **228**:1177–1196.
32. Andrianov, I.V., Danishevskyy, V.V. and Weichert, D. 2010. Continuous model of 2D discrete media based on composite equations. *Acoust. Phys.* **56**(6):807–810.
33. Andrianov, I.V., Danishevskyy, V.V. and Weichert, D. 2008. Simple estimations on effective transport properties of a random composite material with cylindrical fibres. *ZAMP* **59**(5):889–903.
34. Andrianov, I.V., Lesnichaya, V.A. and Manevitch, L.I. 1985. *Homogenization Methods in Statics and Dynamics of Ribbed Shells*. Moscow: Nauka (in Russian).
35. Andrianov, I.V. and Manevitch, L.I. 2002. *Asymptotology: Ideas, Methods, and Applications*. Dordrecht: Kluwer Academic Publishers.
36. Andrianov, I. and Mityushev, V. 2018. Exact and "exact" formulae in the theory of composites. In: *Modern Problems in Applied Analysis*. Drygaś, P., Rogosin, S. (eds.). Birkhäuser, pp. 15–34.

37. Andrianov, I., Starushenko, G., Kvitka, S. and Khajiyeva, L. 2019. The Verhulst-Like equations: Integrable $O\Delta E$ and ODE with chaotic behavior. *Symmetry* **11**:1446.
38. Arvin, H. and Bakhtiari-Nejad, F. 2013. Nonlinear modal interaction in rotating composite Timoshenko beams. *Compos. Struct.* **96**:121–134.
39. Askar, A. 1985. *Lattice Dynamical Foundations of Continuum Theories. Elasticity, Piezoelectricity, Viscoelasticity, Plasticity*. Singapore: World Scientific.
40. Askes, H. and Aifantis, E.C. 2011. Gradient elasticity in statics and dynamics: An overview of formulations, length scale identification procedures, finite element implementations and new results. *Int. J. Solids Struct.* **48**(13):1962–1990.
41. Askes, H. and Sluys, L.J. 2002. Explicit and implicit gradient series in damage mechanics. *Eur. J. Mech. A/Sol.* **21**:379–390.
42. Askes, H. and Metrikine, A.V. 2002. One-dimensional dynamically consistent gradient elasticity models derived from a discrete microstructure. Part 2: Static and dynamic response. *Eur. J. Mech. A/Sol.* **21**:573–588.
43. Askes, H., Metrikine, A.V., Pichugin, A.V. and Bennett, T. 2008. Four simplified gradient elasticity models for the simulation of dispersive wave propagation. *Phil. Mag.* **88**(28-29):3415–3443.
44. Aubry, S. 1995. Anti-integrability in dynamical and variational problems. *Physica D.* 86:284–296.
45. Auffray, N., Le Quang, H. and He Q.C. 2013. Matrix representations for 3D strain-gradient elasticity. *J. Mech. Phys. Sol.* **61**(5):1202–1223.
46. Bacca, M., Bigoni, D., Dal Corso, F. and Veber, D. 2013. Mindlin second-gradient elastic properties from dilute two-phase Cauchy-elastic composites. Part II: Higher-order constitutive properties. *Int. J. Solids Struct.* **50**:4020–4029.
47. Baker, G.A. jr. and Graves-Morris, P. 1996. *Padé Approximants* (2nd ed.). Cambridge, NY: Cambridge University Press.
48. Bakhvalov, N.S. and Eglit, M.E. 1992. Variational properties of averaged equations for periodic media. *Proc. Steklov Inst. Math.* **192**:3–18.
49. Bakhvalov, N.S. and Eglit, M.E. 1993. Averaging of the equations of the dynamics of composites of slightly compressible elastic components. *Comput. Math. Math. Phys.* **33**(7):939–952.
50. Bakhvalov, N.S. and Eglit, M.E. 1993. Homogenization of dynamic problems singularly depending on small parameters. *Proc. Second Workshop on Composite Media and Homogenization Theory*, Trieste, pp. 17–35.
51. Bakhvalov, N.S. and Eglit, M.E. 1994. An estimate of the error of averaging the dynamics of small perturbations of very inhomogeneous mixtures. *Comput. Math. Math. Phys.* **34**(3):333–349.
52. Bakhvalov, N.S. and Eglit, M.E. 1995. The limiting behaviour of periodic media with soft-modular inclusions. *Comput. Math. Math. Phys.* **35**(6):719–729.
53. Bakhvalov, N.S. and Eglit, M.E. 2000. Long-wave asymptotics with dispersion for wave propagation in stratified media. I. Waves orthogonal to the layers. *Russ. J. Num. Anal. Math. Modell.* **15**(1):3–18.
54. Bakhvalov, N.S. and M.E. Eglit. 2000. Long-wave asymptotics with dispersion for wave propagation in stratified media. II. Waves in arbitrary directions. *Russ. J. Num. Anal. and Math. Modell.* **15**(3-4):225–236.
55. Bakhvalov, N. and Panasenko, G. 1989. *Averaging Processes in Periodic Media. Mathematical Problems in Mechanics of Composite Materials*. Dordrecht: Kluwer.

56. Bakhvalov, N.S., Sandrakov, G.V. and Eglit, M.E. 1996. Mathematical study of sound waves propagation progress in mixtures. *Moscow Univ. Math. Bull.* **51**(6):5–7.
57. Bardzokas, D.I., Filshtinsky, M.L. and Filshtinsky, L.A. 2007. *Mathematical Methods in Electro-Magneto-Elasticity*. Springer, Berlin.
58. Bauman, T., Ben Dhia, H., Elkhodja, N., Oden, J.T. and Prudhomme, S. 2008. On the application of the Arlequin method to the coupling of particle and continuum models. *Comp. Mech.* **42**:511–530.
59. Bedford, A. and Drumheller, D.S. 1983. Theories of immiscible and structured mixtures. *Int. J. Eng. Sci.* 21:863–960.
60. Bedford, A. and Drumheller, D.S. 1994. *Introduction to Elastic Wave Propagation*. New York: Wiley.
61. Bedford, A and Stern, M. 1971. Toward a diffusing continuum theory of composite materials. *J. Appl. Mech.* **38**:8–14.
62. Belashov, Yu.V. and Vladimirov, S.V. 2005. *Solitary Waves in Dispersive Complex Media*. Berlin: Springer.
63. Belyaev, A. 2011. On implicit image derivatives and their applications. In: *Proceedings of the British Machine Vision Conference*. Hoey, J., McKenna, St., Trucco, E. (eds.). BMVA Press, pp. 72.1–72.12.
64. Ben Dhia, H. 2006. Global-local approaches: the Arlequin framework. *Europ. J. Comp. Mech.* **15**:67–80.
65. Ben Dhia, H. 2008. Further insights by theoretical investigations of the multiscale Arlequin method. *Int. J. Multisc. Comp. Eng.* **6**:215–232.
66. Ben Dhia, H. and Rateau, G. 2005. The Arlequin method as a flexible engineering design tool. *Int. J. Numer. Met. Eng.* **62**:1442–1462.
67. Benaroya, H. 1997. Waves in periodic structures with imperfections. *Compos. B.* **28**:143–152.
68. Bender, C.M. and Tovbis, A. 1997. Continuum limit of lattice approximation schemes. *J. Math. Phys.* **38**:3700–3717.
69. Benilov, E.S., Grimshaw, R. and Kuznetsova, E.P. 1993. The generation of radiating waves in a singularly-perturbed KdV equation. *Phys. D.* **69**:270–278.
70. Bensoussan, A., Lions, J.-L. and Papanicolaou, G. 1978. *Asymptotic Analysis for Periodic Structures*. Amsterdam: North-Holland.
71. Benveniste, Y. 1985. The effective mechanical behavior of composite materials with imperfect contact between constituents. *Mech. Mater.* **4**:197–208.
72. Benjamin, T.B., Bona, J.G. and Mahoney, J.J. 1972. Model equations for long waves in nonlinear dispersive systems. *Phil. Trans. Roy. Soc. A.* **272**:47–78.
73. Beran, M.J. 1968. *Statistical Continuum Theories*. New York: Wiley Interscience.
74. Berezovski, A., Engelbrecht, J. and Berezovski, M. 2011. Waves in microstructured solids: a unified viewpoint of modeling. *Acta Mech.* **220**:349–363.
75. Berlyand, L. and Mityushev, V. 2001. Generalized Clausius-Mossotti formula for random composite with circular fibres. *J. Stat. Phys.* **102**:115–145.
76. Berlyand, L. and Mityushev, V. 2005. Increase and decrease of the effective conductivity of two phase composites due to polydispersity. *J. Stat. Phys.* **118**:481–509.
77. Bhatnagar, P.I. 1979. *Nonlinear Waves in One-Dimensional Dispersive Systems*. Oxford: Clarendon Press.
78. Bigoni, D. and Drugan, W.J. 2007. Analitical derivation of Cosserat moduli via homogenization of heterogeneous elastic materials. *J. Appl. Mech.* **74**:741–753.
79. Bishop, R.E.D. 1952. Longitudinal waves in beams. *Aeronaut. Quart.* **3**(4):280–293.

80. Blanc, X., Le Bris, C. and Lions, P.-L. 2002. From molecular models to continuum mechanics. *Arch. Rational Mech. Anal.* **164**, 341–381.
81. Blanc, X., Le Bris, C. and Lions, P.L. 2007. Atomistic to continuum limits for computational material science. *ESAIM: Math. Mod. Num. Anal.* **41**(2):391–426.
82. Bloch, F. 1928. Über die Quantenmechanik der Elektronen in Kristallgittern. *Z. Phys.* **52**:555–600.
83. Bobrovnitskii, Yu.V. 2011. Features of normal wave dispersion in periodic structures. *Acoust. Phys.* **57**(4):442–446.
84. Boertjens, G.J. and van Horssen, W.T. 1998. On mode interactions for a weakly nonlinear beam equation. *Nonlin. Dyn.* **17**:23–40.
85. Boertjens, G.J. and van Horssen, W.T. 2000. An asymptotic theory for a weakly nonlinear beam equation with a quadratic perturbation. *SIAM J. Appl. Math.* **60**:602–632.
86. Born, M. and Huang, K. 1988. *Dynamical Theory of Crystal Lattices*. Oxford: Oxford Univ. Press.
87. Born, M. and Wolf, E. 1964. *Principles of Optics: Electromagnetic Theory of Propagation, Interference and Diffraction of Light*. Oxford: Pergamon Press.
88. Boussinesq, J.V. 1872. Théorie des ondes et des remous qui se propagent le long d'un canal rectangulaire horizontal, en communiquant au liquide contenu dans ce canal des vitesses sensiblement pareilles de la surface au fond. *J. Math. Pures Appl.* **17**:55–108.
89. Boutin, C. 1995. Microstructural influence on heat conduction. *Int. J. Heat Mass Transf.* **38**:3181–3195.
90. Boutin, C. 1996. Microstructural effects in elastic composites. *Int. J. Sol. Struct.* 33:1023–1051.
91. Boutin, C. 2000. Study of permeability by periodic and self-consistent homogenization. *Eur. J. Mech. A/Sol.* 19:603–632.
92. Boutin, C. and Auriault, J.L. 1993. Rayleigh scattering in elastic composite materials. *Int. J. Eng. Sci.* **31**:1669–1689.
93. Bowen, R.M. 1976. Theory of mixtures. In: *Continuum Physics: Mixtures and EM Field Theories*. Eringen, A.C. (ed.) New York: Academic Press, pp. 2–129.
94. Braides, A. 2000. Non-local variational limits of discrete systems. *Commun. Contemp. Math.* **2**:285–297.
95. Braun, O.M. and Kivshar, Y.S. 2004. *The Frenkel-Kontorova Model: Concepts, Methods, and Applications*. Berlin, Heidelberg: Springer-Verlag.
96. Brezinski, C. 2004. Extrapolation algorithms for filtering series of functions, and treating the Gibbs phenomenon. *Numer. Algor.* **36**:309–329.
97. Brillouin, L. 1960. *Wave Propagation and Group Velocity*. New York: Academic Press.
98. Brillouin, L. 2003. *Wave Propagation in Periodic Structures: Electric Filters and Crystal Lattices*, 2nd edn. Mineola, New York: Dover Publications.
99. Brocchini, M. 2013. A reasoned overview on Boussinesq-type models: the interplay between physics, mathematics and numerics. *Proc. R. Soc. A.* **469**:20130496.
100. Bryant, J.D., David, T., Gaskell, P.H., King, S. and Lond, G. 1989. Rheology of bovine bone marrow. *Proc. Inst. Mech. Eng.* **203**:71–75.
101. Bykov, V.G. 2015. Nonlinear waves and solitons in models of fault block geological media. *Russ. Geol. Geophys.* **56**(5):793–803.
102. Cai, C.W., Chan, H.C. and Cheung, Y.K. 1997. Localized modes in periodic systems with a nonlinear disorders. *J. Appl. Mech.* **64**:940–945.
103. Cai, C.W., Chan, H.C. and Cheung, Y.K. 2000. Localized modes in a two-degree-coupling periodic system with a nonlinear disordered subsystem. *Chaos, Solitons & Fractals*. **11**:1481–1492.

104. Cai, C.W., Cheung, Y.K. and Chan, H.C. 1995. Mode localization phenomena in nearly periodic systems. *J. Appl. Mech.* **62**:141–149.
105. Cai, G.Q. and Lin, Y.K. 1990. Localization of wave propagation in disordered periodic structures. *AIAA J.* **29**:450–456.
106. Cai, C.W., Liu, J.K. and Chan, H.C. 2003. Exact analysis of localized modes in bi-periodic mono-coupled mass-spring systems with a single disorder. *J. Sound Vib.* **262**:1133–1152.
107. Cao, W. and Qi, W. 1995. Multisource excitations in a stratified biphase structure. *J. Appl. Phys.* **78**:4640–4646.
108. Cao, W. and Qi, W. 1995. Plane wave propagation in finite 2-2 composites. *J. Appl. Phys.* **78**:4627–4632.
109. Carta, G. and Brun, M. 2012. A dispersive homogenization model based on lattice approximation for the prediction of wave motion in laminates. *J. Appl. Mech.* **79**(2):021019-1–021019-8.
110. Carta, G., Brun, M., Movchan, A.B., Movchan, N.V. and Jones, I.S. 2014. Dispersion properties of vortex-type monatomic lattices. *Int. J. Sol. Struct.* **51**:2213–2225.
111. Catheline, S., Gennisson, J.-L. and Fink, M. 2003. Measurement of elastic nonlinearity of soft solid with transient elastography. *JASA* **114**:3087–3091.
112. Cattani, C. and Rushchitsky, J., 2007. *Wavelet and Wave Analysis as Applied to Materials with Micro or Nanostructure*. Singapore: World Scientific.
113. Cencini, M., Cecconi, F. and Vulpiani, A. 2009. *Chaos: From Simple Models to Complex Systems*. Singapore: World Scientific.
114. Chakraborty, G. and Mallik, A.K. 2001. Dynamics of a weakly non-linear periodic chain. *Int. J. Non-Lin. Mech.* **36**:375–389.
115. Chang, C.S., Askes, H. and Sluys, L.J. 2002. Higher-order strain/higher-order stress gradient models derived from a discrete microstructure, with application to fracture. *Engn. Fract. Mech.* **69**:1907–1924.
116. Charlotte, M. and Truskinovsky, L. 2008. Towards multi-scale continuum elasticity theory. *Continuum Mech. Thermodyn.* **20**:133.
117. Charlotte, M. and Truskinovsky, L. 2012. Lattice dynamics from a continuum viewpoint. *J. Mech. Phys. Sol.* **60**(8):1508–1544.
118. Chechin, G.M. and Sakhnenko, V.P. 1998. Interactions between normal modes in nonlinear dynamical systems with discrete symmetry. Exact results. *Physica D* **117**:43–76.
119. Chechin, G.M. and Dzhelauhova, G.S. 2009. Discrete breathers and nonlinear normal modes in monoatomic chains. *J. Sound Vib.* **322**:490–512.
120. Chen, W. and Fish, J. 2001. A dispersive model for wave propagation in periodic heterogeneous media based on homogenization with multiple spatial and temporal scales. *J. Appl. Mech.* **68**:153–161.
121. Cherednichenko, K.D. and Smyshlyaev, V.P. 2004. On full two-scale expansion of the solutions of nonlinear periodic rapidly oscillating problems and higher-order homogenised variational problems. *Arch. Rat. Mech. Anal.* **174**:385–442.
122. Cherkaev, E. and Ou, M.-J.Y. 2008. Dehomogenization: reconstruction of moments of the spectral measure of the composite. *Inver. Probl.* **24**:065008.
123. Christensen, R.M. 2005. *Mechanics of Composite Materials*, 2nd edn. Mineola, New York: Dover Publications.
124. Claude, Ch., Kivshar, Yu.S., Kluth, O. and Spatschek, K.H. 1993. Moving localized modes in nonlinear lattices. *Phys. Rev. B.* **47**(21):14228–14233.

125. Collins, M.A. 1981. A quasi-continuum approximation for solitons in an atomic chain. *Chem. Phys. Lett.* **77**(2):342–347.
126. Colquitt, D.J., Danishevskyy, V.V. and Kaplunov, J. 2019. Composite dynamic models for periodically heterogeneous media. *Math. Mech. Sol.* **24**(9):2663–2693.
127. Colquitt, D.J., Jones, I.S., Movchan, N.V. and Movchan, A.B. 2011. Dispersion and localization of elastic waves in materials with microstructure. *Proc. R. Soc. A* **467**:2874–2895.
128. Conca, C. and Lund, F. 1999. Fourier homogenization method and the propagation of acoustic waves through a periodic vortex array. *SIAM J. Appl. Math.* **59**(5):1573–1581.
129. Cosserat, E. and Cosserat, F. 1909. *Théorie des Corps Déformables*. Paris: Libraire Scientifique A. Hermann et Fils.
130. Craster, R.V., Kaplunov, J. and Pichugin, A.V. 2010. High frequency homogenization for periodic media. *Proc. R. Soc. A.* **466**:2341–2362.
131. Craster, R.V., Kaplunov, J. and Postnova, J. 2010. High-frequency asymptotics, homogenization and localization for lattices. *Q.J. Mech. Appl. Math.* **63**(4):497–520.
132. Dal Maso, G. 1993. *An Introduction to* Γ*-Convergence*. Boston: Birkhäuser.
133. Daya, E.M. and Potier-Ferry, M. 2001. Vibrations of long repetitive structures by a double scale asymptotic method. *Struct. Eng. Mech.* **12**:215–230.
134. Dauxois, T. 1991. Dynamics of breather modes in a nonlinear "helicoidal" model of DNA. *Phys. Lett. A* **159**:390–395.
135. Deift, P. and McLaughlin, K.T.R. 1998. *A Continuum Limit of the Toda Lattice*. Providence: AMS.
136. Del Piero, G. and Truskinovsky, L. 1998. A one-dimensional model for localized and distributed failure. *J. de Physique IY France* **8**(82):199–210.
137. Deymier, P.A. (ed.) 2013. *Acoustic Metamaterials and Phononic Crystals*. Berlin: Springer-Verlag.
138. Dirac, P.A.M. 1929. Quantum Mechanics of many-electron systems. *Proc. R. Soc. London A.* **123**(729):714–733.
139. Dodd, R.K., Eilbeck, J.C., Gibbon, J.D. and Morris, H.C. 1982. *Solitons and Nonlinear Wave Equations*. London: Academic Press.
140. Dreiden, G.V., Khusnutdinova, K.R., Samsonov, A.M. and Semenova, I.V. 2010. Splitting induced generation of soliton trains in layered waveguides. *J. Appl. Phys.* **107**:034909.
141. Dreiden, G.V., Porubov, A.V., Samsonov, A.M., Semenova, I.V. and Sokurinskaya, E.V. 1995. Experiments in the propagation of longitudinal strain solitons in a nonlinearly elastic rod. *Tech. Phys. Lett.* **21**:415–417.
142. Dreiden, G.V., Samsonov, A.M. and Semenova, I.V. 2014. Observation of bulk strain solitons in layered bars of different materials. *Tech. Phys. Lett.* **40**(12):1140–1141.
143. Driscoll, T. and Fornberg, B. 2001. A Padé-based algorithm for overcoming the Gibbs phenomenon. *Numer. Algorithm.* **26**:77–92.
144. Drumheller, D.S. and Bedford, A. 1974. Wave propagation in elastic laminates using a second order microstructure theory. *Int. J. Sol. Struct.* **10**:61–76.
145. Drumheller, D.S. and Sutherland, H.J. 1973. A lattice model for stress wave propagation in composite materials. *J. Appl. Mech.* **40**:157–164.
146. Drygas, P., Mityushev, V., Gluzman, S. and Nawalaniec, W. 2020. *Applied Analysis of Composite Media: Analytical and Computational Approaches for Materials Scientists and Engineers*. New York: Elsevier.

147. Dykhne, A.M. 1971. Conductivity of a two-dimensional two-phase system. *Sov. Phys. JETP* **32**:63–65.
148. Egle, D.M. and Bray, D.T. 1976. Measurement of acoustoelastic and third-order elastic constants for rail steel. *JASA* **60**:741–744.
149. Eglit, M.E. 1998. New models arising in the averaged description of microinhomogeneous media. *Proc. Steklov Inst. Math.* **223**:94–104.
150. Eglit, M.E. 2010. Dispersion of elastic waves in microinhomogeneous media and structures (A review). *Acoust. Phys.* **56**(6):989–995.
151. Ehrlacher, V., Ortner, C. and Shapeev, A.V. 2015. Analysis of boundary conditions for crystal defect atomistic simulations. arXiv:1306. 5334v3.
152. Engelbrecht, J. 1997. *Nonlinear Wave Dynamics: Complexity and Simplicity*. Dordrecht, Boston: Kluwer.
153. Engelbrecht, J. and Braun, M. 1998. Nonlinear waves in nonlocal media. *Appl. Mech. Rev.* **51**:475–488.
154. Eremeyev, V.A. and Sharma, B.L. 2019. Anti-plane surface waves in media with surface structure: Discrete vs. continuum model. *Int. J. Eng. Sci.* **143**:33–38.
155. Eringen, A.C. 1987. Theory of nonlocal elasticity and some applications. *Res. Mech.* **21**:313–342.
156. Eringen, A.C. 1992. Vistas on nonlocal continuum Physics. *Int. J. Eng. Sci.* **30**(10):1551–1565.
157. Eringen, A.C. 1999. *Microcontinuum Field Theories. I. Foundations and Solids*. New York: Springer-Verlag.
158. Erofeev, V.I. 2003. *Wave Processes in Solids with Microstructure*. Singapore: World Scientific.
159. Espinosa, H.D., Dwivedi, S.K. and Lu, H.-C. 2000. Modelling impact induced delamination of woven fibre reinforced composites with contact/cohesive laws. *Comput. Meth. Appl. Mech. Eng.* **183**:259–290.
160. Espinosa, H.D., Zavattieri, P.D. and Dwivedi, S.K. 1998. A finite deformation continuum/discrete model for the description of fragmentation and damage in brittle materials. *J. Mech. Phys. Sol.* **46**:1909–1942.
161. Fanga, X.-Q., Liua, J.-X., Niea, G.-Q. and Hub, Ch. 2010. Propagation of flexural waves and localized vibrations in the strip plate with a layer using Hamilton system. *Eur. J. Mech. A/Sol.* **29**:152–157.
162. Feng, B.F. and Kawahara, T. 2007. Discrete breathers in two-dimensional nonlinear lattices. *Wave Motion* **45**(1-2):68–82.
163. Feng, B.F., Doi, Y. and Kawahara, T. 2004. Quasi-continuum approximation for intrinsic localized modes in Fermi-Pasta-Ulam lattices. *J. Phys. Soc. Jpn.* **73**:2100–2111.
164. Feng, B.F., Doi, Y. and Kawahara, T. 2006. A regularized model equation for discrete breathers in anharmonic lattices with symmetric nearest-neighbor potentials. *Phys. D* **214**:33–41.
165. Filimonov, A.M. 1992. Some unexpected results on the classical problem of the string with N beads. The case of multiple frequencies. *C. R. Acad. Sci. Paris* **1**(315): 957–961.
166. Filimonov, A.M. 1996. Continuous approximations of difference operators. *J. Difference Eq. Appl.* **2**(4):411–422.
167. Filimonov, A.M. and Myshkis, A.D. 1998. Asymptotic estimate of solution of one mixed difference-differential equation of oscillations theory. *J. Difference Eq. Appl.* **4**:13–16.

168. Filimonov, A.M., Mao, X. and Maslov, S. 2000. Splash effect and ergodic properties of solution of the classic difference - differential equation. *J. Difference Eq. Appl.* **6**:319–328.
169. Findlin, A.Ya. 2007. Peculiarities of the use of computational methods in applied mathematics (on global computerization and common sense). In: Blekhman, I.I., Myshkis, A.D., Ya.G. Panovko. 2007. *Applied Mathematics: Subject, Logic, Peculiarities of Approaches. With Examples from Mechanics*. Moscow, URSS: 350–358 (in Russian).
170. Fish, J. and Chen, W. 2001. Higher-order homogenization of initial/boundary-value problem. *J. Eng. Mech.* **127**:1223–1230.
171. Fish, J. and Chen W. 2004. Space-time multiscale model for wave propagation in heterogeneous media. *Comput. Meth. Appl. Mech. Eng.* **193**:4837–4856.
172. Flach, S. and Kladko, K. 1996. Perturbation analysis of weakly discrete kinks. *Phys. Rev. E* **54**:2912–2916.
173. Flach, S. and Willis, C.R. 1993. Asymptotic behavior of onedimensional nonlinear discrete kink-bearing systems in the continuum limit: Problem of nonuniform convergence. *Phys. Rev. E* **47**:4447–4456.
174. Fleck, N.A. and Hutchinson, J.W. 1993. A phenomenological theory for gradient effects in plasticity. *J. Mech. Phys. Sol.* **41**:1825–1857.
175. Fletcher, C.A.J. 1984. *Computational Galerkin Methods*. Berlin: Springer.
176. Floquet, G. 1883. Sur les équations différentielles linéaires á coefficients périodiques. *Ann. sc. l'École Norm. Supér., Sér 2.* **12**:47–88.
177. Flügge, W. 1975. *Viscoelasticity*. New York: Springer.
178. Franzevich, I.N., Voronov, F.F. and Bakuta, S.A. 1982. *Elastic Constants and Modules of Elasticity of Metals and Nonmetals. Reference Book*. Kiev: Naukova Dumka (in Russian).
179. Frederickson, P., Kaplan, J.L., Yorke, E.D. and Yorke, J.A. 1983. The Liapunov dimension of strange attractors. *J. Differ. Eq.* **49**:185–207.
180. Galka, A., Telega, J.J. and Tokarzewski, S. 1999. A contribution to evaluation of effective moduli of trabecular bone with rod-like microstructure. *J. Theor. Appl. Mech.* **37**:707–727.
181. Gambin, B. and Kröner, E. 1989. High order terms in the homogenized stressstrain relation of periodic elastic media. *Phys. Stat. Sol.* **151**:513–519.
182. Gao, Q., Wu, F., Zhang, H.W., Zhong, W.X., Howson, W.P. and Williams, F.W. 2012. Exact solutions for dynamic response of a periodic spring and mass structure. *J. Sound Vib.* **331**(5):1183–1190.
183. Ghavanloo, E., Fazelzadeh, S.A. and Rafii-Tabar, H. 2020. Formulation of an efficient continuum mechanics-based model to study wave propagation in one-dimensional diatomic lattices. *Mech. Res. Comm.* **103**:103467.
184. Giannoulis, J. and Mielke, A. 2004. The nonlinear Schrödinger equation as a macroscopic limit for an oscillator chain with cubic nonlinearities. *Nonlinearity* **17**:551–565.
185. Gibson, J.L. and Ashby, M.P. 1988. *Cellular Solids: Structure and Properties*. New York: Pergamon Press.
186. Gleik, J. 1987. *Chaos: Making a New Science*. New York: Viking Penguin.
187. Golub, M.V., Fomenko, S.I., Bui, T.Q., Zhang, Ch. and Wang, Y.-S. 2012. Transmission and band gaps of elastic SH waves in functionally graded periodic laminates. *Int. J. Sol. Struct.* **49**:344–354.
188. Golub, M.V., Zhang, Ch. and Wang, Y.-S. 2012. SH-wave propagation and scattering in periodically layered composites with a damaged layer. *J. Sound Vib.* **331**:1829–1843.

189. Goupillaud, P.L. 1961. An approach to inverse filtering of near-surface layer effectc from seismic records. *Geophysics* **26**(6):754–760.
190. Grigolyuk, E.I. and Selezov, I.I. 1973. *Nonclassical Theories of Vibrations of Beams, Plates and Shells*. Moscow: VINITI (in Russian).
191. Guckenheimer, J. 1998. Computer simulation and beyond-for the 21st century. *Notices AMS* **45**:1120–1123.
192. Guckenheimer, J. 1998. Numerical computation in the information age. *SIAM NEWS* **31**(5).
193. Guenneau, S., Poulton, C.G. and Movchan, A.B. 2003. Oblique propagation of electromagnetic and elastic waves for an array of cylindrical fibres. *Proc. R. Soc. Lond. A.* **459**:2215–2263.
194. Guo, X.E. 2001. Mechanical properties of cortical bone and cancellous bone tissue. In: Cowin, S.C. (ed.) *Bone Mechanics Handbook*. Boca Raton: CRC Press, pp. 10-1–10-23.
195. Gusev, A.I., Rempel, A.A. and Magerl, A.J. 2001. *Disorder and Order in Strongly Nonstoichiometric Compounds: Transition, Metal Carbides, Nitrides, and Oxides*. Berlin, New York: Springer.
196. Guz, A.N. and Nemish, Yu.N. 1987. Perturbation of boundary shape in continuum mechanics. *Sov. Appl. Mech.* **23**(9):799–822.
197. Guz, A.N. and Shulga, N.A. 1992. Dynamics of laminated and fibrous composites. *Appl. Mech. Rev.* **45**:35–60.
198. Hadjesfandiari, A.R. and Dargush, G.F. 2011. Couple stress theory for solids. *Int. J. Sol. Struct.* **48**:2496–2510.
199. Hadjesfandiari, A.R. and Dargush, G.F. 2013. Fundamental solutions for isotropic size-dependent couple stress elasticity. *Int. J. Solids Struct.* **50**:1253–1265.
200. Hamming, R.W. 1973. *Numerical Methods for Scientists and Engineers*, 2nd ed. New York: McGraw Hill.
201. Hao, H.-Y. and Maris, H.J. 2001. Experiments with acoustic solitons in crystalline solids. *Phys. Rev. B.* **64**:064302.
202. Hashin, Z. 2002. Thin interphase/imperfect interface in elasticity with application to coated fibre composites. *J. Mech. Phys. Sol.* **50**:2509– 2537.
203. Hegemier, G., Gurtman, G.A. and Nayfeh, A.H. 1973. A continuum mixture theory of wave propagation in laminated and fibre reinforced composites. *Int. J. Sol. Struct.* 9:395–414.
204. Hellinger, E. 1914. Die allgemeinen Ansätze der Mechanik der Kontinua. *Enz. Mat. Wiss.* **4**(4):602–694.
205. Herbst, B.M. and Ablowitz, M.J. 1993. Numerical chaos, symplectic intergators, and exponentially small splitting distances. *J. Comput. Phys.* **105**:122–132.
206. Hermann, G., Kaul, R.K. and Delph, T.G. 1978. On continuum modelling of the dynamic behaviour of layered composites. *Archiv. Mech.* **28**(3):405–421.
207. Hersch, J. 1961. Propriétés de convexité du type de Weye pour des problémes de vibration ou d'équilibre. *ZAMP* **12**: 298–322.
208. Hobart, R. 1965. Peierls stress dependence on dislocation width. *J. Appl. Phys.* **36**:1944–1948.
209. Hodges, C.H. and Woodhouse, J. 1986. Theories of noise and vibration transmission in complex structures. *Rep. Prog. Phys.* **49**:107–170.
210. Hollister, S.J., Fyhire, D.P., Jepsen, K.J. and Goldstein, S.A. 1991. Application of homogenization theory to the study of trabecular bone mechanics. *J. Biomech.* **24**:825–839.

211. Hoppensteadt, F.C. and Hyman, J.M. 1977. Periodic solutions of a logistic difference equation. *SIAM J. Appl. Math.* **32**:73–81.
212. Hu, H., Damil N. and Potier-Ferry, M. 2011. A bridging technique to analyze the influence of boundary conditions on instability patterns. *J. Comp. Phys.* **230**:3753–3764.
213. Huang, H.H., Sun, C.T. and Huang, G.L. 2009. On the negative effective mass density in acoustic metamaterials. *Int. J. Eng. Sci.* **47**:610–617.
214. Huges, D.S. and Kelly, I.L. 1953. Second-order elastic deformation of solids. *Phys. Rev*. **92**:1145–1156.
215. Hughes, E.R., Leighton, T.G., Petley, G.M., White, P.R. and Chivers, R.C. 2003. Estimation of critical and viscous frequencies for Biot theory in cancellous bone. *Ultrasonics* **41**:365–368.
216. Hussein, M.I., Hulbert, G.M. and Scott, R.A. 2006. Dispersive elastodynamics of 1D banded materials and structures: analysis. *J. Sound Vib.* **289**:779–806.
217. Hussein, M.I., Leamy, M.J. and Ruzzene, M. 2014. Dynamics of phononic materials and structures: historical origins, recent progress, and future outlook. *Appl. Mech. Rev.* **66**:040802.
218. Indeitsev, D.A., Kuznetsov, N.G., Motygin, O.V. and Mochalova,Yu.A. 2007. *Localization of Linear Waves*. St. Petersburg: St. Petersburg University (in Russian).
219. Jee, W.S.S. 2001. Integrated bone tissue physiology: anatomy and physiology. In: Cowin, S.C. (ed.) *Bone Mechanics Handbook*, Boca Raton: CRC, pp. 51-68.
220. Jerri, A.J. 2011. *Advances in the Gibbs Phenomenon*. New York: ∑ Sampling Publishing.
221. Jiang, M., Shen, Yi. and Liao, X. 2006. Stability, bifurcation and a new chaos in the logistic differential equation with delay. *Phys. Lett. A* **350**:221–227.
222. Joukowsky, N.E. 1937. The work of continuous and non-continuous traction devices in pulling a train from its position and at the beginning of its motion. In: Joukowsky, N.E. *Complete Collected Works, 8: Theory of Elasticity. Railways. Automobiles*, Moscow: ONTI, Kotelnikov, A.P. (Ed.), pp. 221–255 (in Russian).
223. Kafesaki, M. and Economou, E.N. 1999. Multiple-scattering theory for threedimensional periodic acoustic composites. *Phys. Rev. B* 60:11993–12001.
224. Kalamkarov, A.L., Andrianov, I.V. and Danishevs'kyy, V.V. 2009. Asymptotic homogenization of composite materials and structures. *Appl. Mech. Rev.* **62**:030802.
225. Kanaun, S.K. and Levin, V.M. 2003. Self-consistent methods in the problem of axial elastic shear wave propagation through fibre composites. *Arch. Appl. Mech.* **73**:105–130.
226. Kanaun, S.K. and Levin, V.M. 2008. *Self-Consistent Methods for Composites. Vol.2: Wave Propagation in Heterogeneous Materials*. New York: Springer.
227. Kanaun, S.K. and Levin, V.M. 2005. Propagation of shear elastic waves in composites with a random set of spherical inclusions (effective field approach). *Int. J. Sol. Struct.* **42**:3971–3997.
228. Kanaun, S.K., Levin, V.M. and Sabina F.J. 2004. Propagation of elastic waves in composites with random set of spherical inclusions (effective medium approach). *Wave Motion* **40**:69–88.
229. Kantorovich, L.V. and Krylov, V.I. 1958. *Approximate Methods of Higher Analysis*. Groningen: P. Noordhoff.
230. Kaplunov, J.D., Kossovich, L.Ya. and Nolde, E.V. 1998. *Dynamics of Thin Walled Elastic Bodies*. San Diego: Academic Press.

231. Kaplunov, J. and A. Pichugin. 2009. On rational boundary conditions for higher-order long-wave models. In: Borodich, F. (eds.) *IUTAM Symposium on Scaling in Solid Mechanics. IUTAM Bookseries* 10, Dordrecht: Springer, pp. 81–90.
232. Karpov, S.Yu. and Stolyarov, S.N. 1993. Propagation and transformation of electromagnetic waves in one-dimensional periodic structures. *Physics-Uspekhi* **36**(1):1–22.
233. Kasra, M. and Grynpas, M.D. 1998. Static and dynamic finite element analyses of an idealized structurel model of vertebral trabecular bone. *Trans. ASME J. Biomech. Eng.* **120**:267–272.
234. Kawarai, S. 2002. Exact discretization of differential equations by s-z transform. In: *Proceedings of the 2002 IEEE International Symposium on Circuits and Systems*, Phoenix-Scottsdale, AZ, USA, 26-29 May, pp. 461–464.
235. Keller, J.B. 1964. A theorem on the conductivity of a composite medium. *J. Math. Phys.* **5**(4):548–549.
236. Kevrekidis, P.G. 2011. Non-linear waves in lattices: past, present, future. *IMA J. Appl. Math.* **76**(3): 389–423.
237. Kevrekidis, P.G. and Kevrekidis, I.G. 2001. Heterogeneous versus discrete mapping problem. *Phys. Rev. E.* **64**:056624-1 – 056624-8.
238. Kevrekidis, P.G., Kevrekidis, I.G., Bishop, A.R. and Titi, E.S. 2002. Continuum approach to discreteness. *Phys. Rev. E.* **65**(4):46613-1 – 46613-13.
239. Khusnutdinova, K.R. and Samsonov, A.M. 2008. Fission of a longitudinal strain solitary wave in a delaminated bar. *Phys. Rev. E.* **77**:066603.
240. Kissel, G.J. 1988. *Localization in Disordered Periodic Structures*. Massachusetts: Massachusetts Institute of Technology.
241. Kittel, C. 2005. *Introduction to Solid State Physics*. (8th ed.) Hoboken, New York: Wiley.
242. Kivshar, Yu.S. and Salerno, M. 1994. Modulation instabilities in the discrete deformable nonlinear Schrödinger equation. *Phys. Rev. E.* **49**(4):3543–3546.
243. Kliakhandler, I.L., Porubov, A. and Velarde, M.G. 2000. Localized finite-amplitude disturbance and selection of solitary waves. *Phys. Rev. E* **62**:4959–4962.
244. Kochmann, D.M. and Bertoldi, K. 2017. Exploiting microstructural instabilities in solids and structures: from metamaterials to structural transitions. *Appl. Mech. Rev.* **69**:050801.
245. Korn, G.A. and Korn, T.M. 2000. *Mathematical Handbook for Scientists and Engineers: Definitions, Theorems, and Formulas for Reference and Review*. Mineola, New York: Dover.
246. Korteweg, D.J. and de Vries, G. 1895. On the change of the form of long waves advancing in a rectangular channel and on a new type of long stationary waves. *Phil. Mag.* **5**:422–443.
247. Kosevich, A.M. 2005. *The Crystal Lattice: Phonons, Solitons, Dislocations, Superlattices*. Berlin: Wiley.
248. Kosevich, A.M. and Kovalev, A.S. 1975. Self-localization of vibrations in a one-dimensional anharmonic chain. *Sov. Phys.-JETP.* **40**(5):891–896.
249. Kosevich, A.M. and Savotchenko, S.E. 1999. Peculiarities of dynamics of one-dimensional discrete systems with interaction extending beyond nearest neighbors, and the role of higher dispersion in soliton dynamics. *Low Temp. Phys.* **25**(7):550–557.
250. Kosevich, Yu.A. 2003. Nonlinear envelope-function equation and strongly localized vibrational modes in anharmonic lattices. *Phys. Rev. B.* **47**:3138–3152.

251. Kovalev, A.S. and Kosevich, A.M. 1975. Self-localization of vibrations in a one-dimensional anharmonic chain. *Sov. Phys. JETP* **40**:891–896.
252. Kozlov, S.M. 1989. Geometrical aspects of averaging. *Russ. Math. Surv.* **44**:91–144.
253. Kozlov, S.M. 1987. Averagind of difference schemes. *Math. USSR-Sb.* **57**(2):351–369.
254. Kozyrenko, V.N., Kumpanenko, I.V. and Mikhailov, I.D. 1977. Green's function analysis of the vibrational spectra of polymer chain. I. Several approaches to the problem. *J. Polymer Sci.* **15**:1721–1738.
255. Krivtsov, A.M. 2007. *Deformation and Destruction of Microstructured Solids*. Moscow: Fizmatlit (in Russian).
256. Kruzik, J.J. 2009. Predicting fatigue failures. *Science* **325**:156–157.
257. Kunin, I.A. 1966. Model of an elastic medium of simple structure with three-dimensional dispersion. *J. Appl. Math. Mech.* **30**(3):642–652.
258. Kunin, I.A. 1982. *Elastic Media with Microstructure. 1. One-Dimensional Models*. Berlin: Springer.
259. Kunin, I.A. 1983. *Elastic Media with Microstructure. 2. Three-dimensional Models*. Berlin: Springer.
260. Kurchanov, P.F., Myshkis, A.D. and Filimonov, A.M. 1991. Some unexpected results in the classical problem of vibrations of the string with n beads when n is large. *C.R. Acad. Sci. Paris* 1313:961–965.
261. Kurchanov, P.F., Myshkis, A.D. and Filimonov, A.M. 1991. Vibrations of rolling stock and a theorem of Kronecker. *J. Appl. Math. Mech.* **55**(6):870–876.
262. Kushwaha, M.S. 1997. Stop-bands for periodic metallic rods: sculptures that can filter the noise. *Appl. Phys. Lett.* **70**:3218–3220.
263. Kushwaha, M.S. and Halevi, P. 1994. Band-gap engineering in periodic elastic composites. *Appl. Phys. Lett.* **64**:1085–1087.
264. Kushwaha, M.S. and Halevi, P. 1996. Giant acoustic stop bands in twodimensional periodic arrays of liquid cylinders. *Appl. Phys. Lett.* **69**:31–33.
265. Kushwaha, M.S., Halevi, P., Dobrzynski, L. and Djafari-Rouhani, B. 1993. Acoustic band structure of periodic elastic composites. *Phys. Rev. Lett.* **71**:2022–2025.
266. Kushwaha, M.S., Halevi, P., Martinez, G., Dobrzynski, L. and Djafari-Rouhani, B. 1994. Theory of acoustic band structure of periodic elastic composites. *Phys. Rev. B.* **49**:2313–2322.
267. Lagrange, J.L. 1997. *Analytical Mechanics*. Berlin: Springer Nature.
268. Lakes, R.S. 1982. Dynamical study of couple stress effects in human compact bone. *Trans. ASME J. Biomech. Eng.* **104**:6–11.
269. Lamacz, A. 2011. Dispersive effective models for waves in heterogeneous media. *Math. Model. Meth. Appl. Sci.* **21**(9):1871–1899.
270. Landa, P.S. 1996. *Nonlinear Oscillations and Waves in Dynamical Systems*. Dordrecht: Kluwer.
271. Landa, P.S., 2001. *Regular and Chaotic Oscillations*. Berlin, New York: Springer-Verlag.
272. Landa, P.S. and Marchenko, V.F. 1991. On the linear theory of waves in media with periodic structures. *Sov. Phys. Usp.* **34**(9):830–834.
273. Lazar, M., Maugin, G.A. and Aifantis, E.C. 2006. On a theory of nonlocal elasticity of bi-Helmholtz type and some applications. *Int. J. Sol. Struct.* **43**:1404–1421.
274. Le Roux, M.J. 1911. Étude géométrique torsion. *Annales scientifiques de L'École Normale Supérieur* **28**, 523–579.

275. Lebedev, L.P. and Vorovich, I.I. 2002. *Functional Analysis in Mechanics*. New York: Springer.
276. Lenci, S. and Menditto, G. 2000. Weak interface in long fibre composites. *Int. J. Sol. Struct*. **37**:4239–4260.
277. Levy, A.J. and Dong, Z. 1998. Effective transverse response of fibre composites with nonlinear interface. *J. Mech. Phys. Sol*. **46**:1279–1300.
278. Levy, A.J. 1996. The effective dilatational response of fibre reinforced composites with nonlinear interface. *J. Appl. Mech*. **63**:357–364.
279. Levy, A.J. 2000. The fibre composite with nonlinear interface. Part I: Axial tension. *J. Appl. Mech*. **67**:727–732.
280. Li, D. and H. Benaroya. 1992. Dynamics of periodic and near-periodic structures. *Appl. Mech. Rev*. **45**(11):447–459.
281. Li, D. and Benaroya, H. 1996. Vibration localization in multi-coupled and multi-dimensional near-periodic structures. *Wave Motion* **23**:67–82.
282. Liang, B., Yuan, B. and Chen, J.-C. 2009. Acoustic diode: rectification of acoustic energy flux in one-dimensional systems. *Phys. Rev. Let*. **103**:104301.
283. Lin, Y.K. 1996. Dynamics of disordered periodic structures. *Appl. Mech. Rev*. **49**:57–64.
284. Liu, Z., Chan, C.T., Sheng, P., Goertzen, A.L. and Page, J.H. 2000. Elastic wave scattering by periodic structures of spherical objects: theory and experiment. *Phys. Rev. B* **62**:2446–2457.
285. Love, A.E.H. 1927. *A Treatise on the Mathematical Theory of Elasticity*, 4th edn. Cambridge: Cambridge University Press.
286. Lubin, G. 2014. *Handbook of Composites*. Berlin: Springer.
287. Lubineau, G., Azdoud, Y., Han, F., Rey, C. and Askari, A. 2012. A morphing strategy to couple non-local to local continuum mechanics. *J. Mech. Phys. Sol*. **60**:1088–1102.
288. Lucia, U. 2016. Macroscopic irreversibility and microscopic paradox: A Constructal law analysis of atoms as open systems. *Sci. Rep*. **6**:35796.
289. Lur'e, A.I. 1951. *Operational Calculus and its Applications to the Problems of Mechanics*. Moscow: GITTL (in Russian).
290. Lur'e, A.I. 1990. *Nonlinear Theory of Elasticity*. Amsterdam: North-Holland.
291. Malyi, V.I. 1971. About nonlocal theory of elasticity. In: *IV All-Union Conference on Strength and Plasticity*. Moscow: Nauka, pp. 74–78 (in Russian).
292. Malyi, V. 2021. Theoretical determination of the five physical constants of the Toupin-Mindlin gradient elasticity for polycrystalline materials. In: *Nonlinear Dynamics of Discrete and Continuous Systems*, 145–154. Abramian, A.K.,Andrianov, I.V., Gaiko, V.A. (eds). Advanced Structured Matherials 139, Cham: Springer Nature.
293. Manevich, A.I. and Manevitch, L.I. 2005. *The Mechanics of Nonlinear Systems with Internal Resonances*. Singapore: World Scientific.
294. Manevitch, L.I. and Gendelman, O.V. 2011. *Tractable Models of Solid Mechanics. Formulation, Analysis and Interpretation*. Berlin, Heidelberg, London, New York: Springer.
295. Manevitch, L.I. and Pilipchuk, V.N. 1985. Localized vibrations in linear and nonlinear lattices. *Adv. Mech*. **13**:107–134 (in Russian).
296. Manevitch, L.I. and Smirnov, V.V. 2008. *Solitons in Macromolecular Systems*. New York: Nova Science Publishers.
297. Manevitch, L.I., Kovaleva, A.S., Smirnov, V.V. and Starosvetsky, Yu. 2017. *Nonstationary Resonant Dynamics of Oscillatory Chains and Nanostructures*. Singapore: Springer.

298. Manevitch, L.I., Mikhlin, Yu.V. and Pilipchuk, V.N. 1989. *Method of Normal Vibrations for Essentially Nonlinear Systems*. Moscow: Nauka (in Russian).
299. Maradudin, A.A. 1965. Some effects of point defects on the vibrations of crystal lattices. *Rep. Prog. Phys.* **28**:331–380.
300. Maradudin, A.A. 2003. Lattice dynamics. *Ann. Rev. Phys. Chem.* **14**(1):89–116.
301. Maradudin, A.A., Montroll, E.W. and Weiss, G.H. 1971. *Theory of Lattice Dynamics in the Harmonic Approximation*. New York: Academic Press.
302. Martowicz, A., Roemer, J., Staszewski, W.J., Ruzzene, M. and Uhl, T. 2019. Solving partial differential equations in computational mechanics via nonlocal numerical approaches. *ZAMM* **99**(4):e201800342.
303. Maslov, V.P. 1976. *Operational Methods*. Moscow: Mir.
304. Maslov, V.P. and Mosolov, P.P. 2000. *Nonlinear Wave Equations Perturbed by Viscous Terms*. Berlin, New York: de Gruyter.
305. Maslov, V.P. and Omelyanov, G.A. 2001. *Geometric Asymptotic for Nonlinear PDE*. I. Providence: AMS.
306. Maugin, G.A. 1999. *Nonlinear Waves in Elastic Crystals*. Oxford: Oxford University Press.
307. Maugin, G.A. 1999. From Piola's manifold to Cosserats' structure. In: *Geometry, Continua and Microstructure*, Maugin, G.A. (Ed.) Paris: Hermann, pp. 113–120.
308. Maugin, G.A. 2013. *Continuum Mechanics Through the Twentieth Century. A Concise Historical Perspective*. Dordrecht: Springer.
309. May, R. 1976. Simple mathematical models with very complicated dynamics. *Nature* **261**:459–467.
310. McPhedran, R.C. and McKenzie, D.R. 1978. The conductivity of lattices of spheres. I. The simple cubic lattice. *Proc. Roy. Soc. London* **A359**:45–63.
311. McPhedran, R.C., Poladian, L. and Milton, G.W. 1988. Asymptotic studies of closely spaced, highly conducting cylinders. *Proc. Roy. Soc. London* **A415**:185–196.
312. Mencik, J.M. and Ichchou, M.N. 2005. Multi-mode propagation and diffusion in structures through finite elements. *Eur. J. Mech. A/Solids* **24**(5):877–898.
313. Meng, S. and Guzina, B.B. 2018. On the dynamic homogenization of periodic media: Willis' approach versus two-scale paradigm. *Proc. R. Soc. A* **474**:20170638.
314. Metrikine, A.V. 2006. On causality of the gradient elasticity models. *J. Sound Vib.* **297**:727–742.
315. Metrikine, A.V. and Askes, H. 2002. One-dimensional dynamically consistent gradient elasticity models derived from a discrete microstructure. Part 1: Generic formulation. *Eur. J. Mech. A /Sol.* **21**:555–572.
316. Metrikine, A.V. and Askes, H. 2008. An isotropic dynamically consistent gradient elasticity model derived from a 2D lattice. *Phil. Mag.* **86**(21):3259–3268.
317. Maldovan, M. 2013. Sound and heat revolution in phononics. *Nature* **503**:209–217.
318. Mindlin, R.D. 1964. Micro-structure in linear elasticity. *Arch. Rat. Mech. Anal.* **16**:51–78.
319. Mindlin, R.D. and Herrmann, G. 1952. A one-dimensional theory of compressional waves in an elastic rod. *Proc. First U.S. Nat. Congr. Appl. Mech., ASME* 187–191.
320. Montero de Espinosa, F.R., Jiménez E. and Torres, M. 1998. Ultrasonic band gap in a periodic two-dimensional composite. *Phys. Rev. Lett.* **80**:1208–1211.
321. Moon, F.C. 1987. *Chaotic Vibrations. An Introduction for Applied Scientists and Engineers*. Ithaca: Cornell University.
322. Mortell, M.P. and Seymour, R.B. 1972. Pulse propagation in a nonlinear viscoelastic rod of finite length. *SIAM J. Appl. Math.* **22**:209–224.

323. Mortell, M.P. and Varley, E. 1970. Finite amplitude waves in bounded media: nonlinear free vibrations of an elastic panel. *Proc. Roy. Soc. A* **318**:169–196.
324. Movchan, A.B. and Slepyan, L.I. 2007. Band gap Green's functions and localized oscillations. *Proc. R. Soc. A.* **463**:2709–2727.
325. Movchan, A.B., Movchan, N.V., Jones, I.S. and Colquitt, D.J. 2018. *Mathematical Modelling of Waves in Multi-Scale Structured Media*. London: Taylor and Francis.
326. Movchan, A.B., Movchan, N.V. and Poulton, C.G. 2002. *Asymptotic Models of Fields in Dilute and Densely Packed Composites*. London: Imperial College Press.
327. Munkhammar, J. 2013. Chaos in fractional order logistic equation. *Fract. Calc. Appl. Anal.* **16**:511–519.
328. Murnaghan, F.D. 1951. *Finite Deformation of an Elastic Solid*. New York: Wiley.
329. Myshkis, A.D. 2005. Mixed functional differential equations. *J. Math. Sc.* **129**(5):4111–4226.
330. Nayfeh, A.H. 2000. *Perturbation Methods*. New York: Wiley.
331. Nayfeh, A.H. and Mook, D.T. 1979. *Nonlinear Oscillations*. New York: Wiley.
332. Needleman, A. 1990. An analysis of tensile decohesion along an interface. *J. Mech. Phys. Sol.* **38**:289–324.
333. Needleman, A. 1992. Micromechanical modelling of interfacial decohesion. *Ultramicroscopy* **40**:203–214.
334. Nemat-Nasser, S. and Srivastava, A. 2011. Overall dynamic constitutive relations of layered elastic composites. *J. Mech. Phys.* **59**:1953–2258.
335. Nesterenko, V.F. 2018. Waves in strongly nonlinear discrete systems. *Phil. Trans. R. Soc. A* **376**:20170130.
336. Nie, S. and C. Basaran. 2005. A micromechanical model for effective elastic properties of particulate composites with imperfect interfacial bonds. *Int. J. Sol. Struct.* **42**:4179–4191.
337. Ogden, R.W. 1997. *Nonlinear Elastic Deformations*. New York: Dover.
338. Oleynik, O.A., Shamaev, A.S. and Yosifian, G.A. 1992. *Mathematical Problems in Elasticity and Homogenization*. Amsterdam: North-Holland.
339. Osharovich, G.G. and Ayzenberg-Stepanenk, M.V. 2012. Wave localization in stratified square-cell lattices: The antiplane problem. *J. Sound Vib.* **331**:1378–1397.
340. Pagano, S. and Paroni, R. 2003. A simple model for phase transition: from the discrete to the continuum problem. *Q. Appl. Math.* **61**:89–109.
341. Pal, R.K., Rimoli, J.J. and Ruzzene, M. 2016. Effect of large deformation pre-loads on the wave properties of hexagonal lattices. *Smart Mater. Struct.* **25**:054010.
342. Pal, R.K., Ruzzene, M. and Rimoli, J.J. 2016. A continuum model for nonlinear lattices under large deformations. *Int. J. Sol. Struct.* **96**:300–319.
343. Palmov. V.A. 1967. Propagation of vibrations in a nonlinear dissipative medium. *J. Appl. Math. Mech.* **31**:763–769.
344. Palmov, V.A. 2014. *Vibrations of Elasto-Plastic Bodies*. Berlin: Springer.
345. Panasenko, G.P. 2005. *Multi-Scale Modeling for Structures and Composites*. Berlin: Springer.
346. Parks, M.L., Bochev, P.B. and Lehoucq, R.B. 2008. Connecting atomistic-to-continuum coupling and domain decomposition. *Multisc. Mod. Simul.* **7**:362–380.
347. Parnell, W.J. 2007. Effective wave propagation in a prestressed nonlinear elastic composite bar. *IMA J. Appl. Math.* **72**:223–244.
348. Parnell, W.J. 2012. Nonlinear pre-stress for cloaking from antiplane elastic waves. *Proc. R. Soc. A.* **468**:563–580.

349. Parnell, W.J. and Abrahams, I.D. 2006. Dynamic homogenization in periodic fibre reinforced media. Quasi-static limit for SH waves. *Wave Motion* **43**:474–498.

350. Parnell, W.J. and Abrahams, I.D. 2008. Homogenization for wave propagation in periodic fibre-reinforced media with complex microstructure. I-Theory. *J. Mech. Phys. Sol.* **56**:2521–2540.

351. Parnell, W.J. and Abrahams, I.D. 2011. The effective wavenumber of a pre-stressed nonlinear microvoided composite. *J. Phys.: Conf. Ser.* **269**:012007.

352. Parnell, W.J. and Grimal, Q. 2009. The influence of mesoscale porosity on cortical bone anisotropy. Investigations via asymptotic homogenization. *J. Roy. Soc. Interf.* **6**:97–109.

353. Paroni, R. 2003. From discrete to continuum: a Young measure approach. *ZAMP* **54**(2):328–348.

354. Pasternak, E. and Mühlhaus H.-B. 2005. Generalized homogenization procedure for granular material. *J. Eng. Math.* **52**:199–229.

355. Pavlov, I.S. and Potapov, A.I. 2008. Structural models in mechanics of nanocrystalline media. *Dokl. Phys.* **53**(7):408–412.

356. Peerlings, R.H.J., Geers, M.G.D., de Borst, R. and Brekelmans, W.A.M. 2001. A critical comparison of nonlocal and gradient-enhanced softening continua. *Int. J. Sol. Struct.* **38**:7723–7746.

357. Peregrin, D.H. 1966. Calculation of the development of an undular bore. *Fluid Mech.* **25**:321–330.

358. Perrins, W.T, McKenzie, D.R. and McPhedran R.C. 1979. Transport properties of regular arrays of cylinders. *Proc. Roy. Soc.* **A369**:207–225.

359. Petropoulou, E.N. 2010. A discrete equivalent of the logistic equation. *Adv. Differ. Eq.* **2010**:457073.

360. Pfaller, G., Possart, P., Steinmann, M., Rahimi, F., Müller-Plathe and Böhm, M.C. 2012. Acomparison of staggered solution schemes for coupled particle-continuum systems modeled with the Arlequin method. *Comp. Mech.* **49**:565–579.

361. Pichugin, A.V., Askes, H. and Tyas, A. 2008. Asymptotic equivalence of homogenization procedures and fine-tuning of continuum theories. *J. Sound Vib.* **313**:858–874.

362. Picu, R.C. 2002. On the functional form of non-local elasticity kernels. *J. Mech. Phys. Sol.* **50**:1923–1939.

363. Pierre, C. and Dowell, E.H. 1987. Localization of vibrations by structural irregularity. *J. Sound Vib.* **114**:549–564.

364. Pierre, Ch., Castanier, M. and Joe, C.W. 1996. Wave localization in multi-coupled periodic structures. Application to truss beams. *Appl. Mech. Rev*. **49**:65–86.

365. Pilipchuk, V.N. and Starushenko, G.A. 1997. A version of non-smooth transformations of variables for one-dimensional elastic systems of periodic structures. *J. Appl. Math. Mech.* **61**(2):265–274.

366. Pilipchuk, V.N., Andrianov, I.V. and Markert, B. 2016. Analysis of micro-structural effects on phononic waves in layered elastic media with periodic nonsmooth coordinates. *Wave Motion* **63**:149–169.

367. Platts, S.B., Movchan, N.V., McPhedran, R.C. and Movchan, A.B. 2002. Two-dimensional phononic crystals and scattering of elastic waves by an array of voids. *Proc. R. Soc. Lond. A.* **458**:2327–2347.

368. Platts, S.B., Movchan, N.V., McPhedran, R.C., Movchan, A.B. 2003. Band gaps and elastic waves in disordered stacks: normal incidence. *Proc. R. Soc. Lond. A.* **459**:221–240.

369. Porubov, A.V. 2003. *Amplification of Nonlinear Strain Waves in Solids*. Singapore: World Scientific.
370. Porubov, A.V. 2009. *Localization of Nonlinear Strain Waves*. Moscow: FIZMATLIT (in Russian).
371. Porubov, A.V. and Andrianov, I.V. 2013. Nonlinear waves in diatomic crystals. *Wave Motion* **50**:1153–1160.
372. Porubov, A.V. and Velarde, M.G. 2002. Strain kinks in an elastic rod embedded in a viscoelastic medium. *Wave Motion* **35**:189–204.
373. Porubov, A.V., Osokina, A.E. and Antonov, I.D. 2020. Nonlinear dynamics of two-dimensional lattices with complex structure. *Nonlinear Wave Dynamics of Materials and Structures*. Springer Nature: Cham, 309–334.
374. Potapov, A.I., Pavlov, I.S. and Lisina, S.A. 2009. Acoustic identification of nanocrystalline media. *J. Sound Vib*. **322**:564–580.
375. Potapov, A.I., Pavlov, I.S. and Maugin, G.A. 2006. A 2D granular medium with rotating particles. *Int. J. Solids Str.* **43**(20):6194–6207.
376. Poulton, C.G., Movchan, A.B., McPhedran, R.C., Nicorovici, N.A. and Antipov, Y.A. 2000. Eigenvalue problems for doubly periodic structures and phononic band gaps. *Proc. R. Soc. Lond. A*. **456**:2543–2559.
377. Prudnikov, A.P., Brichkov, Yu.A. and Marichev, O.I. 1986. *Integrals and Series. Vol. 1. Elementary Functions*. New York: Gordon & Breach.
378. Psarobas, I.E., Stefanou, N. and Modinos, A. 2000. Scattering of elastic waves by periodic arrays of spherical bodies. *Phys. Rev. B* **62**:278–291.
379. Rayleigh, J.W.S. 1877–1888. *The Theory of Sound*, 1-2. London: Macmillan and Co.
380. Rayleigh, Lord Sec. R.S. 1892. On the influence of obstacles arranged in rectangular order upon the properties of a medium. *The London, Edinburgh, and Dublin Phil. Mag. J. Sci*. **34**(211):481–502.
381. Rednikov, A.Ye., Velarde, M.G., Ryazantsev, Yu.S., Nepomnyashchy, A.A. and Kurdyumov, V.N. 1995. Cnoidal wave trains and solitary waves in a dissipation modified Korteweg - de Vries equation. *Acta Appl. Math*. **39**:457–475.
382. Remoissenet, M. 1986. Low-amplitude breather and envelope solitons in quasi-one-dimensional physical models. *Phys. Rev B* **33**:2386–2392.
383. Ribeiro, P. and Petyt, M. 1999. Non-linear vibration of beams with internal resonances by the hierarchical finite-element method. *J. Sound Vib*. **224**:591–624.
384. Rogers, R.C. and Truskinovsky, L. 1997. Discretization and hysteresis. *Physica B* **233**:370–375.
385. Rosenau, Ph. 1986. Dynamics of nonlinear mass-spring chains near the continuum limit. *Phys. Lett. A* **118**(5): 222–227.
386. Rosenau, Ph. 1987. Dynamics of dense lattices. *Phys. Rev. B* **36**(11):5868–5876.
387. Rosenau, Ph. 1988. Dynamics of dense discrete systems. *Prog. Theor. Phys.* **89**(5):1028–1042.
388. Rosenau, Ph. 1989. Extending hydrodynamics via the regularization of the Chapman-Enskog expansion. *Phys. Rev. A* **40**(12):7193–7196.
389. Rosenau, Ph. 2003. Hamiltonian dynamics of dense chains and lattices: or how to correct the continuum. *Phys. Lett. A* **311**(5):39–52.
390. Rudenko, O.V. 2006. Giant nonlinearities in structurally inhomogeneous media and the fundamentals of nonlinear acoustic diagnostic techniques. *Phys. Uspek.* **49**:69–87.
391. Rushchitsky, J.J. 1999. Interaction of waves in solid mixtures. *Appl. Mech. Rev*. **52**:35–74.

392. Rushchitsky, J. 2014. *Nonlinear Elastic Waves in Materials*. Berlin: Springer.
393. Sachdev, P.L. 1987. *Nonlinear Diffusive Waves*. Cambridge: Cambridge University Press.
394. Sakuma, Y., Païdoussis, M.P. and Price, S.J. 2010. Mode localization phenomena and stability of disordered trains and train-like articulated systems travelling in confined fluid. *J. Sound Vib*. **329**:5501–5519.
395. Salupere, A., Engelbrecht, J. and Maugin, G.A. 2001. Solitonic structures in KdV-based higher-order systems. *Wave Motion* **34**:51–61.
396. Salupere, A., Maugin, G.A. and Engelbrecht, J. 1994. Korteweg-de Vries soliton detection from a harmonic input. *Phys. Lett. A* **192**:5–8.
397. Samarskii, A.A. and Mikhailov, A.P. 2002. *Principles of Mathematical Modeling. Ideas, Methods, Examples*. London: Taylor and Francis.
398. Samsonov, A.M. 2001. *Strain Solitons in Solids and How to Construct Them*. Boca Raton: CRC Press.
399. Santosa, F. and Symes, W. 1991. A dispersive effective medium for wave propagation in periodic composites. *SIAM J. Appl. Math*. **51**:984–1005.
400. Schrödinger, E. 1914. Zur Dynamik elastisch gekoppelter Punktsysteme. *Annalen der Physik* **349**(14):916–934.
401. Scott, A. 2003. *Nonlinear Science: Emergence and Dynamics of Coherent Structures*. New York: Oxford University Press.
402. Seeger, A. 2010. On the simulation of dispersive wave propagation by elasticity models. *Philos. Mag*. **90**(9):1101–1104.
403. Sen, A. and Mukherjee, D. 2009. Chaos in the delay logistic equation with discontinuous delays. *Chaos Soliton Fract*. **40**:2126–2132.
404. Seymour, R.B. and Varley, E. 1970. High frequency, periodic disturbances in dissipative systems. I. Small amplitude, finite rate theory. *Proc. Roy. Soc. A* **314**:387–415.
405. Shamrovskyi, A.D. 2004. *Discrete Approaches in Economics*. Zaporozhye: Zaporozhye State Engineering Academy (in Russian).
406. Shodja, H.M., Zaheri, A. and Tehranchi, A. 2013. Ab initio calculations of characteristic lengths of crystalline materials in first strain gradient elasticity. *Mech. Mater*. **61**:73–78.
407. Shokin, Y.I. 1983. *The Method of Differential Approximation*. Berlin: Springer.
408. Shulga, N.A. 2003. Propagation of elastic waves in periodically inhomogeneous media. *Int. Appl. Mech*. **39**:763–796.
409. Shulga, N.A. 2003. Propagation of coupled waves interacting with an electromagnetic field in periodically inhomogeneous media. *Int. Appl. Mech*. **39**:1146–1172.
410. Sievers, A.J. and Page, J.B. 1995. Unusual local mode systems. In: *Dynamical Properties of Solids*, Vol 7. *Phonon Physics the Cutting Edge*, Horton, G.K., Maradudin, A.A. (eds.), pp. 137–256.
411. Sigalas, M.M. and Economou, E.N. 1992. Elastic and acoustic wave band structure. *J. Sound Vib*. **158**:377–382.
412. Sigalas, M.M. and Economou, E.N. 1993. Band structure of elastic waves in two dimensional systems. *Sol. Stat. Comm*. **86**:141–143.
413. Sigalas, M.M. and Economou, E.N. 1994. Elastic waves in plates with periodically place inclusions. *J. Appl. Phys*. **75**:2845–2850.
414. Sigalas, M.M. and Economou, E.N. 1996. Attenuation of multiple-scattered sound. *Europhys. Lett*. **36**:241–246.

415. Slepyan, L.I. 1972. *Non-steady-state Elastic Waves*. Leningrad: Sudostroyenie (in Russian).
416. Slepyan, L.I. 1981. Crack propagation in high-frequency lattice vibration. *Sov. Phys. Dokl.* **26**:900–902.
417. Slepyan, L.I. 1981. Dynamics of a crack in a lattice. *Sov. Phys. Dokl.* **26**:538–540.
418. Slepyan, L.I. 1982. The relation between the solutions of mixed dynamic problems for a continuous elastic medium and a lattice. *Sov. Phys. Dokl.* **27**:771–772.
419. Slepyan, L.I. 1984. Dynamics of brittle fracture in media with a structure. *Mech. Sol.* **19**(6):114–115.
420. Slepyan, L.I. and Troyankina, L.V. 1984. Fracture wave in a chain structure. *J. Appl. Mech. Techn. Phys.* **25**(6):921–927.
421. Slepyan, L.I. and Troyankina, L.V. 1988. Impact waves in a nonlinear chain. In: *Plasticity and Fracture of Solids*, Gol'dstein, R.V. (ed.) Moscow: Nauka, pp. 175–186 (in Russian).
422. Smith, B., Bjorstad, P. and Gropp, W. 1996. *Domain Decomposition*. Cambridge: Cambridge University Press.
423. Smyshlyaev, V.P. 2009. Propagation and localization of elastic waves in highly anisotropic periodic composites via two-scale homogenization. *Mech. Mat.* **41**:434–447.
424. Strogatz, S.H. 2018. *Nonlinear Dynamics and Chaos with Applications to Physics, Biology, Chemistry, and Engineering*. Boca Raton: CRC Press.
425. Sun, C.T., Achenbach, J.D., Herrmann, G. 1968. Continuum theory for a laminated media. *J. Appl. Mech.* **35**:467–475.
426. Tadmor, E.B., Ortiz, M. and Phillips, R. 1996. Quasicontinuum analysis of defects in crystals. *Phil. Mag. A* **73**:1529–1563.
427. Tamura, S., Shields, J.A. and Wolfe, J.P. 1991. Lattice dynamics and elastic phonon scattering in silicon. *Phys. Rev. B* 44:3001–3011.
428. Tan, H., Liu, C., Huang, Y. and Geubelle, P.H. 2005. The cohesive law for the particle/matrix interfaces in high explosives. *J. Mech. Phys. Sol.* **53**:1892–1917.
429. Tarasov, V.E. 2017. Discrete wave equation with infinite differences. *Appl. Math. Inf. Sci. Lett.* **5**(2):41–44.
430. Tarasova, V.V. and Tarasov, V.E. 2017. Logistic map with memory from economic model. *Chaos Solit. Fract.* **95**:84–91.
431. Ting, T.C.T. 1980. Dynamic response of composites. *Appl. Mech. Rev.* **33**:1629–1635.
432. Tokarzewski, S., Telega, J.J. and Galka, A. 2000. Prediction of torsional rigidities of bone filled with marrow: The application of multipoint Padé approximants. *Acta Bioeng. Biomech.* **2**:560–566.
433. Torres, M., Montero de Espinosa, F.R., Garcia-Pablos, D., Garcia, N. 1999. Sonic band gaps in finite elastic media: surface states and localization phenomena in linear and point defects. *Phys. Rev. Lett.* **82**:3054–3057.
434. Toupin, R.A. 1964. Theories of elasticity with couple-stresses. *Arch. Ration. Mech. Anal.* **17**(2):85–112.
435. Truskinovsky, L. and Vainchtein, A. 2005. Quasicontinuum modelling of short-wave instabilities in crystal lattices. *Phil. Mag.* **85**(33-35):4055–4065.
436. Tvergaard, V. 1995. Fibre debonding and breakage in a whisker reinforced metal. *Mater. Sci. Eng. A* **90**:215–222.
437. Tvergaard, V. and Hutchinson, J.W. 1993. The influence of plasticity on mixed-mode interface toughness. *J. Mech. Phys. Sol.* **41**:1119–1135.

438. Tzirakis, N. and Kevrekidis, P.G. 2005. On the collapse arresting effects of discreteness. *Math. Comput. Simul.* **69**:553–566.
439. Ulam, S.M. 1960. *A Collection of Mathematical Problems*. New York: Interscience Publishers.
440. Vakakis, A.F., King, M.E. and Pearlstein, A.J. 1994. Forced localization in a periodic chain of nonlinear oscillators. *Int. J. Non-Lin. Mech.* **29**:429–447.
441. Vakakis, A.F., Manevitch, L.I., Mikhlin, Yu.V., Pilipchuk, V.N. and Zevin, A.A. 1996. *Normal Modes and Localization in Nonlinear Systems*. New York: Wiley.
442. Vakakis, A.F., Nayfeh, A.H. and King, M.E. 1993. A multiple scales analysis of nonlinear localized modes in a cyclic periodic system. *J. Appl. Mech.* **60**:388–397.
443. Van Dyke, M. 1975. *Perturbation Methods*. Stanford: The Parabolic Press.
444. Van Rietbergen, B. and Huiskes, R. 2001. Elastic constants of cancellous bone. In: Cowin, S.C. (ed.) *Bone Mechanics Handbook.* Boca Raton: CRC Press, 15-1 – 15-24.
445. Varley, E. and Rogers, T.G. 1967. The propagation of high frequency, finite acceleration pulses and shocks in viscoelastic materials. *Proc. Roy. Soc. A* **296**:498–518.
446. Vasiliev, V.V. and Lurie, S.A. 2016. Correct nonlocal generalized theories of elasticity. *Phys. Mesomech.* **19**(3):269–281.
447. Vasiliev, A.A., Dmitriev, S.V. and Miroshnichenko, A.E. 2010. Multi-field approach in mechanics of structural solids. *Int. J. Sol. Struct.* **47**:510–525.
448. Vedenova, E.G., Manevich, L.I. and Pilipchuk, V.N. 1985. Normal oscillations of a string with concentrated masses on non-linearly elastic supports. *Appl. Math. Mech.* **49**:153–159.
449. Verhulst, P.F. 1845. Recherches mathématiques sur la loi d'accroissement de la population. *Nouv. Mém. del' Académie Royale des Sci. et Belles-Lett. de Brux.*, **18**:1–42.
450. Verhulst, P.F. 1847. Deuxiéme mémoire sur la loi d'accroissement de la population. *Mém. de l'Académie Royale des Sci. des Lett. et des Beaux-Arts de Belg.* **20**:1–32.
451. Vishik, M.I. and Lyusternik, L.A. 1960. The asymptotic behaviour of solutions of linear differential equations with large or quickly changing coefficients and boundary conditions. *Russ. Math. Surv.* **15**(4):23–91.
452. Wagner, G.J. and Liu, WK. 2003. Coupling of atomistic and continuum simulations using a bridging scale decomposition. *J. Comput. Phys.* **190**:249–274.
453. Wattis, J.A.D. 1993. Approximations to solitary waves on lattices, II: quasi-continuum approximations for fast and slow waves. *J. Phys. A: Math. Gen.* **26**:1193–1209.
454. Wattis, J.A.D. 2000. Quasi-continuum approximations to lattice equations arising from discrete nonlinear telegraph equation. *J. Phys. A: Math. Gen.* **33**:5925–5944.
455. Weinan, E. and Engquist, B. 2003. The heterogeneous multiscale methods. *Comm. Math. Sci.* **1**(1):87–132.
456. Weissert, T.P. 1997. *The Genesis of Simulations in Dynamics: Pursuing the Fermi-Pasta-Ulam Problem*. Berlin: Springer.
457. Wilcox, D.C. 1995. *Perturbation Methods in the Computer Age*. La Canada, Ca.: DCW Industries Inc.
458. Williams, J.L. and Lewis, J.L. 1982. Properties and an anisotropic model of cancellous bone from the proximal tibial epiphysis. *Trans. ASME J. Biomech. Eng.* **104**(1):50–56.
459. Willis, J.R. 2009. Exact effective relations for dynamics of a laminated body. *Mech. Mater.* **41**:385–393.
460. Willis, J.R. 2011. Effective constitutive relations for waves in composites and metamaterials. *Proc. R. Soc. A* **467**(2131):1865–1879.

461. Willis, J.R. 2012. A comparison of two formulations for effective relations for waves in a composite. *Mech. Mater.* **47**:51–60.
462. Wolfe, J.P. 1998. *Imaging Phonons: Acoustic Wave Propagation in Solids*. Cambridge: Cambridge University Press.
463. Xiao, S.P. and Belytschko, T. 2004. A bridging domain method for coupling continua with molecular dynamics. *Comp. Met. Appl. Mech. Eng.* **193**:1645–1669.
464. Yan, Z.-Z. and Zhang, Ch. 2012. Band structures and localization properties of aperiodic layered phononic crystals. *Phys. B: Phys. Condens. Matter.* **407**:1014–1019.
465. Yan, Z.-Z., Zhang, Ch. and Wang, Y.-S. 2009. Analysis of wave propagation and localization in periodic/disordered layered composite structures by a mass-spring model. *Appl. Phys. Lett.* **94**:161909.
466. Yan, Z.-Z., Zhang, Ch. and Wang, Y.-S. 2010. Wave propagation and localization in randomly disordered layered composites with local resonances. *Wave Motion* **47**:409–420.
467. Zabusky, N.J. and Deem, G.S. 1967. Dynamics of nonlinear lattices. I. Localized optical excitations, acoustic radiation, and strong nonlinear behavior. *J. Comput. Phys.* **2**:126–153.
468. Zabusky, N.J. and Kruskal, M.D. 1965. Interaction of "solitons" in a collisionless plasma and the recurrence of initial states. *Phys. Rev. Lett.* **15**:240–243.
469. Zaitsev, V.Yu., Nazarov, V.E. and Talanov, V.I. 2006. @Nonclassical" manifestations of microstructure-induced nonlinearities: new prospects for acoustic diagnostics. *Phys. Usp.* **49**:89–94.
470. Zakharov, V.E. 1974. On stochastization of one-dimensional chains of nonlinear oscillators. *Sov. Phys. JETP* **38**(1):108–110.
471. Zakharov. Yu. 2019. On physical principles and mathematical mechanisms of the phenomenon of irreversibility. *Phys. A: Statist. Mech. Appl.* **525**:1289–1295.
472. Zalipaev, V.V., Movchan, A.B., Poulton, C.G., McPhedran, R.C. 2002. Elastic waves and homogenization in oblique periodic structures. *Proc. R. Soc. Lond. A.* **458**:1887–1912.
473. Zevin, A.A. 1991. Symmetrization of functionals and its applications in mechanics problems. *Mech. Sol.* **26**(5):109–118.
474. Zevin, A.A. 1996. Localization of periodic oscillations in discrete non-linear systems. *J. Sound Vib.* **193**(4):847–862.
475. Zevin, A.A. 1998. Estimates of eigenvalue of self-adjont boundary-value problems with periodic coefficients. *Ukr. Math. J.* **50**(5):719–722.
476. Zheludev, N.I. and Kivshar, Yu.S. 2012. From metamaterials to metadevices. *Nat. Mater.* **11**:917–924.
477. Zhikov, V.V. 1991. Estimates for the averaged matrix and the averaged tensor. *Russ. Math. Surv.* **46**(3):65–136.
478. Zivieri, R., Garesci, F., Azzerboni, B., Chiappini, M. and Finocchio, G. 2019. Nonlinear dispersion relation in anharmonic periodic massspring and mass-in-mass systems. *J. Sound Vib.* **462**:114929

Index

Printed in the United States
by Baker & Taylor Publisher Services